Contemporary Strategies and Approaches in 3-D Information Modeling

Bimal Kumar
Glasgow Caledonian University, UK

A volume in the Advances in Civil and Industrial Engineering (ACIE) Book Series

IGI Global
DISSEMINATOR OF KNOWLEDGE

Published in the United States of America by
IGI Global
Engineering Science Reference (an imprint of IGI Global)
701 E. Chocolate Avenue
Hershey PA, USA 17033
Tel: 717-533-8845
Fax: 717-533-8661
E-mail: cust@igi-global.com
Web site: http://www.igi-global.com

Library of Congress Cataloging-in-Publication Data

Names: Kumar, B. (Bimal), editor.
Title: Contemporary strategies and approaches in 3-D information modeling / Bimal Kumar, editor.
Description: Hershey, PA : Engineering Science Reference, [2018] | Includes bibliographical references.
Identifiers: LCCN 2017049732| ISBN 9781522556251 (hardcover) | ISBN 9781522556268 (ebook)
Subjects: LCSH: Building information modeling. | Three-dimensional display systems. | Visual analytics.
Classification: LCC TH438.13 .C66 2018 | DDC 720.285--dc23 LC record available at https://lccn.loc.gov/2017049732

This book is published in the IGI Global book series Advances in Civil and Industrial Engineering (ACIE) (ISSN: 2326-6139; eISSN: 2326-6155)

British Cataloguing in Publication Data
A Cataloguing in Publication record for this book is available from the British Library.

Advances in Civil and Industrial Engineering (ACIE) Book Series

ISSN:2326-6139
EISSN:2326-6155

Editor-in-Chief: Ioan Constantin Dima, University Valahia of Târgovişte, Romania

MISSION

Private and public sector infrastructures begin to age, or require change in the face of developing technologies, the fields of civil and industrial engineering have become increasingly important as a method to mitigate and manage these changes. As governments and the public at large begin to grapple with climate change and growing populations, civil engineering has become more interdisciplinary and the need for publications that discuss the rapid changes and advancements in the field have become more in-demand. Additionally, private corporations and companies are facing similar changes and challenges, with the pressure for new and innovative methods being placed on those involved in industrial engineering.

The **Advances in Civil and Industrial Engineering (ACIE) Book Series** aims to present research and methodology that will provide solutions and discussions to meet such needs. The latest methodologies, applications, tools, and analysis will be published through the books included in **ACIE** in order to keep the available research in civil and industrial engineering as current and timely as possible.

COVERAGE

- Production Planning and Control
- Hydraulic Engineering
- Transportation Engineering
- Ergonomics
- Earthquake engineering
- Quality Engineering
- Structural Engineering
- Materials Management
- Engineering Economics
- Operations research

The Advances in Civil and Industrial Engineering (ACIE) Book Series (ISSN 2326-6139) is published by IGI Global, 701 E. Chocolate Avenue, Hershey, PA 17033-1240, USA, www.igi-global.com. This series is composed of titles available for purchase individually; each title is edited to be contextually exclusive from any other title within the series. For pricing and ordering information please visit http://www.igi-global.com/book-series/advances-civil-industrial-engineering/73673. Postmaster: Send all address changes to above address.

Titles in this Series

For a list of additional titles in this series, please visit:
https://www.igi-global.com/book-series/advances-civil-industrial-engineering/73673

Handbook of Research on Perception-Driven Approaches to Urban Assessment and Design
Francesco Aletta (University of Sheffield, UK) and Jieling Xiao (Birmingham City University, UK)
Engineering Science Reference • ©2018 • 641pp • H/C (ISBN: 9781522536376) • US $295.00

Convective Heat and Mass Transfer in the Free Atmosphere Emerging Research...
Kaliyeva Kulyash (Lorraine University, France)
Engineering Science Reference • ©2018 • 127pp • H/C (ISBN: 9781522530503) • US $135.00

Innovative Applications of Big Data in the Railway Industry
Shruti Kohli (University of Birmingham, UK) A.V. Senthil Kumar (Hindusthan College of Arts and Science, India) John M. Easton (University of Birmingham, UK) and Clive Roberts (University of Birmingham, UK)
Engineering Science Reference • ©2018 • 395pp • H/C (ISBN: 9781522531760) • US $235.00

Dynamic Stability of Hydraulic Gates and Engineering for Flood Prevention
Noriaki Ishii (Osaka Electro-Communication University, Japan) Keiko Anami (Osaka Electro-Communication University, Japan) and Charles W. Knisely (Bucknell University, USA)
Engineering Science Reference • ©2018 • 660pp • H/C (ISBN: 9781522530794) • US $245.00

Handbook of Research on Trends and Digital Advances in Engineering Geology
Nurcihan Ceryan (Balikesir University, Turkey)
Engineering Science Reference • ©2018 • 765pp • H/C (ISBN: 9781522527091) • US $345.00

Intelligent Vehicles and Materials Transportation in the Manufacturing Sector...
Susmita Bandyopadhyay (University of Burdwan, India)
Engineering Science Reference • ©2018 • 230pp • H/C (ISBN: 9781522530640) • US $155.00

Recent Advances in Applied Thermal Imaging for Industrial Applications
V. Santhi (VIT University, India)
Engineering Science Reference • ©2017 • 306pp • H/C (ISBN: 9781522524236) • US $205.00

For an entire list of titles in this series, please visit:
https://www.igi-global.com/book-series/advances-civil-industrial-engineering/73673

IGI Global
DISSEMINATOR OF KNOWLEDGE

701 East Chocolate Avenue, Hershey, PA 17033, USA
Tel: 717-533-8845 x100 • Fax: 717-533-8661
E-Mail: cust@igi-global.com • www.igi-global.com

Table of Contents

Detailed Table of Contents

Since it spread on the market, the professionals consider building information modelling as one of the most efficient methods to handle and manage a building and its entire life cycle, including costs, energy simulation, construction production data, and more. Thanks, also, to the legislation of the past few years, it is possible to say that this approach is known almost by everyone in the new construction field, and its employment is growing. The current use of BIM software is mostly referred to new buildings, made of regular elements and standard parameters. Is it possible to use the BIM process for the maintenance and the conservation of cultural heritage? The only way to answer this question is to research the academic environment, starting to train the professionals of tomorrow earlier, and proposing interesting cases studies on the subject. This chapter explores BIM and cultural heritage.

The authors investigated issues of geometric interoperability for reusable BIM components across multiple platforms using industry foundation classes (IFCs) which many proprietary BIM software platforms claim to fully support. These assertions were tested, first in 2012 and then in 2017 to assess the state and evolution of interoperability in the industry. A simple test model was created representing significant types of geometry encountered in component libraries, which were then expressed in IFC files. In the 2012 study, 11 commonly used BIM tools showed a dramatic failure to process the geometries as intended, indicating that the authoring tools, whilst technically capable of supporting required component geometric representations, were constrained from doing so by their conversion interfaces with IFC geometries. In the 2017 tests, improvements were observed though there were still significant processing failures that could result in serious errors; particularly in the case of the BIM library components imported into project design models.

A novel approach to higher dimensional spatial database design is introduced by replacing the canonical solid–face–edge–vertex schema of topological data by a common type SpatialEntity, and the individual "bounded-by" relations between two consecutive classes by one separate binary relation BoundedBy on SpatialEntity defining an Alexandrov topology. This exposes mathematical principles of spatial data design. The first consequence is a mathematical definition of topological "dimension" for spatial data. Another is that every topology for spatial data is an Alexandrov topology. Also, version histories have a canonical Alexandrov topology, and generalizations can be consistently modeled by continuous foreign keys between LoDs. The result is a relational database schema for spatial data of dimension 6 and more, seamlessly integrating space-time, LoDs, and version history. Topological constructions enable queries across these different aspects. Giving points coordinates amounts can give rise to topological inconsistencies which can be measured with topological invariants.

Despite significant progress for the adoption of BIM in AEC, currently its adoption for FM has been sparse, scarce, and extraneous. There are few cases in the world where robust adoption has taken place that are able to demonstrate success and are willing to disseminate the positive impact of BIM FM on sustainability, operational efficiency, and cost reduction. To date, there is no approach, motivation, or support in place to enable the extensive adoption of BIM for FM worldwide. In the UK, for instance, the UK BIM initiative, mandate, and the Digital Built Britain cannot count on the participation of FM stakeholders; the government has only started promoting initiatives that could trigger an extensive BIM approach, generating benefits for organizations and more importantly, society as a whole. In this chapter, data from authors' various research projects has been put together to generate an agenda for BIM FM implementation. The findings reveal that unless an intervention, such as a mandate for FM services suppliers, is put in place, very little will happen with regards to BIM FM.

The 3DIR project investigated the use of 3D visualization to formulate queries, compute the relevance of information items, and visualize search results. Workshops identified the user needs. Based on these, a graph theoretic formulation was created to inform the emerging system architecture. A prototype was developed. This enabled relationships between 3D objects to be used to widen a search. An evaluation of the prototype demonstrated that a tight coupling between text-based retrieval and 3D models could enhance information retrieval but add an extra layer of complexity.

Building information modelling is further globalizing architecture, engineering, and construction (AEC) professional partnerships. However, little is known on the effect of cultural and human factors on BIM-enabled visualization applications. This desktop study examined the extant literature on factors relating to application of BIM-enabled visualization technologies as a process that can improve, leverage, and conduct visual communication for coordination during implementation of global projects. It identifies BIM-enabled visualization having the capability in facilitating knowledge flows in complex discontinuous working environment of a property development's life cycle, and supports designers' understanding in its early working phases. This chapter presents the development of a theoretical proposition for embedding local

Preface

BACKGROUND

This book summarises some key research efforts in an area which is quite topical now and strives to bring together historically two distinct areas, i.e. 3D Drawing and Information Modelling. 3D traces its origins to the developments in Computer-aided Drawing/Draughting (CAD) whereas Information Modelling has been a long-established field in information management and computing. 3D modelling was an exciting development back in the mid-70s when solid modelling emerged almost simultaneously out of research at Cambridge University in the UK and Stanford in the USA. This was about a decade after Ivan Sutherland at MIT had invented Sketchpad for 2D CAD in 1963. Right through the 70s, several research projects were undertaken around the world to make the 2D CAD robust enough for commercialisation and it was in 1982 that Autodesk was founded which still remains the market leader in 2D and 3D CAD and 3D Information modelling (more commonly called Building Information Modelling in the AEC sector). The latest and the popular incarnation of 3D Solid Modelling is BIM which currently is extremely topical and is taking the whole AEC world by storm around the world. BIM is the culmination of intensive research efforts in the visual 2D and 3D modelling and Product Modelling through the 90s which addressed the standardisation of building element schemas to facilitate interoperation between different systems. BIM is essentially an implementation of *object-based parametric modelling* approach to system design and implementation. Object-based parametric modelling was first commercialised in the field of mechanical systems design. So, several parallel developments from the 60s onwards converged to result in 3D information modelling that is so ubiquitous today. The whole idea behind 3D information modelling has been to fuse information with visualisation resulting in quite powerful approaches to eliciting, understanding and exchanging information.

BIM has become a major topic of interest for the construction industry around the world in recent years. There are several high-profile examples of BIM-driven project delivery that different organisations have claimed to have delivered over the recent past. At the same time, there has been a considerable amount of misunderstanding about what exactly a BIM-driven project implies. The clear majority of practitioners appear to take the usage of BIM software in some aspects of the lifecycle of built asset procurement for it to qualify to be a BIM-enabled asset delivery. Lately, it has been pointed out by several organisations and experts that whilst the 'lonely' use of BIM technologies does accrue benefits largely to the organisations using them, it does not necessarily benefit the whole project and the end client organisation. To achieve benefits at the project level and to the end clients over the entire lifecycle of the built asset, *Collaborative BIM* needs to be implemented with appropriate processes, information exchange standards and contractual protocols in place. The implications of *Collaborative BIM* are far-reaching and could potentially alter the relationships between the key stakeholders profoundly. There are also potential implications for the government bodies as well as the society at large. As increased amounts of digital information get captured by utilising a collaborative BIM approach, several hitherto unknown inferences could be drawn about the performance of built assets including their users' behaviour patterns giving rise to several policy level and strategic decisions about their procurement, operation and maintenance and future capital planning processes. It is, therefore, no surprise that governments around the world are actively adapting their procurement processes to suit a Collaborative BIM-enabled approach. In the UK, an entire set of processes and standards for information exchanges, generally referred to as Level 2 BIM have been made mandatory in all publicly procured projects from 2016. 3D Information Modelling comprises of different components and the following paragraphs will shed some light on some of these.

INFORMATION MODEL

An information model in software engineering is a representation of concepts and the relationships, constraints, rules, and operations to specify data semantics for a chosen domain of discourse. Typically, it specifies relations between kinds of *things* or objects, but may also include relations between these individual *things*. The term *information model* in general is used for models of individual things, such as facilities, buildings, process plants, etc. In those cases the concept is specialised to facility information model, building information model, plant information model, etc. Such an information model is an integrated *view* of the facility that the data and documents about the facility relate to.

GEOMETRIC AND BUILDING DATA MODEL

In order to appreciate the power of 3D information models it is important to understand the difference between the traditional geometric data model as contained in a 2D-CAD drawing as opposed to a building model typically found in BIM models.

The main difference is the ability to 'model' concepts like *space* in addition to physical objects. This requires the ability to include information regarding topology (i.e. connectivity between different elements of a building). For example, a *wall* may enclose a *space* and may be connected to other *walls*. In addition, even a concept like *space* can be dealt with and reasoned with by storing and manipulating various *parameters* that define its major characteristics like its function (office space, circulation space etc.), its area, its occupants etc. This is exactly what a BIM system does in order to model *spaces*.

3D INFORMATION MODELLING-BASED DESIGN

A brief discussion of information model as well as geometric and building data models, can now be followed up with a more practical application of these approaches called Model-Based Design (MBD). MBD is a mathematical and visual method of addressing problems associated with designing complex systems like buildings, plants or indeed any other product benefits of this approach over the more traditional drawing-based design. Among other things, model-based design provides an efficient and cost-effective approach for facilitating seamless communication and hence integration throughout the design and construction processes. Latterly, the developments in internet technologies has facilitated model-based design approaches by allowing retrieval of design related information available on the web and seamlessly integrating distributed applications as web-based services within CAD systems. This approach allows for incorporating the most up to date information in a design whether that relates to individual elements of the design or any other aspect of design such as perhaps latest regulatory information. Model-based design can also be used to link a design system with an online *catalogue* of products which can be simply *dropped* into the design model with some simple mouse clicks. Once the products or elements are *dropped* into the models, they can then be manipulated as if they were an integral part of the original model itself. Several design software vendors have produced such technologies (like Autodesk's i-Drop or Google's 3D Warehouse). This makes the communication and exchange of information between different designers and manufacturers so much more seamless, timely and cost-effective. Further downstream, it is then possible to carry out parametric and other analyses by manipulating these products by interchanging them with perhaps other manufacturers and choose the best possible solution for the client.

BUILDING INFORMATION MODELLING (BIM)

So, with the background provided in the preceding paragraphs and a very brief introduction to model-based design earlier, it is now an opportune time to introduce BIM per se. However, before that it should be mentioned that one of the problems with the whole BIM 'thing' is that BIM technology hit the scene way before BIM processes, protocols and standards. The upshot of this is that someone familiar with BIM technologies takes it as such and no more. In fact, it would arguably be more appropriate if the 'B' in BIM should stand for 'Building' as a verb rather than a noun implying its scope to be vertical (building) structures as well as horizontal (civil infrastructure) structures like bridges, tunnels etc. Similarly, the 'IM' in BIM should more appropriately stand for 'Information Management' rather than 'Information Modelling'. The wider connotation of BIM is much bigger than just modelling and has huge implications for overall information management for the entire asset lifecycle from inception to demolition. This includes modelling but is not restricted to it. In fact, a lot of the misconceptions about BIM arise because of this historical fact than anything else. Therefore, BIM normally means not just the technology but also the associated set of processes, standards and protocols for information exchange.

Finally, a brief word about distinguishing BIM from 3D CAD. In CAD, the core entity is a *drawing*. However, in BIM the core entity is *building objects* with attached information as parameters as well as rules about their behaviour. BIM is based on *Object-based parametric modelling* as briefly described before. This is the key difference between BIM and CAD.

OUTSTANDING CHALLENGES AHEAD

In the AEC industry, BIM was probably implemented in new build projects in the USA before most other countries. However, it would appear that for most part this was driven by using BIM technologies without much insistence on information management processes and other softer aspects of BIM implementation like information management processes and contractual protocols. This led to some issues (like sharing of liability due to shared ownership of 3D information models) in some cases that prompted some other countries, most prominently the UK, to develop structured information management processes including contractual diktats before mandating the use of BIM in projects. The UK approach, commonly known as Level 2 BIM, promises to deliver significant benefits to all stakeholders in the industry although jury is still out as robust cost-benefit studies are yet to be carried out on large projects.

Despite the popularity of 3D Information Modelling (or BIM) and a global acceptance of their robustness and utility as the preferred platform for information management throughout the lifecycle of an asset, there remain a number of challenges still to be resolved. These challenges could be grouped into two major categories. There are significant technical challenges as well as arguably the more intractable problems relate to the *softer*, non-technical aspects of their implementation like information management processes, information exchange standards and contractual issues. Although not exhaustive by any means, some of these challenges are briefly discussed next.

Technical Challenges

As is well-established, existing buildings constitute majority of the building stock in any country. If the entire sector is supposed to be implementing BIM, it is imperative that BIM-based 3D models should be implemented for the entire building stock. Clearly, this is a mammoth task not just for the volume of effort required but also due to technical challenges. So far, a fully or even semi-automatic method of converting a scan (e.g. 3D laser scan) of an existing asset into a full-blown object-based parametric model does not exist. This is an even bigger challenge for older, historical buildings due to lack of standard components and obscure construction methods used in the past. Therefore, creating retro-fit 3D information models for existing buildings in an efficient manner is still a major challenge although there are several research projects underway to address this problem.

Another technical issue that has plagued the application of BIM technologies has been issues surrounding re-usability of building components. One of the main perceived benefits of using these technologies is seamless exchange of information. This obviously relies on smooth, error-free interoperation between systems. However, when one attempts to use a BIM component across different platforms, one still encounters issues like loss of information and erroneous translation of information. Intensive academic and commercial research over the years have resulted in significant improvements. However, there still remains more work to be done.

Besides, these near-term challenges mentioned above, there are several medium to long term challenges that the research community needs to address so the industry can realise the full potential of 3D information modelling. Some of these include

Big Data analytics, Internet of Things (IoT) and Cyber Physical Systems. It will be considerable length of time before all these challenges reach a level of maturity and commercial robustness that would benefit the industry and society at large.

Non-Technical Challenges

Arguably the most significant potential benefits from the implementation of BIM can accrue beyond the construction phase of an asset. It is well known that a huge (up to 70%) majority of lifecycle costs of an asset transpires during the operation and maintenance phases of an asset. Despite this well-established fact, adoption in FM (Facilities Management) remains relatively low compared to design and construction phases. This is clearly an important challenge that needs to be addressed urgently.

Another major non-technical challenge related to the value created by BIM. Many critics are not convinced if BIM provides any significant benefits in relation to costs incurred. Unfortunately, there is little (if any) evidence provided so far to conclusively demonstrate the monetary benefits of implementing BIM over the lifecycle of an asset. There are some empirical, anecdotal figures mentioned by some organisations but a major longitudinal study is needed to address this problem properly.

Finally, it is generally accepted that certain procurements routes and methods are more suited to BIM-based project implementation than others. This has been an active area of research and several approaches have been proposed. However, for a truly collaborative and shared ownership of models to be successfully implemented there may well be a requirement for more innovative procurements methods yet to be developed.

ORGANISATION OF THE BOOK

This book presents ten contributions from some of the leading researchers from 3D information modelling world. The contributors are drawn from authors who have already published in *International Journal of 3D Modelling* (IJ3DIM) and in most cases their contributions are significantly enhanced versions of their earlier publications in IJ3DIM. These contributions have been carefully chosen so each one of them addresses one or more challenges mentioned before.

In Chapter 1, Achille et al. contend that the professionals consider Building Information Modelling as one of the most efficient methods to handle and manage a building and its entire life cycle, including costs, energy simulation, construction production data, and more. They further suggest that due to the BIM-related legislation in recent years in many countries its implementation is growing fast.

However, vast majority of the current use of BIM software is mostly confined to new buildings comprising of regular elements and standard parameters. They raise the question whether it is possible to use the BIM process for the maintenance and the conservation of Cultural Heritage and provide some interesting answers.

In Chapter 2, Benghi and Greenwood present an investigation into issues of geometric interoperability for reusable BIM components across multiple platforms using Industry Foundation Classes (IFCs). They tested the reusability of some components, first in 2012 and then in 2017 to assess the state and evolution of interoperability in the industry. They created a simple test model representing significant types of geometry encountered in component libraries, which were then expressed in IFC files. They point out that in the 2012 study eleven commonly-used BIM tools showed a dramatic failure to process the geometries as intended, indicating that the authoring tools, whilst technically capable of supporting required component geometric representations, were constrained from doing so by their conversion interfaces with IFC geometries. In the most recent set of tests they carried out in 2017, they observed improvements though there were still significant processing failures that could result in serious errors; particularly in the case of the BIM library components imported into project design models. The chapter provides important insights into this very important area of research of practical significance and application.

Bradley and Paul propose a novel approach in Chapter 3 for the design of higher dimensional spatial databases by replacing the canonical Solid–Face–Edge–Vertex schema of topological data by a common type SpatialEntity, and the individual "bounded-by" relations between two consecutive classes by one separate binary relation BoundedBy on SpatialEntity defining an Alexandrov topology. A highly technical and fundamental contribution, the chapter explicates on mathematical principles of spatial data design. The authors claim several contributions of their work, the first being a mathematical definition of topological "dimension" for spatial data as well as their important conclusion that every topology for spatial data is an Alexandrov topology. They conclude that one important upshot of this conclusion is that topological constructions enable queries across different aspects of the stored spatial data.

Codinhoto et al., in Chapter 4, contend that despite significant progress in the adoption of BIM in AEC, currently its adoption for FM has been sparse, scarce and extraneous. In their opinion, there are few cases in the world where robust adoption of BIM in FM has taken place demonstrating success and positive impacts of BIM FM on sustainability, operational efficiency and cost reduction. They claim that in the UK, the UK BIM initiative mandate and the Digital Built Britain, cannot count on the participation of FM stakeholders. The authors present data from their various research projects to generate an agenda for BIM FM implementation. In final conclusion, they suggest that unless an intervention, such as a mandate for FM services suppliers is put in place, very little will happen with regards to BIM FM.

In Chapter 5, Demian et al. describe their 3DIR project which investigated the use of 3D visualisation to formulate queries, compute the relevance of information items, and visualize search results. Based on the user needs elicited from many workshops, they created a graph theoretic formulation to inform the emerging system architecture and a prototype was developed. The main objective of the work was to enable relationships between 3D objects to be used to widen a search. They conclude that their prototype demonstrated that a tight coupling between text-based retrieval and 3D models could enhance information retrieval but also add an extra layer of complexity.

In Chapter 6, by Ghafar et al. suggest that despite little understanding of the effect of cultural and human factors on BIM enabled visualisation applications, BIM is globalising AEC professional partnerships. They describe their desktop study which examined the extant literature on factors relating to application of BIM enabled visualisation technologies as a process that can improve, leverage and conduct visual communication for coordination during implementation of global projects. They claim that their study identifies BIM enabled visualisation as having the capability for facilitating knowledge flows in complex discontinuous working environments of a property development's life cycle and supporting designers' understanding in its early working phases. This chapter presents the development of a theoretical proposition for embedding local work culture etiquette in BIM enabled visualisation application for augmenting dynamic knowledge transfer among discontinuous members in a building project. The authors believe that the results of their study are expected to benefit rapidly developing countries, e.g. Malaysia, in enabling successful partnerships with counterparts from developed countries.

In Chapter 7, Jowett et al. address the significant issue of mapping BIM-enabled workflows to appropriate procurement routes and discuss its implications on value creation for the stakeholders. The authors present an overview of BIM tools and workflows, new procurements methods, and emerging management practices and how they are being adopted on projects to overcome collaboration barriers to improve project performance within the Architecture, Engineering, Construction and Operation (AECO) sector. They claim that although the use of collaborative procurement methods such as Design and Build (DB) procurement and Integrated Project Delivery (IPD) when adopting BIM workflows are recommended by various people, there is little operationalization and empirical evidence of the value realization potential when using BIM in conjunction with these procurement methods. Their study draws upon five case studies of BIM-based DB projects to analyse and quantify the potential of value realization. The authors conclude that their results reveal potential hurdles inhibiting BIM from reaching its full potential and they propose changes to the current processes to facilitate BIM in enhancing value on DB projects.

Chapter 8 by Rossman et al. is the only contribution in the book from outside the AEC domain. The authors present how simulation applications benefit from realistic and exact forest models. They contend that these benefits range from training simulators like flight or harvester simulators to economic and ecological simulations for tree growth or succession. Their nD forest simulation and information system integrates the necessary methods for data extraction, modelling and management of highly realistic models. The authors present the use of semantic world modelling and show that tree data can efficiently be extracted from remote sensing data even for very large areas. In their approach, data is modelled using a GML-based modelling language and an integrated, flexible data management approach provides caching, persistence, a central communication hub and a versioning mechanism. The authors show that by combining various simulation techniques and data versioning, their nD forest simulation and information system can provide applications for historic 3D data in multiple time dimensions (hence nD) as well as with predicted data based on simulations.

Saleeb, in Chapter 9, provides an interesting account of research tests and experiments that have provided evidence for the disparity between human perception of space in the physical and 3D virtual environment. The author suggests that this disparity could have dire effects on the decision-making process throughout the whole construction lifecycle of an asset due to non-precision of perceived spaces. Quite alarmingly, some results have shown an infidelity in displaying the actual dimensions of the space in the 3D virtual environment and previous research by the author has identified the magnitude of this disparity. Despite the apparent importance of this issue of disparity, it would appear that there is little consensus and conclusive reasoning proposed by anyone behind the causes for this disparity. This chapter presents an investigation of this issue and highlights different psychophysical factors that might cause this difference in perception and compares these factors with previous research.

The final chapter by Underwood et al. reports on a practical application of BIM by providing an account of a project between the University of Salford and Links FF&E - a design, manufacture and fit-out company in the UK. The reported project is aiming to implement BIM as a catalyst for a Lean transformation to streamline processes and operations through the adoption of a case study methodology on a Design for Manufacture and Assembly (DfMA) BIM implementation at the company. The authors present the key findings of this work which highlight that the challenges for smaller companies adopting disruptive technologies could be mitigated with a business case that considers the changes on organisational processes and workflows by embedding technologies within the company with the focus on eliminating waste in the processes and thereby adding value.

I hope readers will find this state-of-the-art book stimulating and a valuable resource for their research. I believe this book will be a valuable reference material for academics and practitioners alike interested in 3D information modelling in one way or another.

Bimal Kumar
Glasgow Caledonian University, UK

Chapter 1
BIM and Cultural Heritage: Compatibility Tests on Existing Buildings

Cristiana Achille
Politecnico di Milano, Italy

Nora Lombardini
Politecnico di Milano, Italy

Cinzia Tommasi
Politecnico di Milano, Italy

ABSTRACT

Since it spread on the market, the professionals consider building information modelling as one of the most efficient methods to handle and manage a building and its entire life cycle, including costs, energy simulation, construction production data, and more. Thanks, also, to the legislation of the past few years, it is possible to say that this approach is known almost by everyone in the new construction field, and its employment is growing. The current use of BIM software is mostly referred to new buildings, made of regular elements and standard parameters. Is it possible to use the BIM process for the maintenance and the conservation of cultural heritage? The only way to answer this question is to research the academic environment, starting to train the professionals of tomorrow earlier, and proposing interesting cases studies on the subject. This chapter explores BIM and cultural heritage.

DOI: 10.4018/978-1-5225-5625-1.ch001

INTRODUCTION

The aim of the research, in progress today, is to investigate, with real data and real case studies, the BIM process and highlight the difficulties to improve it. In particular, the target is to test different methods and software, which satisfy the requirements of the new buildings, and see if they can also be successfully used in complex situations. The restoration work on Cultural Heritage, where it is necessary to consider the unicity of the architectural and archaeological elements, the construction systems and materials (not standardizable as the new ones), the damage to the structures and the environmental conditions, represents definitively, a prominent example.

Nowadays, there is a constant diffusion of BIM systems among different fields of application. Building Information Model is not a single software, but it is a process that supports both the information sharing and the maintenance of buildings, during their life-cycle. Harpaceas (2017) states that the BIM modules are three: "Authoring", including all the software that build the 3D model; "Tools", used for the computes and the information about the materials in the construction site; "Review", checking the final model implemented by all the information, ensuring that the structural, architectural and plant models correspond.

It is clear that the potential which, until now, has been used mostly in the new constructions field, can be adapted and refined in the CH area.

The restoration work aims at the conservation of the authenticity of the object, handing down the object as it has come to us, in its aesthetic, artistic and historical reach. It treats the built as a historical document, carriers of information that, even if not recognised, cannot be lost. As with a historical record, it must be guaranteed readability to future generations as well.

Ultimately, the conservation project is a sustainable project, considering that the preservation of the material is obtained by keeping it in situ, with the use of eco-friendly materials and techniques, entail the reduction of the risks (including environmental) and the costs associated with the demolition and the management of the wastes.

The maintenance work could be considered as an epistemological problem: during its development it is necessary to choose how deeply must be its knowledge, starting with the ontological question of the object itself. It means that, from one side, it is necessary to define how and how much deeper this object can be described. On the other hand, if there is a need to examine it in its entirety or considering different levels of knowledge for each portion or element, (planning survey activities according to the aim of the project).

Any conservation project, as well as a maintenance project, requires the knowledge of the object and requests specific competencies and equipment, whose application needs a schedule to optimising resources, including the financial ones.

An indispensable element of this process of knowledge is the graphic representation (plans, sections, profiles, and more) of the building. It is possible to consider this aspect as an “a posteriori” description of the structure, while the BIM process for the new constructions feels an “a priori” model of the building. The purpose of the BIM process it to represent the new architectures “as built”. The care in describing each element, in its geometry and shapes, and even in its manifest degradation or deformation, offers the possibility to come into “a confidential level” (borrowing the concept from the NTC – Technique Norms for Construction - 2008) with the building or the single built element. The “confidential level” invites to define the level of knowledge arose among architects and engineers during the structural evaluation of existing buildings. From this “level” depends the safety factor assumed for the structural assessment. The “confidential level” concept developed inside the Italian Code is as useful draft or outline to determine the “level of knowledge” inside the BIM process for the conservation project.

Precisely, because it refers to something that is already built and it is tangible, with its history, the conservation project, which also includes reuse, cannot suggest ideal or “typed” representations. The analysis is, in fact, always directed to something that, in its seriality (for example, repeated elements, hand-made or not) is recognised as singular, because it is affected by the temporal dimension.

The conservation work, based on respect for the historical value of each element of the building, sets itself as the historical stage of the building and its life cycle. It has, therefore, its documentary value in the historical sense and cannot be assumed only as a functional instrument for the study and execution of the project.

The BIM process for the new building contemplates all project and executive activities, providing a parameterization of the object itself. Parameterization facilitates the interoperability of the process.

In conservation, however, the main difficulty lies in the initial acquisition of information and data use to describe the actual state of the object.

It is easier, however, is the phase when the conservation work is developing the project of the utilisation of the old building, considering new technologies to conserve in situ the existing materials and to enhance the building techniques, when it is possible to find standardised solutions of which a metric estimation can be made.

In fact, the descriptive standards adopted for new products cannot be applied to existing ones: the processes that characterise new materials and technologies cannot be transposed into the study of old equipment and techniques, of which there is no standardised description.

The stndard for existing constructions is defined by specific international and national norms such as RIEM, ISO and UNI-EN and UNI-Cultural Heritage (ex Nor.ma.L, UNI Normal Cultural Heritage Code, 2006).

The difficulty of describing the existing building emerges when it comes to dealing with its problems and unknowns, and with the training of the specialists in charge to develop these studies.

The research starts from the necessity to apply the BIM's process on an existing building that could be or not be Cultural Heritage.

The interesting on the BIM procedures is connected with a changing scenario involving the construction labour market and the roles of all the actors participating in the building process, as owners, contractors, public administration, architects, engineers, professionals in general.

The main question is developed mainly from the methodological and disciplinary point of view and to reinforce the practice in this process, considering not only the new buildings but, especially, the existing ones.

In Italy, one of the main questions is, according to the Italian law, to deliver the restoration work to specific companies, legally recognised and licensed by the government, and hired through national /international public tenders.

Given that the Italian market trend prevails the intervention on the existing building rather than the construction of the new one, it becomes essential to understand how the BIM can work on the conservation/restoration project without losing the consolidated know-how and methodologies in the preservation fields. This necessity is linked to - the urgent need of anti-seismic security, which requires awareness of the state of conservation and the construction techniques employed (as required by NTC 2008); - the implementation of the National standards for energy certification of buildings (DM 26/06/2009). This fact implies that the object is detected and returned-in from the geometric consistency, including cracking and deformations, and from the technological and construction characteristics, which are strictly necessary for the required evaluations.

It is also wondered how the project on the existing building, can be transposed by an ICT (Information and Communication Technologies) method such as BIM, conceived for the design of new Buildings, at a time when:

- This method is made mandatory by the international and Italian rules for public contracts;
- The method is a substitute and, only partially, supplementary to the "traditional" design methods;
- An important training activity at a national and international level has begun, involving the adoption of this approach by professionals in the field (designers, technicians, public administrators, contractors and architects, builders);
- The assumption of this procedure implies a change in collaboration in the project phases, imposing a reorganisation that starts from the assumption

of responsibility of the project itself (which must, for the restoration, be delegated, necessarily to the architect);
- It is necessary to rethink methodologies and approaches to teaching, training and learning at the levels of school, university, and lifelong learning.

What is attractive to the potentiality offered by the BIM technology is the methodical approach not only for the interventions needed but also for the management of the cultural Heritage, with the activities to monitor its health condition and the maintenance.

The working method realised using these instruments would allow the achievement of these tasks:

- The digital representation of the existing building;
- An informative database about all the different elements of the heritage;
- A recurring checking of the health condition of the legacy;
- The evaluation of the effects of the different typology of intervention.

The management of existing site has different needs from the new buildings. Therefore it is important to know the precise information that the model must contain:

- Geo localization
- Shape of the element
- Materials
- Different layers of materials
- State of decay
- Historical information
- Environmental conditions

The "retrospective reading" of the object is always seen as something particularly burdensome and, unfortunately, not fundamental. A standard or canon do not define "Retrospective reading". The protection of built cultural heritage, existing buildings in general, and old centres, with their urban features, follow the indications provided by the Restoration Charters, with particular reference to the Italian one, edited in 1972 (Carta Italiana del Restauro 1972).

In fact, the Restoration Charter, as well as Manuals of Restoration and researchers based on the material-energetic-structural performance of existing buildings, show that it is hard to define a standard for materials and techniques.

For example, masonry walls or wooden structures are non-specific definitions unable to describe their construction or their arrangement and of their mechanical (and energetic) behaviour.

Only the survey procedure can be standardised. Designations or terminologies can be standardised, but parameters, especially those related to performance, require interpretation by interpolating an uncountable number of information, also obtained by indirect survey. Interpolation is possible only by the knowledge of the object and by the awareness of the limits of that knowledge and "confidential level" (Lombardini & Cantini, 2017).

The challenge is to maintain the representation (in all its forms) of the cultural process underlying the conservation project, to reduce the risk of misunderstandings of the object's shape and its material reach.

The search for completeness in the representation allows creating an interpretative model of reality that can denounce the "conscious" level of knowledge the designer has come to.

The objects constituting the existing construction, on the other hand, cannot be typed and normalised, not even in the stage of representation: every element, though repeatedly, has its originality and component of authenticity that is in its being man's work in his *Hic* et *Nunc*. These two aspects are indispensable and compel (they must force us) to assume nothing of this knowledge in a priori manner.

The BIM methodologies and the same software are designed for new constructions. Parameterization, LOD process and interoperability are analysed and standardised concerning the construction of new standardised facilities.

For the conservation project, the existing object is not the point of arrival but the starting point. In its existence, then, this represents the unacknowledged to be detected and studied.

His relief and his research can only come to a model that, though graphically detailed, needs continuous refinement in:

- Graphic information (thinking only to inaccessible structures, such as foundations or section descriptions of structures: wood, m properties of materials and structural functioning (useful for assessing building behaviour, even in case of earthquake);
- Parameters, both equipment and environments, for an energy assessment of structures.

The representation of the existing object is always a model and does not mean that the level of detail of the graphic design has to correspond to a certain degree of knowledge.

BACKGROUND

The current legislation and the continuous updating of the rules testify the great interest in this argument and the need to find a correct BIM methodology to apply to CH.

On January 15th, 2014, the European Parliament made official the adoption of BIM methodology in the public contracts thanks to the European Procurement Directive (EUPPD, 2014). In particular, they expressed the will of opening the public competition to innovative methodologies, as they wrote in the art.22 c.4: "For public works contracts and design contests, Member States, may require the use of specific electronic tools, such as of building information electronic modelling tools or similar." (EUPPD, 2014).

The Directive underlines many times the desire of producing research and pushing the market towards BIM methodology, where it is possible to cut the costs and the times and to control the project from the early phase of the design.

Speaking about the countries involved, Europe is now hosting the greatest concentration of government-led BIM programs in the world. In fact, Finland and Norway were first to set standards, followed by procurement policies from the UK, Netherlands and Italy; and most recently joint government and industry initiatives from France, Germany and Spain (Table 1).

What is certain so far, is that within five years the European Industry will move from digitally naïve to "digital natives"15 (EU BIM Task Group 2017). The primary target in Europe is to end the waste of public resources (money, materials, professionals, and others) for automatic procedures. This need was one of the leading causes of BIM approach spreading in all over our continent and the rest of the world.

Table 1. Some of the most developed BIM Guides and their latest updates (Cobim – Common BIM Requirements, 2012; Stastbygg BIM Manual (2013); British Standard Institution BSI – PAS (UK Government, 2016); UNI 11337, 2017; GSA – U.S. General Administration National BIM Standards, 2017).

State	Regulations	Latest Update
Finland	Cobim – Common BIM Requirements	2012
Norway	Stastbygg BIM Manual	2013
United Kingdom	British Standard Institution BSI - PAS	2016 (in development 2017)
Italy	UNI 11337	2017 (in development)
USA	GSA – U.S. General Administration National BIM Standards, National BIM Guide for Owners	2016 2017

These modern Information Technologies (IT) tools, applied to Cultural Heritage, offer several possibilities to organise the knowledge of the artefacts, to conserve, protect and valorise them. The general trend is to build dynamic and suitable tools to handle the information about the restoration and management process.

The digital archiving techniques, analysis and administration of the data, use the 3D model as a support where it is possible to georeference the material and topological information, the transformations in time, the construction and technical installations (as climate services, utility services, lighting, etc.). In this way, if all the archive is continually updated, it contributes to the life cycle management and the plan of work of the ancient building or archaeological site.

Many European countries create informative platforms that try to collect all the data about their CH. Some examples are the Kist o Riches by the School of Scottish Studies (University of Edinburgh, 2015); the CultureSampo developed by the Helsinki University of Technology (2015); the Italian SICaR (SiCAR, Sistema Informatico per la documentation e la Progettazione dei Cantieri di Restauro).

In particular, the Italian experience organises the historical and scientific literature to evaluate the interventions needed, the execution, the maintenance and the monitoring. This approach is a practical example of a system not only used for valorisation and tourism purposes but also useful for restorers and operators of the construction yards (Tommasi et al., 2016).

Many types of research are focusing their attention on building a system that can be adapted to all Cultural Heritage cases studies, even to the archaeological areas (Scianna et al., 2014). However, it is still tough to find it because each situation is very different from the others, having particular and unique needs and history. For this reason, the hard structured information system cannot be used in Cultural Heritage field, which requires more dynamic systems, capable of self-adapting to every case for covering all possible needs. Nowadays, it is possible to distinguish two different categories of software related to 3D models. So far, the BIM software are not able to solve all the issues that come from the modelling of the Cultural Heritage, but they need to be joined the integration with other instruments to reach a satisfactory result: 3D modelling software and BIM software. The first category includes the modelling software (as 3DStudio, AutoCAD, Rhinoceros etc.) that can manage real based 3D models. These have high complexity, great accuracy, high-resolution and heterogeneous features (both line based and surface based at the same time), represented through Mesh and NURBS. These models are directly created from 3D dense point clouds, coming from the survey.

On the other hand, speaking about BIM software, it is important to know that the most significant letter of the word BIM is the "I"- Information - as they add to the 3D buildings models an information system, namely connecting digital models with a different kind of information useful for building management.

Revit, for example, is one of best known "building design software specifically built for Building Information Modelling (BIM), with features for architectural design, (MEP and structural engineering, and construction" based on "parametric components". BIM software offer hard structured information system can connect data and information to the objects, with the possibility to perform queries and simulations on them. Inside the universe BIM, there is an experimental process called HBIM, explained by Dore and Murphy (Dore, Murphy 2013), that represents an example of BIM specifically directed to Digital Heritage. It is a new prototype library of parametric objects based on historic architectural data and a system of cross platform programs for mapping parametric objects into a point cloud and image survey data (Fai et al., 2011; Dore et al., 2012). From all the case studies considered emerged the fundamental features that all the BIM software ought to operate in Cultural Heritage:

- To be able to collect historical-cultural information to understand where to act;
- To guarantee the possibility of managing the survey data (CAD, point clouds), to ease the operation of virtual reconstruction of the building;
- To foresee the possibility of modifying or implementing, anytime and by any operator, to ensure both the update of all the information linked to the different elements and the validity of the evaluations on the present state of the heritage;
- To let the construction of an informative and open database, containing the indications about the materials and the constructive methodologies of the building;
- To ensure the exchange of different data between the various parametric or non-parametric software, thanks to the identification of some format (IFC, XLM, etc.).
- To make the BIM systems convenient in Cultural Heritage instead of using regular 3D modelling software, some operative issues needed to be resolved:
- Lack of interoperability between BIM and the technologies for the topographical survey; hard management of heavy data coming from surveying instrument (e.g. laser scanner, photogrammetry, etc.) such as massive point clouds;
- Simplification of the geometry of the monument;
- Complexity in the restitution of the unicity and the details of the historical elements;
- Inability to assign punctually accurate data to points placed on the surface of the object.

The need to ease the process of 3D model production and graphical elaborations for the CH became a goal not only for a correct reproduction of the existing monument but also for a complete comprehension of the hierarchical logic among its parts.

Nowadays the use of BIM for new constructions requires the development of platforms containing National database with technical, financial and scientific information useful for the civil construction process (as buildings, bridges, roads, all the facilities of the urban planning and others). As well known, the public construction goes from the ideation of the object through its execution and its management up to its "demolition" and renovation according to the concept of "life cycle" as previewed in the BIM methodology. The Cultural Heritage, like the monument, is included in this process but, according to the conservation approach, it is possible only considering the restoration of it, bypassing the demolition phase, saving the whole construction from the technical, constructive, material point of view, (bearing in mind, also, the existing utility services as water, electrical, gas supplies, drainages, etc.). Talking about HBIM (Dore, Murphy, 2013), the life cycle process going from the design to the end of the construction gets altered because it is necessary to substitute the demolition or the renovation phase with the restoration or conservation phase. Defining renovation as the possibility to change the existing materials, structures and any utility services with new ones, considering the conservation project as the saving of each existing elements of the construction itself using all the methodologies or materials able to avoid their substitution or the replacement.

The introduction of HBIM leads to a reflection on the BIM life cycle, dealing with BIM for the heritage implies to do maintenance and restoration on something which is heritage, as part of the widest range, that can be defined as the existing. In the life cycle of the monument (as the archaeological ruins) it is not anymore considered the end.

Thinking about sustainability and resources management, it is necessary to seek the conservation of the existing building, which could also be, a monument, extending its life under a strict and conscious control. As for the BIM for new constructions, also the BIM for the existing buildings requires the creation of a database that allows an immediate management of the project for the maintenance and the restoration through the BIM procedure. As already said (starting from a full range that is the existing, in which is pointed out the smaller range of older constructions and therefore even that of ancient buildings, called cultural heritage) the issue is to build an accurate database of the cultural heritage.

How is it possible to create this database starting from old constructive systems (structures and process plant engineering) and old materials that are not certified, unlike new ones? What is it possible to do, today? To assign to each drawn element, through a BIM software, a schedule that describes the object, or the portion of the subject matter under exam? This procedure may only work through an absolute standardisation of the terminology. Even if Italian and International legislation have codified the typologies of degradation of stone materials, there is no precise codification of the ancient materials and capable systems. In the same time, alphanumeric codes are not assigned to the decays listed in the Italian technical rules UNI (UNI Normal Cultural Heritage Code, 2006), or in the ICOMOS ISCS Glossary (ICOMOS, 2008). What does codification mean? It gives to each element an unequivocal denomination that corresponds to an alphanumeric code, which makes it possible to identify all the surveyed items in the existing construction (as old buildings, monuments, archaeological ruins, etc.).

In this sense, the schedules developed by the Central Institute for the Cataloguing and Documentation of the Italian Ministry may offer us a huge help so as it was for stones deterioration patterns, it is important that the materials and the efficient systems are defined by a group of experts unambiguously. What it is possible to do today is to build the object using a BIM program and assign it a schedule that has to be standardised, where are located standardised information and, also, personal comments about the project. In this way, it is possible to support the communication and the interdisciplinary process. This approach aims to optimise the duration of the design process that, in case of the existing, goes from the geometrical survey to the historical analysis, to the materials and construction techniques analysis, to the deterioration diagnosis, up to the intervention of conservation and restoration. There is, of course, a substantial difference between the process of restoration that can be done on a construction which has been built through a BIM process, which has already been computerized with all its elements and all the attached information, and the restoration of a building which has been done before the BIM introduction, which needs to be studied, analysed and surveyed in all its parts: all these parts have to be defined "a posteriori".

Why transposing the BIM in Cultural Heritage? One of the reasons, but not the only one, may be found in the fact that the spreading diffusion of this software will lead to having a lot of specialised professionals, able to use different kinds of instruments or methods. It is necessary to reorganise the information known today, to create this new alphanumeric database. The updating, into the BIM database, of the materials and techniques for the intervention project according to what has been said so far, could be simplified by the fact that they represent standardised products (Figure 1). Of course, the schedule may also contain other information, but it could not disregard the exact cataloguing of the object (ICOMOS, 2008; UNI Normal Code, 2006).

Figure 1. Workflow of the management of the data related to the construction materials in order of their implementation and their use by the BIM software: how the model is built

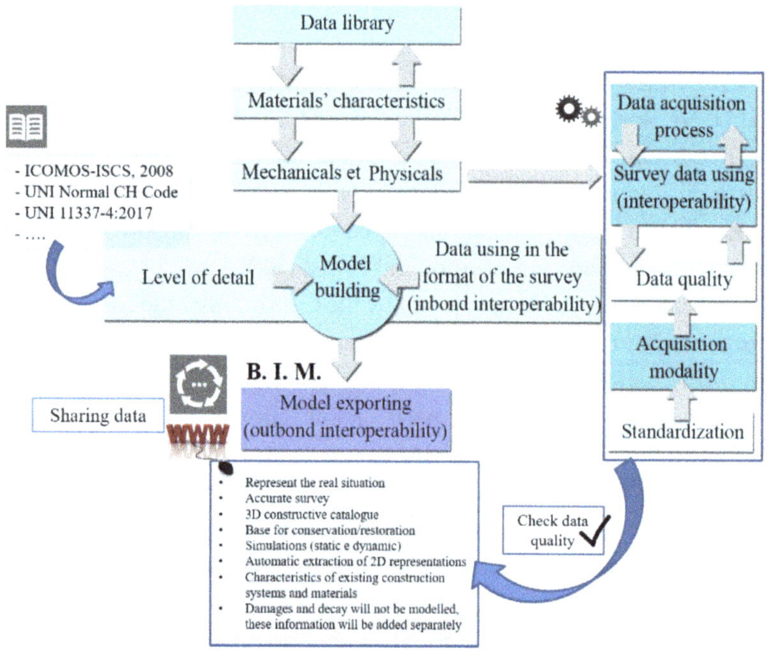

MAIN FOCUS OF THE CHAPTER

Issues, Controversies, Problems

The knowledge of CH needs to be accurate and detailed to ensure the preservation and the transmission to the future generations the historical values that they bring with them. For this reason, the survey plays a crucial role, as it is a knowledge process that helps to document the object, and it is the basis for a conservation project and further analyses. In fact, it will contribute to growing the archives of the available documentation.

A 'good survey' requires an awareness of the instruments and techniques, familiarity with software for managing and processing data and the clarity of purposes.

The availability of more refined survey technologies and methods allows you to choose which is 'the best' and most suited. The operators adopt different approaches and instruments depending on the peculiarities of the object to survey and the amount of information required.

The aims of these activities are the knowledge of size data, the geometrical characteristics, the materials, and the construction techniques used. The geometric knowledge of the objects investigated is the first step of the survey process, which today is possible with excellent results. In fact, the market is plenty of a vast number of remote sensing sensors (photogrammetric and laser scanner), available for mapping purposes and digital recording of Cultural Heritage. Regardless of the techniques or instrument to choose, the result of a survey with remote sensors is always a point cloud. This element is a rough data that can be very big and has to be processed before importing it into a modelling software. The BIM software is just a category of programs that can use these data, but there are a lot of different solutions, according to the needs of the operator (navigation and visualise, editing and modelling of point clouds).

If the aim of the researcher or the professional is to create a BIM model of an ancient building or site, the starting point will not derive by a new project, but it will be a survey data (Figure 2). For this reason, it is necessary for this kind of program to be able to handle these input correctly.

Until a few years ago, the management of point clouds was one of the biggest problems of BIM applied to Cultural Heritage. The most common problems that came across opening the point clouds in Graphisoft Archicad and Autodesk Revit (2012-2013) were:

- Impossibility to import big point clouds (with millions point), making it impossible to reach a high level of details;
- Inability to recognise the points as snap point and redraw the profiles;
- Failure to edit the surfaces created with the triangulation.

Figure 2. Comparison between the workflows for the modelling of a new building and an ancient one

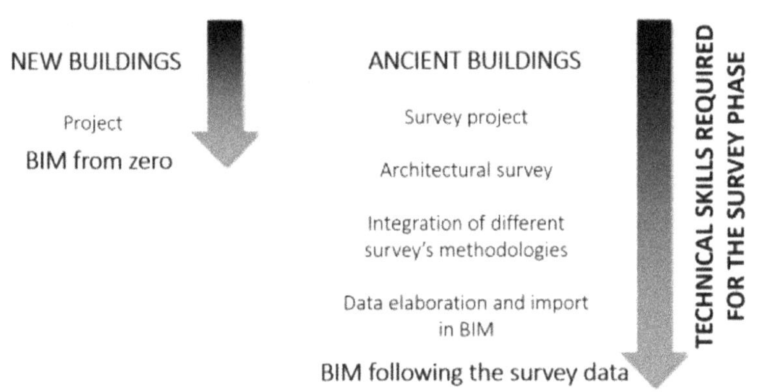

To try to overcome these problems, some plugins were available: for example, Green Spider for Revit developed by Garagnani (2013), and Cadimage for Archicad, which improved the functionalities of the original programs as the recognition of the points as a snap. Given that BIM is not a single software, but a process, it is impossible to have one unique program able to solve all the matters. For this reason, some software mediated between the instrument for surveying (e.g. laser scanner) and the BIM modeller (e.g. Archicad, Revit, Allplan, and others), e.g. one program created with this aim is Scalypso.

Today –2017/2018- on the other hand, BIM software can import point clouds, and the technical problem is almost overcome, but it's important to underline that is not easy to manage them. The focus now is on studying the best pipeline to handle the cloud, implementing the visualisation function given by the software. In fact, there are the different combination to use the point cloud in this kind of program, which leads to different results, and the choice of the methodology has to consider the needs of the building, the operator, the time at disposal and the budget involved.

The employment of the survey data is only part of the problem: "BIM" is a bond between the model and its information. The issues related to the management of

Figure 3. Changing of point cloud's import pipeline through the years. The number of phases is considerably reduced, and the result is improved. On the right different kind of visualisation of point cloud through the years and the different level of detail: from a tiny and not precise point cloud (2012/2014) to a highly detailed and dense point cloud with RGB values (2015/2016), without the need to intermediate with plug-ins

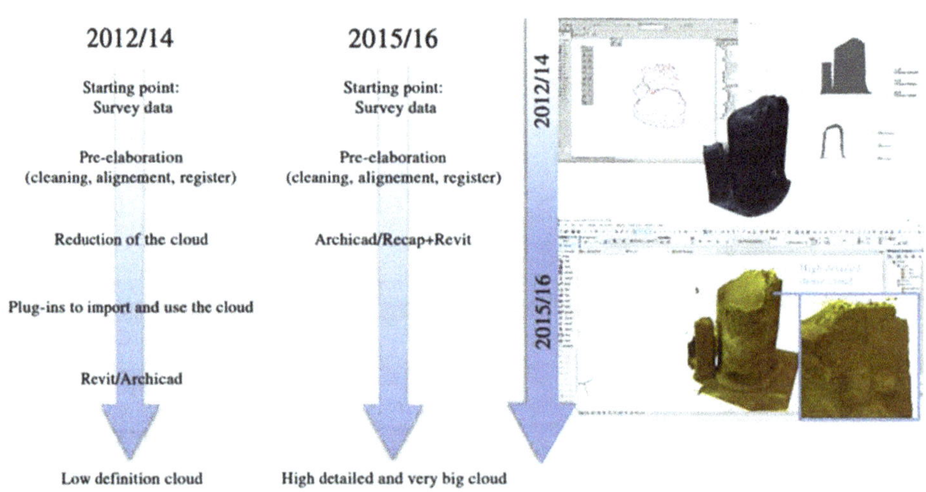

point cloud try to solve only the problems linked to the construction of the model, but they cannot define the features and the properties of the materials of the object.

The resolution of technical problems (as the ones related to the point clouds), contributes to the application of BIM process also for conservation projects, even if they differ from the new construction's ones.

The availability of tools with improved performance does not correspond to the correct solution of the problem. To operate correctly on the architectural and archaeological heritage, it is necessary to own an accurate preparation. There is a need to have a real competence of the theoretical principles that norms the conservation, the functioning of the tools (hardware and software) and, the ability to elaborate and handle the data.

The passage from the rough data to the elaborated one is not automatic. The elaboration is not easy because each case is unique. In particular, in the CH field, the introduction of information technologies for the managing of these problems has transformed the figure of the "classic" operator. Today, the network of men-instruments and workers has taken off the auto-sufficient and individual nature to the surveyor-conservator, who is now inserted in a team. Now the trend is to work in-group, together with other figures highly qualified belonging to other scientific fields, as it should happen in the BIM process, that, for this reason, seems to respond to the necessities of this area.

SOLUTIONS AND RECOMMENDATIONS

To operate in Cultural Heritage field, as in the past, also today the staff training is a crucial role:

"Safeguarding and valorising cultural heritage require specialised professional figures with a wide spectrum of knowledge and skills. In Italy, schools and universities provide the basic training for operators safeguarding and valorizing cultural heritage. Several universities offer three-year plus degree courses (3+2), possibly accompanied by further specialization. Though still based on the historical-critical and technical structure of traditional postgraduate courses and postgraduate schools in history of art, archaeology, restoration, the new schools and training possibilities offered in degree courses, now also foreseen for other subjects, pay greater attention to the legal and economic aspects and have a more incisive "in the field" training. Italian universities deal with several subjects linked to managing and using CH and also offer specific training courses for those wanting to handle aspects of historical, architectural and protected landscape Heritage in-depth. The really wide CH horizon, especially in Italy, justifies this ample teaching spectrum and variety of teaching structures. For this, Universities offer, like in this case, training courses to be done in addition to ordinary courses and which represent a real theoretical-practical, hands-on moment for students on specific subjects" (Achille et al. 2011)

The Universities are currently beginning to dispense curricular lessons about BIM to the students. However, this phase hasn't produced "BIM" graduates, yet, but the trend is to create post-graduate courses focused on BIM topics (e.g. "BIM Manager. Methods, Models and Applications" at Politecnico di Milano, or BIM Master at University of Pisa). The training plans that the professional field is obliged to hold, require a money and time waste (even if they are part of the mandatory credits for the professional updating, DPR 137/2012).

To avoid the post-graduate training and to transmit before an awareness about the BIM theme to the "professionals of tomorrow", the education has to begin during the university career, taking advantage of the academic case study offered by the professors. In this article are presented three (Figure 4) various examples of bachelor degree thesis, where the topic was BIM applied to Cultural Heritage, including the early stages of the process (survey, elaboration of the data and import in BIM software).

In particular, the modelling of different cases was aimed to test the difficulties associated with their representation, considering their differences and irregularities. The objects are: the ruins of Nemi's Shrine, Rome (Lombardini et al. 2011; Achille et al. 2012; Di Benedetto & Gozzo, 2012; Lombardini, N., 2015), also appeared in 2015; the small Lombard church of Santa Maria Delle Grazie in Cesate, Milan (Figure 4), sixteenth century, attributed to architect Francesco Maria Richini (Camastra & Innocenti 2014); and a portion of the San Michele Basilica in Pavia (De Dartein, 1865; Degasperi, 2014).

Students elaborated data and information in a BIM process, starting from the data surveyed on the sites,

Regarding the modelling part, once, working on these platforms (2012-2014), there was a simplification caused by the parameterisation of the elements, which had to become more flexible and embraced the irregularities and the unicity of a historic building or an archaeological area.

The modelling phase was complicated and (at that moment) time-consuming, but there were significant advantages in using it, as having a model of an information system to customise following the site's needs (Figure 5). The simplification decreased going forward through time, and today it is possible to reach easily the LOD (Level of Development) desired (usually the reference scale is 1:50).

The 3D modelling of the archaeological area was made in different ways, which led to different results and built the final model. The modelling of the coverage project (existing and a new one) reached a very high level of detail (for 2014), while the modelling of the ruins, created many difficulties and the result was satisfactory enough.

Figure 4. The archaeological area of Nemi (Rome, Italy); the Lombard Church of Santa Maria delle Grazie in Cesate (Milan, Italy); and the Church of S. Michele Maggiore in Pavia (Pavia, Italy)

Figure 5. Realisation of ad-hoc material sheets as they were not available in the standard BIM libraries, with the insertion of the elements and the state of conservation

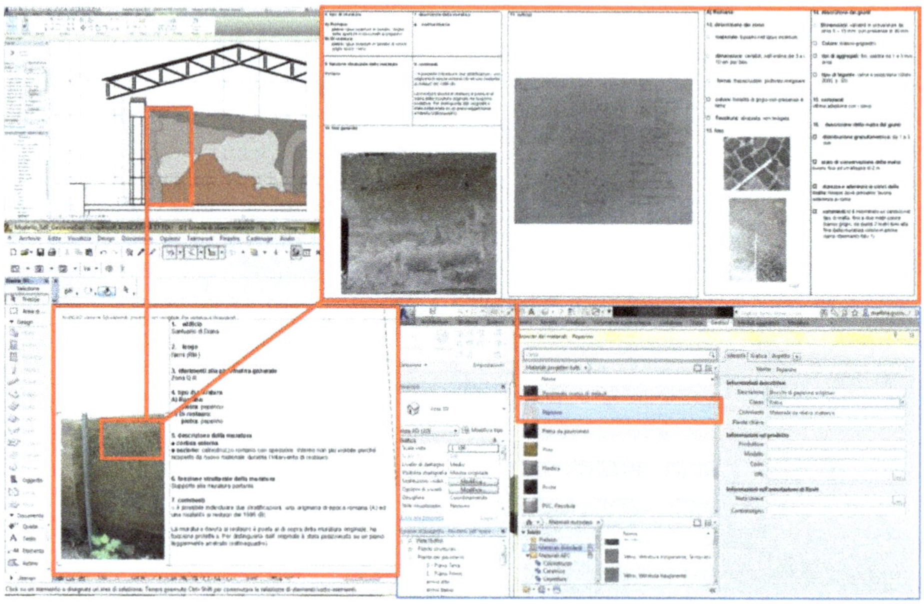

On the other hand, the modelling of the church (2015/2016), as it happened for the ruins of Nemi (Figure 5), has shown some issues. They were linked precisely to the fact that the software worked on the single blocks of the church (and therefore struggling to reproduce irregular shapes and to conceive of inhomogeneity from the point of Constructive view). Even more complex has been the design of the capitals (Figure 6), which are very articulated in the decorative apparatus (far from the classic canonization formalised by the historical treatises).

In both the case studies, it was wondered how to represent the objects to a given LOD (not necessarily coinciding with those set by US law), and which non-graphic information hooked up, making them interactive (Defined as LOI – Level of Information - by British and Italian law).

The information added to the models were: text, CAD, photos, link to web pages and others.

The text is the specific information about the materials, the decay, micro environmental, etc. What is required by a BIM software is the freedom to assign specific characteristics to specific points, but this software work for elements and components, and they do not recognise the single points on their surface.

This is a problem for the Cultural Heritage field, where potentially every point of the surface can have different parameters, considering not only the particular

Figure 6. Top line: recognition and insertion of materials and state of decay, inserting-implementing data into specific sheets made for the 3D elements. Bottom line: modelling of the structural elements and decorations

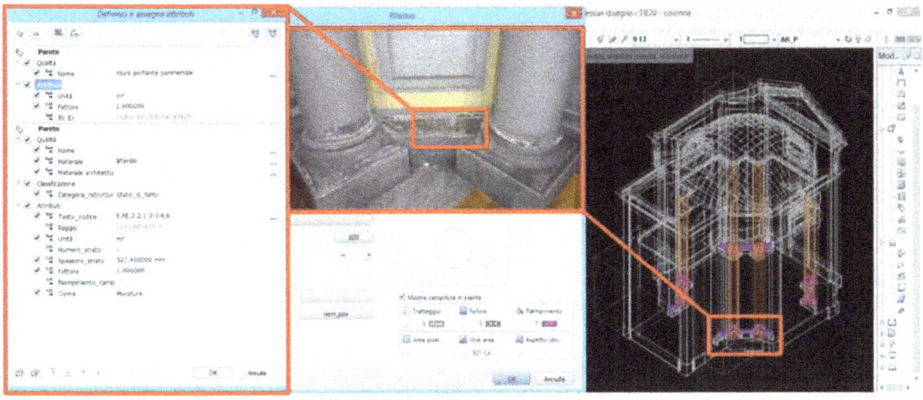

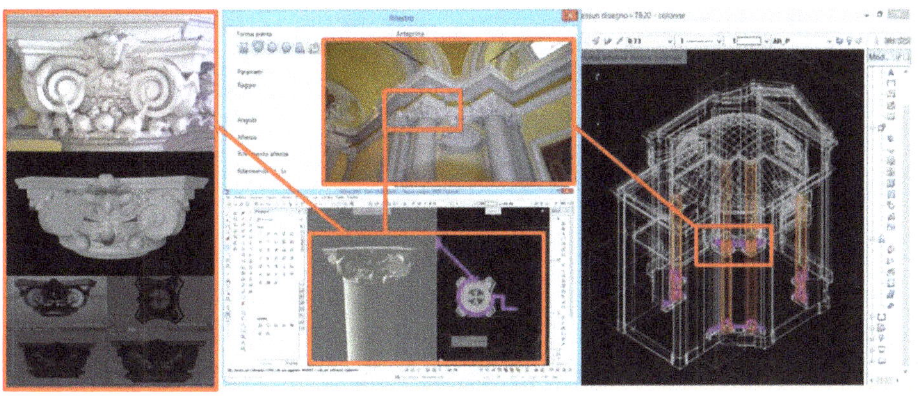

characteristics that are not standardised but also their damages. This problem was partially resolved to put some tags in selected points. Now, with the last releases of the software, it is possible to divide an element into several parts and assign to every block (Figure 7) accurate information about it.

The CAD file can be a product of direct survey, so it is important to be able to handle it easily with the BIM software. In general, in 2013-14, while in Archicad there were no issues in managing them, in Revit the format could be imported and correctly viewed, but the files could not be edited, and they worked only as tracing path. On the other hand, the point clouds are the product of laser scanner survey or photogrammetric survey. The management of this type of data is more complex, and it is one of the biggest problems of BIM technology in Cultural Heritage field.

Figure 7. Building one of the pillars of San Michele, with the new material inserted and modelling by blocks

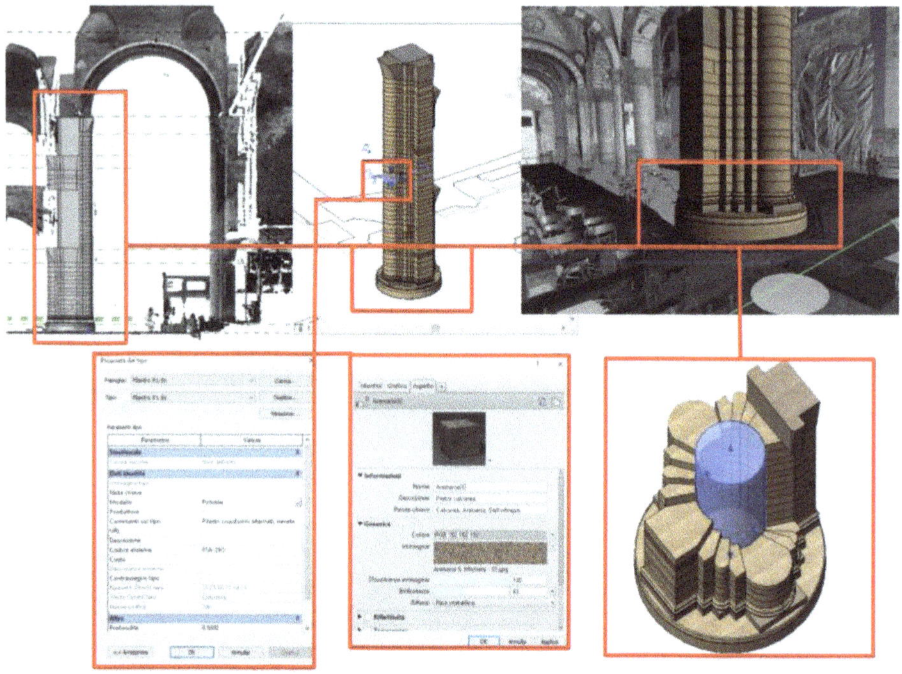

The tests, run with the former versions of the software (Revit 14 and Archicad 17) highlighted the limitations of the programs: in Revit 14, the points were not considered as snaps, so it was possible to correctly view the point cloud, without using it to precisely define the shape of the elements. On the other hand, Archicad 17 worked in the entirely different way: it is not able to import a big cloud, so before starting that operation, it was necessary to reduce it. Once you have a smaller point cloud, you can import it and thank the add-on Archisuite, and in particular, with Architools, it is possible to recognise the points as 3D elements, using them to trace the shape. However, it is clear that using this method; you are going to lose the level of detail because of the previous reduction of the cloud.

Today, most of the issues that the students met in the phases of their thesis are mostly overcame, thanks also to the interest among the researchers on the topic about BIM and Cultural Heritage. It is clear that there are still some issues to solve, but considering the fast progress of the information technologies, the functions of these software grow more and more every year, making also easier to use this methodology in the academic environment.

FUTURE RESEARCH DIRECTIONS AND CONCLUSION

The use of the BIM methodology in Italy is strongly recommended for the public contracts starting from 2016 (ANIMA, n.d.), and it speeded up the changing process in the construction field since the technological innovation and the digitalisation are defining the general economic scenario.

BIM is destined to become a diffuse process, but the transition from the traditional to the innovative methodology is not easy; in fact, the conversion requires not only an enormous economic and time investment but also a cultural change.

As told before, the University begins to dispense curricular lessons about BIM to the students, even if this phase hasn't already produced certified specialist. For this reason, the professional field is obliged to a training period that requires high time and costs waste (even if this kind of courses are part of the credits needed for the professional updating, DPR August 7th, 2012).

This paper shows how the BIM process can be fundamental also in the Cultural Heritage field: having a smart model that is possible to interrogate in every element (Figure 8). This actual and particular state of the development, at the same time, of the BIM process and the building industry, is able, and must demonstrate the necessity of the enhancement and improvement of the course in the new labor market.

This kind of digital competences approach, starting with the basic introduction of the interoperability, is imposing itself inside a transformed and widespread employment, demolishing a lot of skills' limits and geographical boundaries. The Information and Communication Technology (ICT) is raising the necessity of sharing common competencies and methodologies inside different cultures and diverse training approach in the field of building construction and, especially, of the care of Architectural Heritage. Of course, despite the significant progress of the technology applied to existing buildings, there are still some difficulties to overcome.

For example, one of the issues to improve is the interoperability, i.e. in the exchange of information and parameters from one sector to another one.

The BIM manager can assume a pivotal role in interoperability coordination, in the case of design on the existing one. In fact, he not only can synchronise and interrelate the BIM process, but also ensure that the various stages of the project are respecting the expected level of knowledge, or, as was above described, "confidential level".

Another aspect to enhance is the representation of the building that must provide support for the information (parameters) and subsequently serve as a basis for a project that involves reuse, structural consolidation, intervention on existing materials, respecting the compatibility and reversibility.

The object detected with any instrument, and more specifically with photogrammetric and 3D laser-scanner methods, even at high detail levels is an "empty" envelope. Unlike the new design, the BIM for the existing building is

Figure 8. The BIM model meets the archaeological heritage, containing drawings, ad-hoc conservation sheets, structural analyses, photos, material analyses, walls' stratification, and more

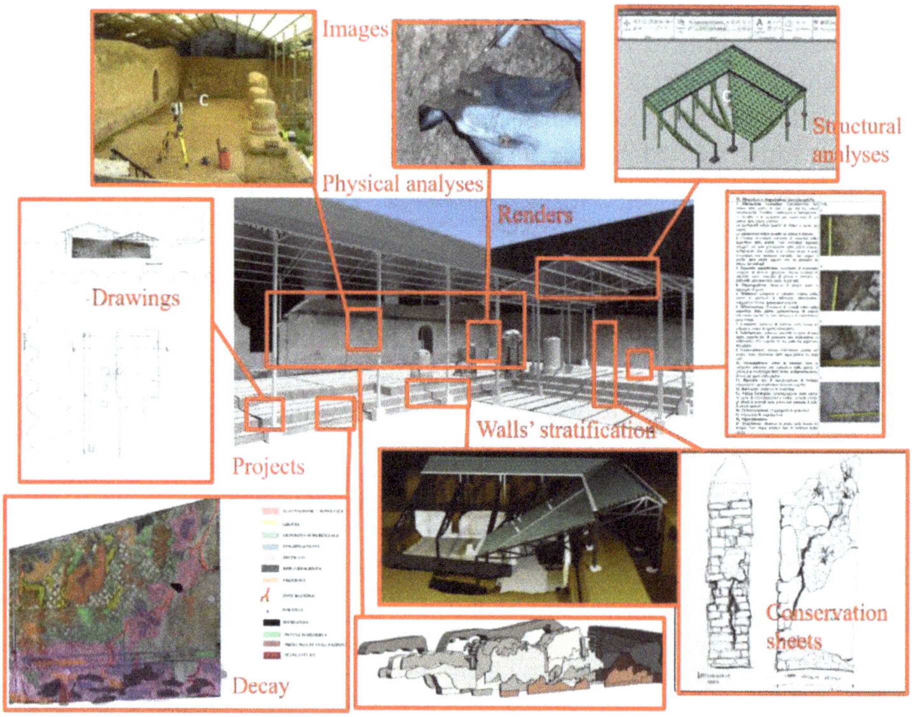

based on a detailed survey of the exterior surfaces but without material consistency, structural connections, construction of the current systems. All that is tangible and invisible (and which it can be hypothesised, though only indicatively, using other ways and forms of knowledge other than those of the geometric survey).

At the particular levels of knowledge that the conservation work refers to, it should be considered that the BIM, as conceived, adds additional levels of simplification that do not help the development of the preservation project itself. The general representation of complex or not well-understood elements may result in erroneous knowledge of the artefact at the expense of a good conservation project.

Buildings and, more generally, the constructions with which the conservation approach comes into contact, do not carry with us the standard information used to define all the components. Even so, these documents, though relevant, could not exhaust the necessary experience of the authenticity of the object. As has already been stated, the impossibility of establishing the certifiable parameters of existing materials is replaced by the standardisation of processes and survey tools, which can guarantee the reliability of the collected information which, in turn, is defined when intersected with other data.

Furthermore, there is no an unambiguously and shareable lexicon useful to describe all the elements of an existing building, as nowadays international rules had established, in different languages, for alterations and degradations.

In a word, now (2017-18) there are still some problems regarding the modelling, the interoperability and the information adding, but the new releases are more and more flexible and able to overcome the issues.

The problem is not technical anymore, but is more theoretical: now operators can create well-defined models, complete with a lot of data, but the question is: how can we use them? When the data, a rough entity, become information? The focus is on display, manage, promotion and valorisation of the massive amount of data that are produced with the most advanced techniques.

Moving forward, it is possible to say that there is a uniformity in the BIM software skills, which are overcoming their differences, shaping the more and more advanced performance on the specific needs of Cultural Heritage world.

REFERENCES

Achille, C., Fassi, F., Lombardini, N., Gaudio, & Galbusera. (2011). Survey of the archaeological site of Nemi. A training experience. In *Proceedings of XXIIIrd International CIPA Symposium*. Prague, Czech Republic: Academic Press.

Achille, C., Lombardini, N., & Valentini, M. (2012). Aree archeologiche e sistemi di protezione: modelli di valutazione di compatibilità, in 'La conservazione del patrimonio architettonico all'aperto. Superfici, strutture, finiture e contesti'. In *Proceedings of the XXVIII Congress Science and Cultural Heritage* (pp. 269-277). Bressanone, Italy: Academic Press.

ANIMA. (n.d.). *Legislative Decree 18 2016 April, n. 50*. Retrieved from http://en.anima.it/contenuti/10654/appalti-pubblici

Autodesk Revit. (2012-2013). *Auto Desk*. Retrieved from http://www.autodesk.it

Building Smart. (2012). *Common BIM Requirements*. Retrieved from http://www.en.buildingsmart.kotisivukone.com/3

Camastra, C., & Innocenti, A. (2015). *Dal dato storico-materico alla modellazione: il metodo BIM per i Beni Culturali nella Chiesa di Santa Maria delle Grazie a Cesate* (Unpublished thesis dissertation). Politecnico di Milano, Italy.

Carbonara, G. (2012). An Italian contribution to architectural restoration. *Frontiers of Architectural Research, 1*(1), 2–9. doi:10.1016/j.foar.2012.02.007

Codice dei Beni Culturali e del Paesaggio. (2014). *Gazzetta Ufficiale n°45.*

De Dartein, F. (1865). *Étude sur l'Architecture Lombarde et sur les origins de l'architecture Romano-Byzantine.* Paris, France: Dunod Éditeur.

Degasperi, M. (2015). *Il progetto della conservazione nel confronto con la metodologia BIM: problemi di modellazione e di 'linguaggio' nello studio di una porzione della Basilica di San Michele Maggiore a Pavia* (unpublished thesis). Politecnico di Milano, Italy.

Della Torre, S. (2016). *Beni Culturali e antisismica: il commento di Stefano Della Torre sull'annullamento della Circ. 18 del MiBACT.* Retrieved September 2, 2016, from http://www.ingenio-web.it/

Di Benedetto, C., & Gozzo, M. (2013). *Aree Archeologiche e Bim: Verifiche di compatibilita* (Unpublished master's thesis dissertation). Politecnico di Milano, Italy.

Dore, C., & Murphy, M. (2012). Integration of Historic BIM (HBIM) and 3D GIS for Recording and Managing Cultural Heritage Sites. *Proc. of the 18th Int. Conf. on Virtual Systems and Multimedia (VSMM)*, 369–376.

Dore, C., & Murphy, M. (2013). Semi-automatic modelling of building façades with shape grammars using historic building information modelling. *Proc. of the Int. Conf, on 3D ARCH 2013 - 3D Virtual Reconstruction and Visualization of Complex Architectures*, 57-64. 10.5194/isprsarchives-XL-5-W1-57-2013

EU BIM Task Group. (2017). *EUBIM Task Group.* Retrieved from http://www.eubim.eu

EUPPD. (2014). *D.L. n° 42 January 22nd.* EUPPD.

European Procurement Directive. (2014). *Official Journal of the European Union.* Art.22, c.4.

Fai, S., Graham, K., Duckworth, T., Wood, N., & Attar, R. (2011). Building Information Modeling and heritage documentation. *Proc. of the 23rd Int. CIPA Symposium*, 43–48.

Garagnani, S., & Manferdini, A. M. (2013). Parametric accuracy: building information modeling process applied to the cultural heritage preservation. *Proc. of the Int. Conf, on 3D ARCH 2013 - 3D Virtual Reconstruction and Visualization of Complex Architectures*, 87-92. 10.5194/isprsarchives-XL-5-W1-87-2013

Graphisoft Archicad. (n.d.). *Graphisoft*. Retrieved from https://www.graphisoft.com

Harpaceas. (2017). *Harpaceas the BIM Expert*. Retrieved from http://www.harpaceas.it

ICOMOS. (2008). *Illustrated glossary on stone deterioration patterns*. Retrieved from https://www.icomos.org/publications/monuments_and_sites/15/pdf/Monuments_and_Sites_15_ISCS_Glossary_Stone.pdf

Lombardini, N. (Ed.). (2015). Un approccio interdisciplinare al progetto di conservazione. La valorizzazione dell'area archeologica dedicata al culto della Diana Nemorense. Firenze, Italy: Altralinea edizioni.

Lombardini, N., & Cantini, L. (in press). Non-standardized data in the bim process. the management of construction systems data in the cultural heritage conservation. In *Proceedings of the 3rd International Conference on Protection of Historical Constructions*. Lisbon, Portugal: Academic Press.

Ministry of the public education. (1972). Carta italiana del restauro. *Circular n° 117*, April 6th, 1972.

National Institutes of Building Sciences. (2018). *National BIM Guide for Owners*. Retrieved from https://www.nibs.org/?nbgo

Normal Cultural Heritage Code, U. N. I. (2006). *UNI 11182, 01.020 91.100.15*. retrieved from http://www00.unibg.it/dati/corsi/20067/70707-UNI11182.pdf

NTC. (2008). *Norme Tecniche per le Costruzioni, D.M., January 14th*. Retrieved from http://www.ingegneriasoft.com/NTC2008_Norme_tecniche_per_le_costruzioni.htm

Politecnico di Milano, B. I. M. Manager. Methods, Models and Applications. III Edition. (n.d.). Retrieved from https://www.polimi.it

Scianna, A., Gristina, S., & Paliaga, S. (2014). Experimental BIM Applications in Archaeology: A Work-Flow. *Proc. Of the 5th Int. Conf. on Digital Heritage. Progress in Cultural Heritage: Documentation, Preservation, and Protection*, 490–498. 10.1007/978-3-319-13695-0_48

SiCAR. (n.d.). *Sistema Informatico per la documentazione e la Progettazione dei Cantieri di Restauro*. Retrieved from http://www.infobc.uniud.it/progetto/attivita/ricerca/sicar-sistema-informatico-per-i-cantieri-di-restauro

Stastbygg. (2013). *Stastbygg BIM Manual 1.2.1*. Retrieved from http://www.statsbygg.no/files/publikasjoner/manualer/StatsbyggBIM-manual-ver1-2-1eng-2013-12-17.pdf

Tommasi, C., Achille, C., & Fassi, F. (2016). From point cloud to BIM: a modelling challenge in the cultural heritage field. The International Archives of the Photogrammetry, Remote Sensing and Spatial Information Sciences, XLI-B5, 429-436.

UK Government. (2016). *PAS 2016: 2010: next generation access for new build homes guide*. Retrieved from https://www.gov.uk/government/publications/pas-2016-2010-next-generation-access-for-new-build-homes-guide

UNI 11337-4:2017. (2017). *Edilizia e opere di ingegneria civile - Gestione digitale dei processi informativi delle costruzioni - Part 4: Evoluzione e sviluppo informativo di modelli, elaborati e oggetti*. Retrieved from http://www.uni.com/index.php?option=com_content&view=article&id=5753:gestione-digitale-dei-processi-informativi-delle-costruzioni-pubblicate-le-parti-1-4-e-5-della-uni-11337&catid=170:commerciale&Itemid=2612

University of Edinburgh. (2015). *Kist o Riches*. Retrieved from http://www.tobarandualchais.co.uk

University of Helsinki. (2015). *CultureSampo*. Retrieved from http://www.kulttuurisampo.fi

University of Pisa. (n.d.). *BIM Master 2017. II Edition*. Retrieved from http://www.unibim.it/

U.S. General Services Administration. (n.d.). Retrieved from https://www.gsa.gov/

Chapter 2
Constraints in Authoring BIM Components:
Results of Longitudinal Interoperability Tests

Claudio Benghi
Northumbria University, UK

David Greenwood
Northumbria University, UK

ABSTRACT

The authors investigated issues of geometric interoperability for reusable BIM components across multiple platforms using industry foundation classes (IFCs) which many proprietary BIM software platforms claim to fully support. These assertions were tested, first in 2012 and then in 2017 to assess the state and evolution of interoperability in the industry. A simple test model was created representing significant types of geometry encountered in component libraries, which were then expressed in IFC files. In the 2012 study, 11 commonly used BIM tools showed a dramatic failure to process the geometries as intended, indicating that the authoring tools, whilst technically capable of supporting required component geometric representations, were constrained from doing so by their conversion interfaces with IFC geometries. In the 2017 tests, improvements were observed though there were still significant processing failures that could result in serious errors; particularly in the case of the BIM library components imported into project design models.

DOI: 10.4018/978-1-5225-5625-1.ch002

INTRODUCTION

As Building Information Modelling (BIM) becomes more prevalent in the construction industries of the world some of the practical problems of authoring and sharing models are also becoming more evident. It is generally accepted that BIM-related technologies offer considerable advantages to many, perhaps most, participants in the construction sector (Eastman et al., 2011). In the UK, for example, the Government has mandated the use on all its projects of "fully collaborative 3D BIM" (Cabinet Office, 2011) and supported the development of standards to support the industrial uptake (BIM Task Group et al., 2016).

A fundamental limiting factor in the uptake of BIM is the issue of interoperability, defined by the International Alliance for Interoperability (IAI, now 'buildingSMART') as 'an environment in which computer programs can share and exchange data automatically, regardless of the type of software or where the data may be residing' (Fischer and Kam, 2002, p. 14). Currently, there exists a whole range of commercially available BIM software platforms that have specialised to suit the functional needs of their main users (architects, structural engineers, services engineers, constructors, and so on) and consequently differ structurally and semantically. The aspirational ideal of fully collaborative BIM presumes a single model, allowing the full integration of all aspects of the design and further, for the same information to be re-used in the delivery and operation of the constructed facility (UK Treasury and Cabinet Office, 2016. p. 7). To do this effectively, secure and reliable exchange of data is essential. It is this requirement that underlies the concept of 'interoperability' (Yang and Zhang, 2006). As Cerovsek points out a key issue has been how to achieve 'inter-operability between multiple models and multiple tools that are used in the whole product lifecycle' (Cerovsek, 2011, p. 224) and BIM usage is still largely restricted to coordinated models that relate to the contribution of each of the disciplines involved. Currently, as evidenced by NBS's National BIM Report (NBS, 2017) full collaboration is still not a reality. For some time the recognised basis of BIM interoperability has been the system of Industry Foundation Classes (IFCs) designed by the International Alliance for Interoperability and maintained by buildingSMART (Tolman, 1999; Fischer and Kam, 2002). However, the mere presence of IFC is not sufficient for overcoming the problems of interoperability, and, for some critics, data exchange remains 'unreliable and unpredictable' (e.g. Sacks et al., 2010, p. 420). Fischer and Kam (2002, p. 40) identify such problems, particularly when they result in 'geometric misrepresentations across different software packages reading the same IFC source file'.

The Problem of Standard Component Libraries

The problem is perhaps most acute when it comes to the effective authoring and use of standard building component libraries for BIM; a development that has, for some time been seen as to have significant potential for improving the productivity of designers and specifiers (Howard and Björk, 2008). Demand for library components from the industry and construction product manufacturers is high, with 93% of practitioners stating they "need access to well-structured generic digital objects" (NBS, 2017, p. 20) and there are organisations currently attempting to author library components that can be delivered across multiple BIM platforms with the minimum amount of re-authoring. Buildings contain numerous components – windows, doors, etc. - that are standardly procured from their manufacturers. However, when they exist within a proprietary BIM tool, these components are native to that particular tool and not easily shared between different BIM platforms; for example, a boiler component authored in one tool is unlikely to be useable in another BIM tool. This leads to a major weakness in data sharing and restricts any efficiency gains that the industry may make through data reuse, both in terms of content creation and subsequent delivery and management. Worse, when the transfer of data appears to work, but is incorrect, the reliability of the whole process is undermined. This has arguably contributed to the relatively slow uptake of BIM standards to date and hampered the efforts of buildingSMART to support the development of library standards through IFC. The current releases of the buildingSMART data model (Ifc2x3 and Ifc4) do not support explicit parametric objects and therefore do not allow the exchange of object's geometric constraints. Although some of these features are under development in the upcoming Ifc5 standard (buildingSMART, 2017), it is likely that for the next few years the industry will need to create library objects within the constraints of the current buildingSMART data standards. In these, whilst explicit parametric behaviour is not supported, implicit parametric behaviour can be approximated by careful selection of the available geometric objects. For example, the IFC-Extruded-Area-Solid defines a solid in terms of a footprint, an extrusion direction and a depth of extrusions, as illustrated in Figure 1.

Essentially the profile, depth and direction allow the object to behave in a simple parametric way by determining the shape of the resulting object on the single values, as opposed to the use of geometries such as IfcTriangulatedFaceSet, which would require the changing of several individual vertex points to determine its shape. As reported by Fischer and Kam (2002, p. 25) considerable efficiency gains can be made by authoring library components through these implicitly parametric geometric objects.

Figure 1. Pseudo-parametric vs. meshed geometric entities in IFC (buildingSMART, 2013a, 2013b)

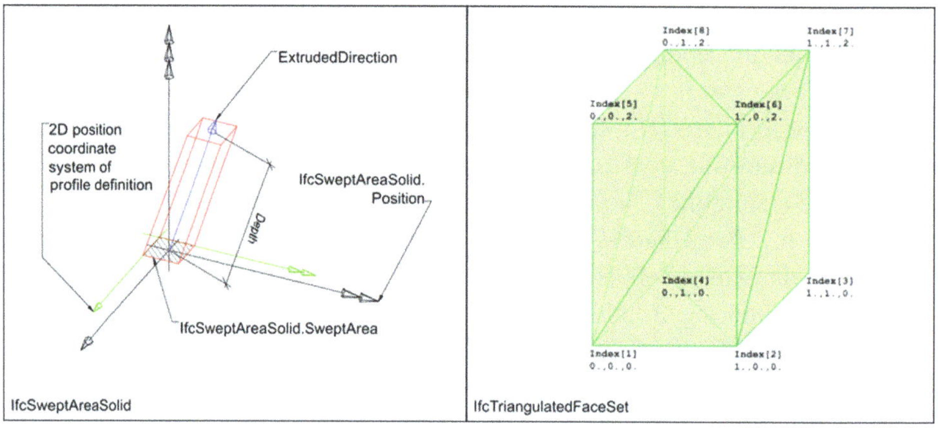

RESEARCH AIMS AND METHOD

This study investigated the issues around authoring standard reusable BIM components that can be delivered across multiple platforms using IFC as a single source of content. The primary aim was to determine what constraints exist for modelling library components to support current BIM platforms, and to illustrate their effects.

Selecting the Software Applications

In 2012 we reviewed the BIM software applications that claimed support for the Ifc2x3 standard and these claimants were then filtered to include only those that had some form of geometric presentation and/or authoring engine. These are identified in Table 1. Whilst other platforms exist, it is suggested that these are representative and typical of those currently available for use in the Construction Industry.

Alongside the commercially-available BIM software applications, listed in Table 1, the open-source software IFC viewer XbimXplorer (xBimTeam, 2017), was adopted as a reference test, as the authors had previously found it to have performed well in trial tests of this nature and its open source nature provided transparency into its geometric modelling operations. A reference test suite has been published along with the source of a geometry engine that correctly processes these geometries and a viewer to examine the results. This is available at http://docs.xbim.net/research/ifc2x3-test-harness.html.

Table 1. Selection of BIM platforms

Author	BIM Tool	2012 test version	2017 test version	Editing
Bentley	AECOsim	AECOsim v8i	View v. 8i	x
Graphisoft	ArchiCAD	ArchiCAD 16	ArchiCAD 20	x
Nemetcheck	DDS Cad Viewer	V 8.0	V 13.0	
Karlsruhe Institute of Technology	FZKViewer	V 2.2	V 4.8	
	GTDS Viewer	v 0.6	replaced by FZKViewer	
	IfcStorey Viewer	v2.2b		
Autodesk	Revit	v2012	v2017	x
Nemetcheck	Solibri Model Viewer	v7.1	v9.7	
Trimble	Tekla Structures	v18.1	v2017sp2	x
	Tekla BIMsight	v2012	v1.99	
Nemetcheck	VectorWorks	v2012	v2017	x

Selection of Relevant Geometry Cases

The focus of this study is the authoring of building components. These can be defined as discrete compound geometric solid objects that are inserted from a library into a construction project model. Typically these are items which are manufactured off-site and installed during construction; examples would be a boiler, light fitting, door or window. Their three-dimensional visual representation would be described using a combination of the classes represented in Figure 2, which also shows the changes occurred between the schemas of Ifc2x3 and Ifc4.

Figure 2. IFC solid model representations

Since the test model was developed ahead of the Ifc4 release, only solid model representation entities defined in buildingSMART data standard Ifc2x3 were included in the tests. In order to develop a representative test we investigated over 200 IFC files produced in professional BIM workflows; this resulted in a clear prevalence of geometries defined with a simplified form of Boundary Representation (*Brep*) where all faces are planar and every face bound is defined as a polyline. This type of solid is defined as *IfcFacetedBrep* or *IfcFacetedBrepWithVoids*, (which rely on the same underlying geometric functions). Since not a single instance was found of a geometry using the format of *IfcSweptDiskSolid* in the sample, this representation was excluded from the tests as it was prima facie unlikely to give any interoperability concerns in actual BIM workflows. Therefore, five solid geometry representations (*IfcCsgSolid, IfcExtrudedAreaSolid, IfcRevolvedAreaSolid, IfcFacetedBrep* and *IfcSurfaceCurveSweptAreaSolid*) were used to create a simple test reference model that was designed to validate the correct interpretation of geometry and topology. At the time of authoring all of their respective topological shapes were supported as output format by the majority of the available BIM tools.

Design of the Geometric Test Model

A single compound geometric element was constructed that embodied each of these geometries in a form that assisted visual assessment (as well as providing a pleasing irony). This is illustrated in Figure 3 with annotation to indicate the underlying IFC geometries being tested.

The objective was to define individual parts of the model that could provide evidence for correct support for the following six IFC solid cases:

- *IfcCsgSolid* (the letter “B”)
- *IfcExtrudedAreaSolid* (the body of the letter “i”)
- *IfcRevolvedAreaSolid* (the dot of the letter “i”)
- *IfcSurfaceCurveSweptAreaSolid* (the letter “M”)
- *IfcFacetedBrep* (the body of the exclamation mark)
- *IfcRevolvedAreaSolid* (the dot of the exclamation mark)

The difference between the two *IfcRevolvedAreaSolid* instances stands in central void of the dot of the letter “i” due to the offset position of the revolution axis. The underlining block upon which the model sits was included only to assist visual comparison, and was developed with the typical geometries of a standard wall (an *IfcExtrudedAreaSolid* of an *IfcRectangleProfileDef* lying on the XY plane) to maximize the likelihood of correct interpretation; its rendering was excluded from

Figure 3. Geometry test model

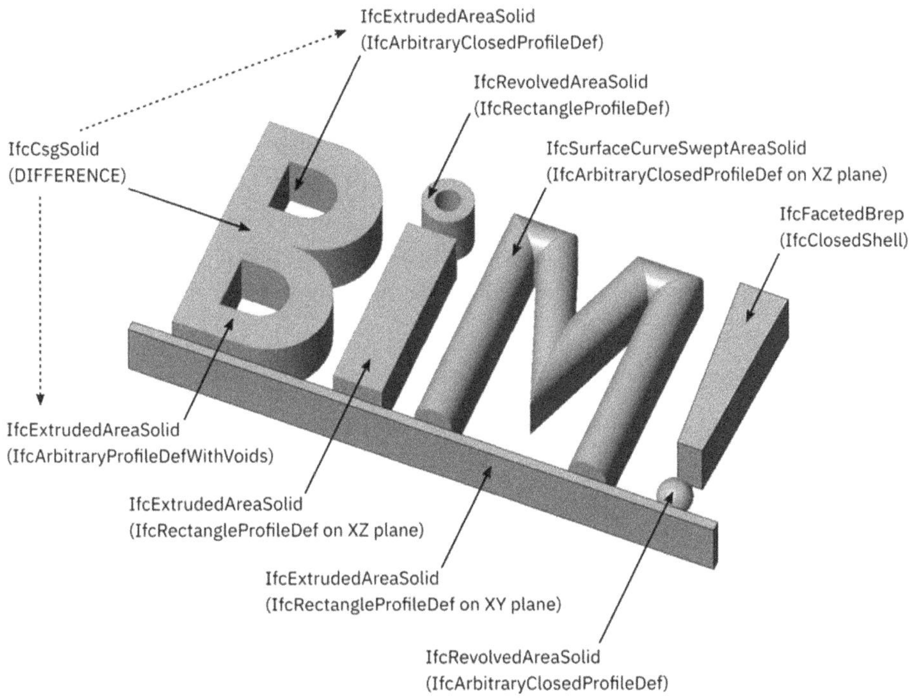

the analysis of results. The construction of each of the solid geometries has been designed to validate the following measures of fidelity:

- Visual Presence
- Geometric Equivalence (Cartesian composition)
- Topological Equivalence (Relational composition)

In this test case, measurement of Semantic Fidelity (Element, Material, Properties etc.) and Presentation (Texture, Colour, Line Style etc.) have been excluded (to be the subject of further separate investigation).

Results

The tests were performed by observing the results of importing the IFC files containing all of the composite test model's geometric entities into each of the listed BIM platforms. A 'visual presence' test was performed on the results and a pass was given where any geometry was visible and appeared to be the same as that of the test case. The following Table 2 summarizes the outcomes of each platform's ability to interpret the IFC geometries correctly. Screenshots of resultant representations of the test model are presented (as Figures 4-23) in an appendix to the text of the paper with Figure 24 showing the correct representation from XbimXplorer, which was used to create the reference test suite (available at http://docs.xbim.net/research/ifc2x3-test-harness.html) and accordingly passed all tests. The supplier names of the products listed in Table 2 have been omitted, but can be seen by referring to the earlier, Table 1.

As Table 12 shows, in the 2012 tests all the tools accurately displayed the body of the "i" in BIM (*IfcExtrudedAreaSolid*). This was not initially the case with ArchiCAD

Table 2. Pass/Fail Summary for all geometries of the test model (x signifies a pass)

Product	B	i body	i dot	M	! body	! dot	Pass count and (%)	Entire BiM!	Screen shot
AECOsim v8i		x			x		2 (33%)		Figure 4
View v. 8i 2017		x	x	x	x	x	5 (83%)		Figure 5
ArchiCAD 16		x	x				2 (33%)		Figure 6
ArchiCAD 20		x	x	x	x		4 (67%)		Figure 7
Cad Viewer 2012		x			x		2 (33%)		Figure 8
Cad Viewer 2017		x			x	x	3 (50%)		Figure 9
FZKViewer v2.2		x	x		x	x	4 (67%)		Figure 10
FZKViewer v4.8	x	x	x	x	x	x	7 (100%)	x	Figure 11
GTDSViewer 0.6	x	x					2 (33%)		Figure 12
	Discontinued software has not been re-tested in 2017								
IfcStorey Viewer[4]		x					1 (17%)		Figure 13
	Discontinued software has not been re-tested in 2017								
Revit 2012		x			x		2 (33%)		Figure 14
Revit 2017		x			x		2 (33%)		Figure 15
Solibri SMV 7.1		x	x				2 (33%)		Figure 16
Solibri SMV 9.7	x	x	x	x	x	x	7 (100%)	x	Figure 17
Tekla Structures 18.1	x	x	x				3 (50%)		Figure 18
Tekla Structures 2017	x	x	x		x	x	5 (83%)		Figure 19
Tekla BIMsight 2012	x	x	x				3 (50%)		Figure 20
Tekla BIMsight 2017	x	x	x		x	x	5 (83%)		Figure 21
VectorWorks 2012		x					1 (17%)		Figure 22
VectorWorks 2018	x	x					2 (33%)		Figure 23

16, but on further examination it was observed that there had been a total failure in reading the IFC file as a compound entity, resulting in all six geometries being rejected as a single entity due to an internal error in processing. Further compound geometry testing indicated the possibility that a failure in processing any single geometric entity would lead to a total failure to process the compound geometry: when tests were repeated using individual geometry test objects all platforms reported the same results as before, with the exception of ArchiCAD 16, in which the "i"-body was accurately displayed, thus passing the "i"-body geometry tests. (This issue did not occur in the 2017 test with ArchiCAD 20, where the geometries that successfully displayed had all been correctly parsed from the single test file as with the other platforms.)

Analysis of Results

Whilst in 2012 none of the tools processed all six of the geometries correctly, and the results of the tests indicated a general inherent misinterpretation of modelling of Sweeps and Boolean operations, the situation has improved in the latest (2017) tests with two of the tools correctly interpreting all the geometries. In the 2012 tests all tools consistently failed to render the "M" in BIM. Further inspection showed clear diversity in their interpretation of the buildingSMART standard. The most common problem was lack of adherence to interpreting the normal of the reference plane defined in *IfcSurfaceCurveSweptAreaSolid*. This resulted in the "M" regularly being mirrored or rotated. Several tools also failed to relocate the profile so that its origin was aligned with the first point of the directrix and its "Z" direction was tangential to that of the directrix. Whilst the standard documentation is reasonably clear it appears to have been misinterpreted. In the tests on the 2017 versions of the tools, four presented a correct rendering of the "M" in BIM.

The remaining geometries all showed some form of support in each of the tools but there was little consistency. The Coordination View 2.0 specification requires that all solid geometries are transmitted as some form of *IfcSolidModel*. Boolean operations (*IfcBooleanResult*) do not fall into this scope. In order to transfer these operations (such as cutting the hole in the "B") the standard requires an *IfcCSGSolid* to be used rather than an *IfcBooleanResult*. It was noted that more tools would support the "B" when the *IfcBooleanResult* was incorrectly used rather than when *IfcCSGSolid* was correctly used. As the geometry operation is the same this is clearly an indication that there were minor (and therefore, simple to remedy) processing errors in most of the BIM IFC readers. There was a general problem with the use of compound geometries. An *IfcShapeRepresentation* can only have one representation type; these are defined by the Standard as the types of *IfcSolidModel* shown in Table 3.

Table 3. IfcSolidModel types

Type	Description
Solidmodel	Including swept solid, Boolean results and Brep bodies more specific types are:
Sweptsolid	Swept area solids, by extrusion and revolution
Brep	Faceted Brep with and without voids
Csg	Boolean results of operations between solid models, half spaces and Boolean results
Clipping	Boolean differences between swept area solids, half spaces and Boolean results
Advancedsweptsolid	Swept area solids created by sweeping a profile along a directrix

There is some ambiguity as to whether *IfcSolidModel* can be used as a declaration for mixed types. This approach was adopted for the purposes of the tests, but it should be noted that some tools interpret this differently. For example, Solibri would not show an object of type *Brep* unless it was explicitly declared as a *Brep*. This causes problems when authoring complex components where the entity may be modelled as a set of different geometries; a practice that is very normal for BIM authors. One alternative - to provide different representations for each type of geometry - is an impractical one to impose on authors. The other alternative, providing a compound representation and a *Brep* representation for visual integrity, works for the viewing-only scenario but does not improve support for data reuse. In the 2012 tests all but one of the tools failed to report any errors to the user when processing the IFC file, even in the case of total failure. This is of particular concern for reasons of model validation, as those tools which aim to support model checking should consider failure to read geometry as a valid accuracy test. When the tests were repeated in 2017, with the same software platforms and with Ifc2x4, there was clear evidence that some of these problems had been overcome: all but two of the nine software tools that were re-tested in 2017 (two of the tools had been discontinued, as shown in Table 2) showed improved processing of the geometries, with two attaining a 100% result.

CONCLUSION

Between 2012, when the tests were first conducted, and 2017 the IFC standard itself had evolved. In 2012, the then current IFC2x3 specification had support for various kinds of solid geometry, and these definitions have been improved and enhanced in the IFC2x4 standard. Any software implementation of the standard is according to a conformance class, called a "Model View Definition" (MVD) that restricts support to a defined subset. The IFC implementations in software tools adhered to the MVD subset of the IFC2x3 specification, called "IFC2x3 Coordination View 2.0". Not all of the solid geometry types were included. For example, the *IfcSurfaceCurveSweptAreaSolid*, the CSG primitives and the *IfcCSGSolid* root were not part of the IFC2x3 Coordination View 2.0 and therefore not readily supported by most applications. As we move from the coordination view to supporting library views these geometries have become important and need to be considered.

The inconsistent support for these relatively non-complex solid geometry representations, as demonstrated by the 2012 tests, represents a problem that was likely to be compounded by an increase in the prevalence of these geometries over time. This indicated the urgency in the need to address such issues within the scope of the then current Ifc2x3 standards. The Ifc2x3 standard defined "types" of shape representation and enforced all *IfcShapeRepresentation* instances to declare their representation type. It was unclear from the standard how to interpret this representation type. Some BIM tools enforced compliance with the type rigorously: thus if the type was specified as a *Brep* but the actual object was a swept solid, the object geometry was ignored, and vice versa. However, the standard implies that if the type is specified as a *SolidModel* then all subtypes shown in Table 4 are valid. Some BIM tools obeyed this rule; others did not. It was therefore recommended that all BIM tools observe a representation type of *SolidModel* to support geometry objects that are either *SweptSolid, CSG, Brep, Clipping* or *AdvancedSweptSolid.* Where they cannot process the specified geometry a bounding box representation should be substituted with an appropriate error message. The use of *IfcCsgSolid* was found not to be correctly supported by most BIM platforms; this was the only way to create a Boolean result as a valid single solid geometry. It was therefore recommended that either *IfcBooleanResult* be a permitted *SolidModel* type within the then current Ifc2x3 standard or that implementers' agreements require support for *IfcCSGSolid. IfcRevolvedSolids* appeared to be variably implemented and it was unclear what geometric constraints existed to creating these solids. The Ifc2x3 standard states "the axis shall not intersect the interior of the swept area"; this left a grey area relating to an axis which intersects the boundary of the swept area - a construction often used to create simple solids such as spheres which contain no holes. It was recommended that this be clarified in the documentation and that coincidental axis and boundary edge permitted, as this was the only way to create a sphere without the use of the unsupported CSG primitives.

Although *IfcExtrudedSolid* was shown to be generally well supported there are some aspects of the IFC documentation which could be improved to avoid problems with profiles of type *IfcArbitraryClosedProfileDef* and *IfcArbitraryClosedProfileDefWithVoid* - a geometry used to define a 2D profile that has an outer wire bound and zero or more inner loops or holes. Convention in standards such as OpenGL defines that the winding of the outer bound should be counter clockwise and the inner bounds clockwise. Whilst this is not strictly necessary to define geometric objects which have no topology (such as *IfcArbitraryClosedProfileDef*) it was considered to be useful for BIM tools to obey this convention and avoid corrections during processing. Also implicit in the documentation is that the normal of the profile shape that is to be extruded points away from the extrude direction. If this were not the case the result would be a negative solid. It was concluded that this could have been made more explicit in the technical documentation.

In summary, the consistency of support for appeared to have been poor across all BIM platforms. Relatively simple IFC geometries have been tested (there are, in fact, over 90 geometric classes that could be involved in the full execution of a compared format) and none of the BIM tools tested in 2012 gave complete support. The most likely reasons for failure were connected with incorrect interpretation of the Standard. There was demonstrable evidence that all authoring tools could natively support the tested geometries, visible through the user interface. It was thus considered that this should be a matter of software programme correction by the BIM tool author to significantly improve the level of support. The 2012 tests had revealed that model checking tools that ignored any geometry that is not correctly processed left the possibility of serious errors if they are relied upon to test actual compliance within designs for construction projects. For example, the failure to show a pipe in a clash detection check could have serious ramifications. It was therefore concluded that Bounding Box proxies should be substituted where failure-to-process occurred. Greater clarity was necessary as to which geometric representation contexts were chosen to depict an object and its geometry type and these needed to be consistent or explicit across BIM tools. In view of the critical nature of these issues, the test model used in this investigation was circulated to experts in the area, including members of BuildingSMART, and have, in part, resulted in the improved modifications to their products that have been found to be present in the 2017 tests. The tests in 2017 have shown improvements in a number of areas, with two of the BIM tools attaining 100% results. However, the fact remains that confidence in BIM modelling tools will be undermined unless interoperability improves, especially if tangible objects disappear from a project merely because a particular BIM tool has failed to process them. It is therefore further recommended that all BIM tools should indicate failure to correctly process a geometry object either in part or in full. The expectations for BIM library components within the construction industry must be supported by the functionality of existing BIM tools. These tools embody proprietary parametric components that greatly reduce geometric authoring time; however, these components have to be specifically authored for each BIM platform.

For a construction product manufacturer wishing to deliver BIM library components for their product range this greatly increases the cost of authoring and maintaining those components.

ACKNOWLEDGMENT

The authors would like to acknowledge the contribution of their former colleagues, Professor Stephen Lockley (Emeritus Professor of Building Modelling at Northumbria University, UK) and Dr. Jane Matthews (of Curtin University, Perth, Australia) in co-designing the test model and the contributing to the first software tests.

REFERENCES

BIM Task Group. (n.d.). *British Standards Institution, Department of Business, Energy & Industrial Strategy, 2016. BIM Level 2 Standards*. Retrieved from: http://bim-level2.org/en/standards/

buildingSMART. (2013a). *IfcExtrudedAreaSolid specification*. Retrieved from: http://www.buildingsmarttech.org/ifc/IFC4/final/html/schema/ifcgeometricmodelresource/lexical/ifcextrudedareasolid.htm

buildingSMART. (2013b). *IfcTriangulatedFaceSet specification*. Retrieved from: http://www.buildingsmarttech.org/ifc/IFC4/final/html/schema/ifcgeometricmodelresource/lexical/ifctriangulatedfaceset.htm

buildingSMART. (2017). *Summary of IFC Releases*. Retrieved from: http://www.buildingsmart-tech.org/specifications/ifc-releases/summary

Cabinet Office. (2011). *Government Construction Strategy*. Retrieved from https://www.gov.uk/government/uploads/system/uploads/attachment_data/file/61152/Government-Construction-Strategy_0.pdf

Cerovsek, T. (2011). A review and outlook for a "Building Information Model" (BIM): A multi-standpoint framework for technological development. *Advanced Engineering Informatics, Information mining and retrieval in design, 25*, 224–244.

Eastman, C. M., Eastman, C., Teicholz, P., & Sacks, R. (2011). *BIM Handbook: A Guide to Building Information Modeling for Owners, Managers, Designers, Engineers and Contractors*. Hoboken, NJ: John Wiley & Sons.

Fischer, M., & Kam, C. (2002). *PM4D final report, CIFE technical report*. CIFE. Retrieved from https://web.stanford.edu/group/4D/download/PM4D_Final_Report.pdf

Howard, R., & Björk, B.-C. (2008). Building information modelling – Experts' views on standardisation and industry deployment. *Advanced Engineering Informatics, Network methods in engineering, 22*, 271–280.

NBS. (2017). *National BIM Report 2017*. Retrieved from https://www.thenbs.com/knowledge/nbs-national-bim-report-2017

Sacks, R., Kaner, I., Eastman, C.M., & Jeong, Y.-S. (2010). The Rosewood experiment — Building information modeling and interoperability for architectural precast facades. *Automation in Construction, Building information modeling and interoperability, 19*, 419–432.

Tolman, F. P. (1999). Product modeling standards for the building and construction industry: Past, present and future. *Automation in Construction*, *8*(3), 227–235. doi:10.1016/S0926-5805(98)00073-9

Treasury and Cabinet Office. (2016). *Government Construction Strategy*. Retrieved from https://www.gov.uk/government/publications/government-construction-strategy-2016-2020

xBimTeam. (2017). *XbimXplorer Download*. Xbim Toolkit Documentation. Retrieved from: http://docs.xbim.net/downloads/xbimxplorer.html

Yang, Q. Z., & Zhang, Y. (2006). Semantic interoperability in building design: Methods and tools. *Computer Aided Design*, *38*(10), 1099–1112. doi:10.1016/j.cad.2006.06.003

APPENDIX

Representation of the Testing Model in the Selected Applications

The following figures are Screenshots of the representations of the test model as computed by each of the BIM software platforms tested.

Figure 4. AECOsim from Bentley (tested 2012)

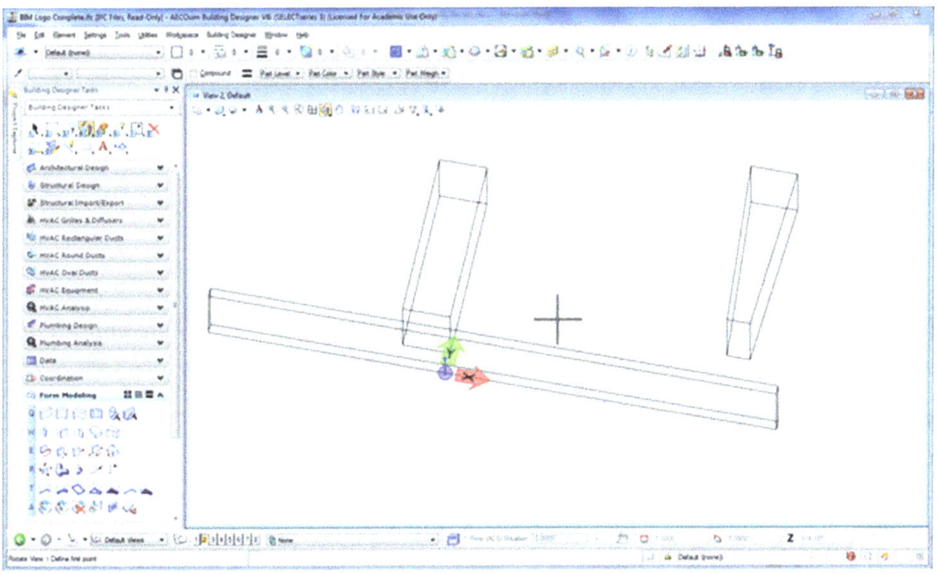

Figure 5. View v. 8i The current viewer from Bentley has been evaluated in 2017 as AECOsim was not available for testing

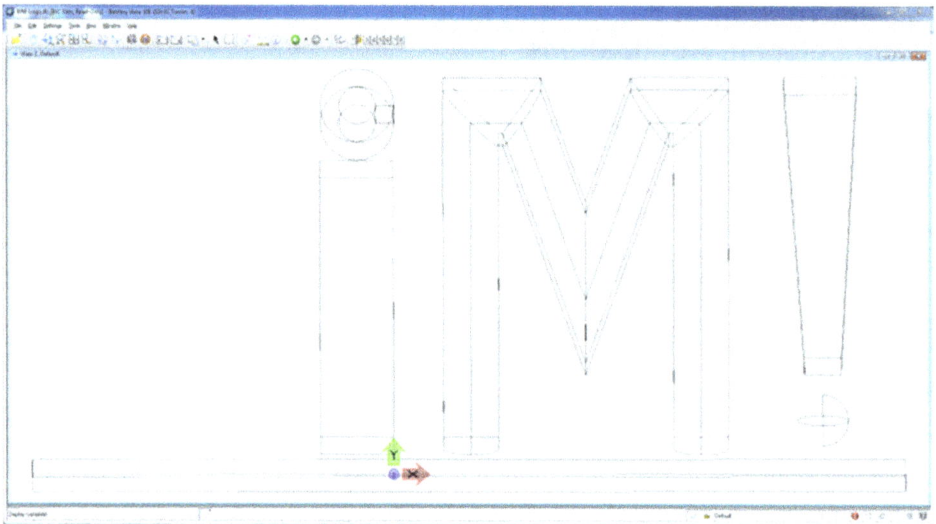

Figure 6. ArchiCAD version 16, the failure of one part affected the whole import, but more geometries were visible when primitives were loaded individually (tested 2012)

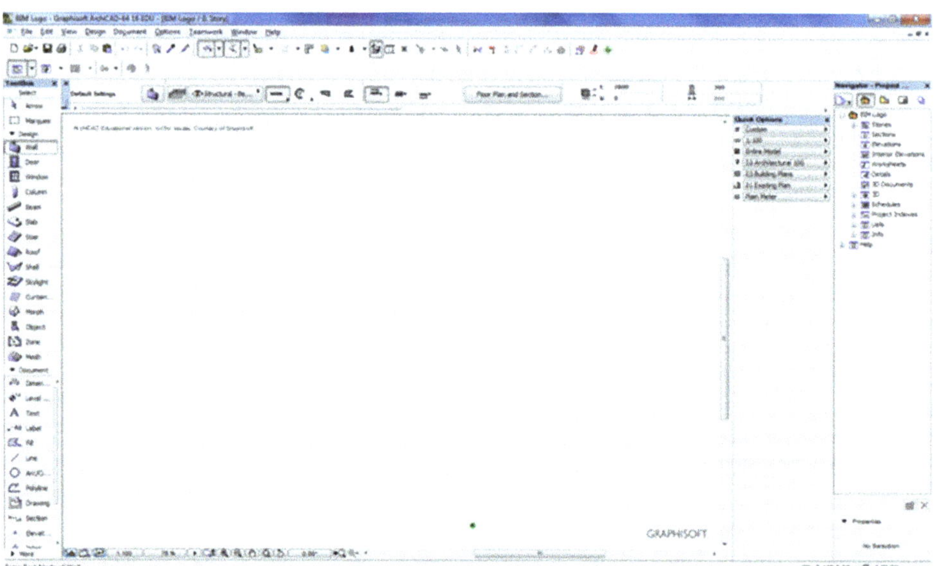

Figure 7. ArchiCAD v. 20 (tested 2017)

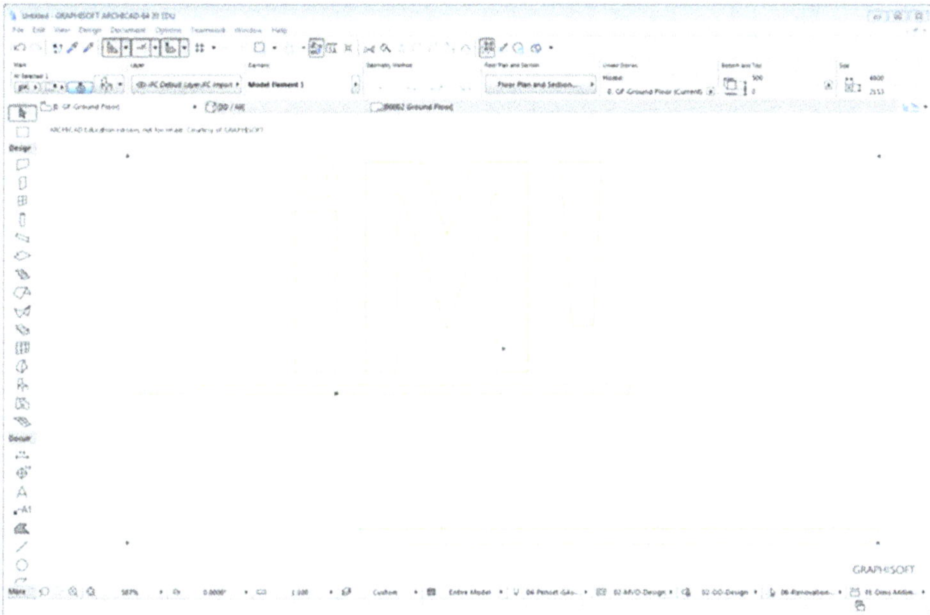

Figure 8. DDS CAD 2012 (tested 2012)

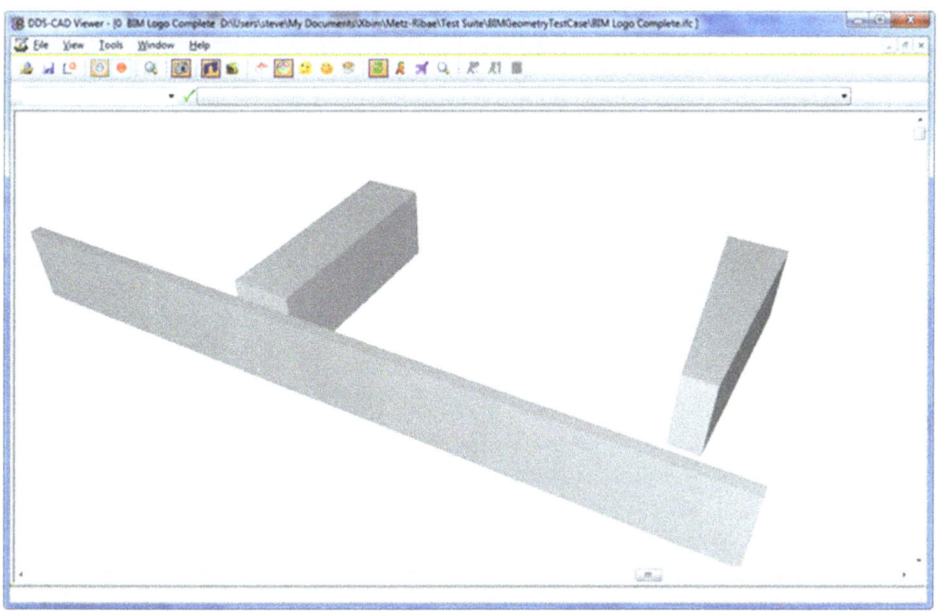

Figure 9. DDS CAD 2017(tested 2017)

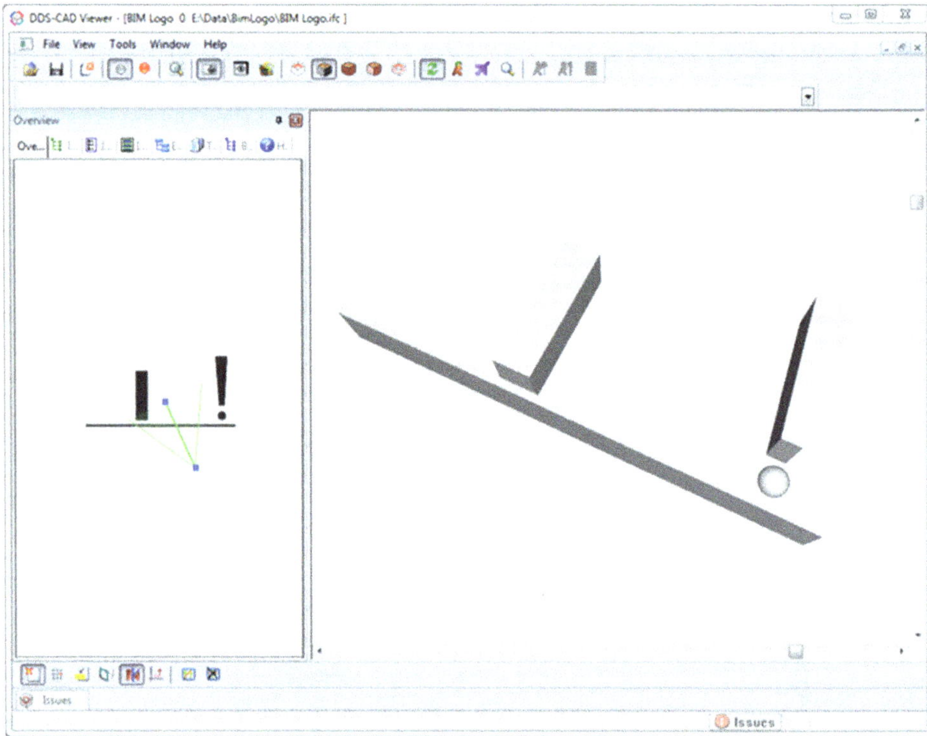

Figure 10. FZKViewer V2.2 (tested 2012)

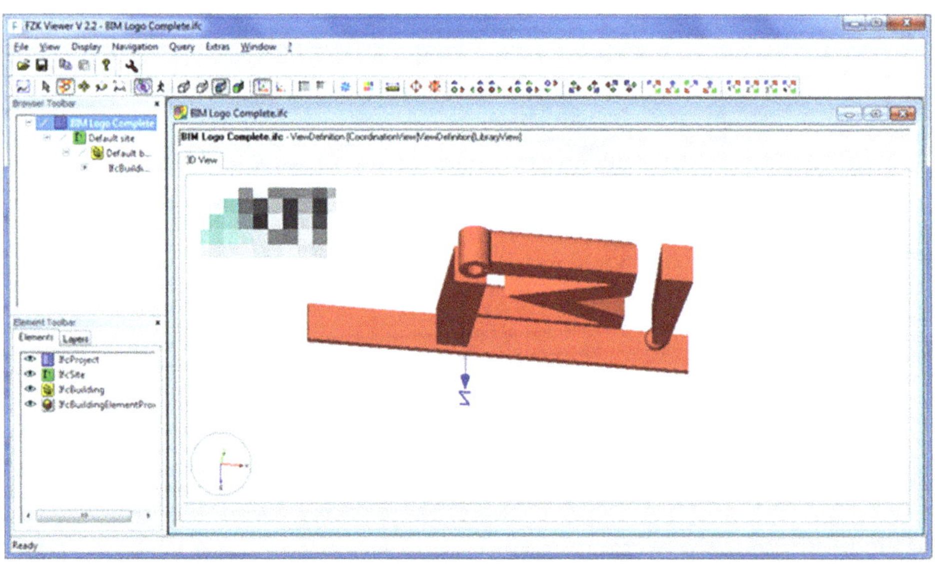

Figure 11. FZKViewer V4.8 (tested 2017)

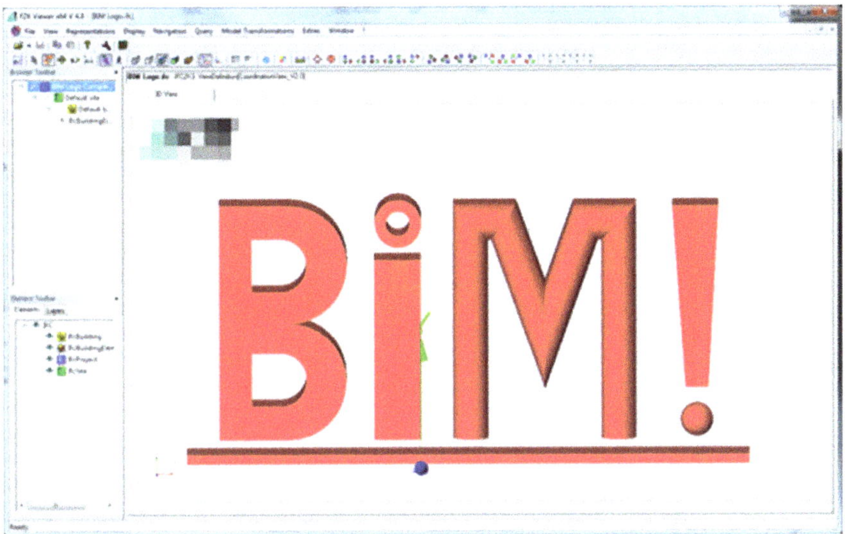

Figure 12. GTDS Viewer 0.6 (tested 2012: product discontinued before 2017)

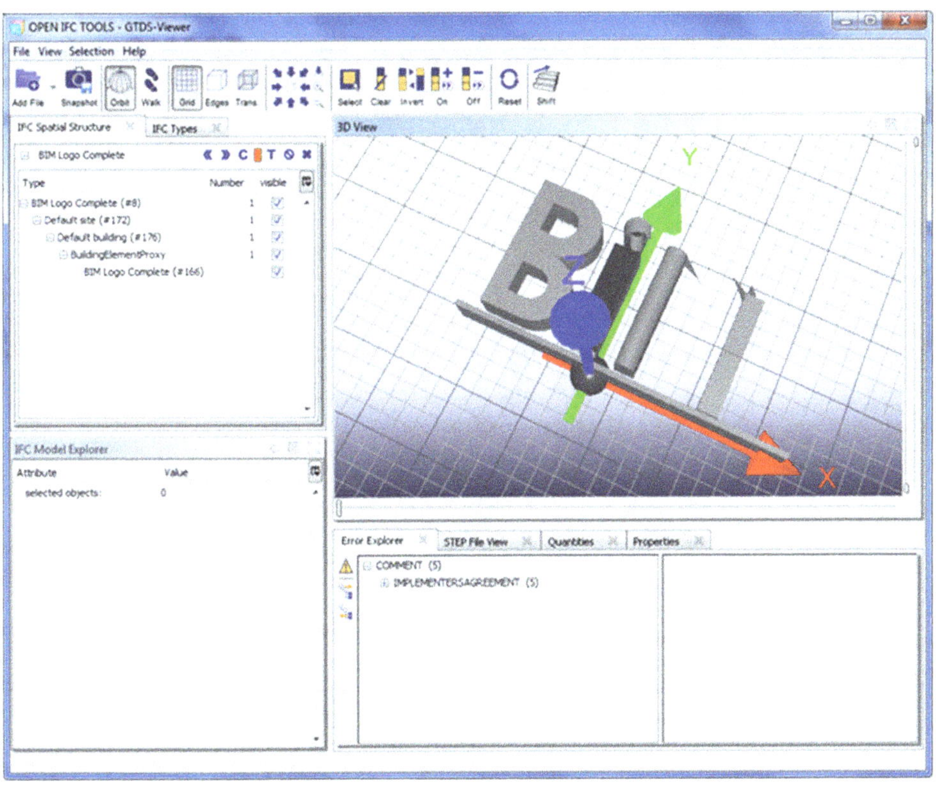

Figure 13. IfcStoreyViewer (tested 2012: product discontinued before 2017)

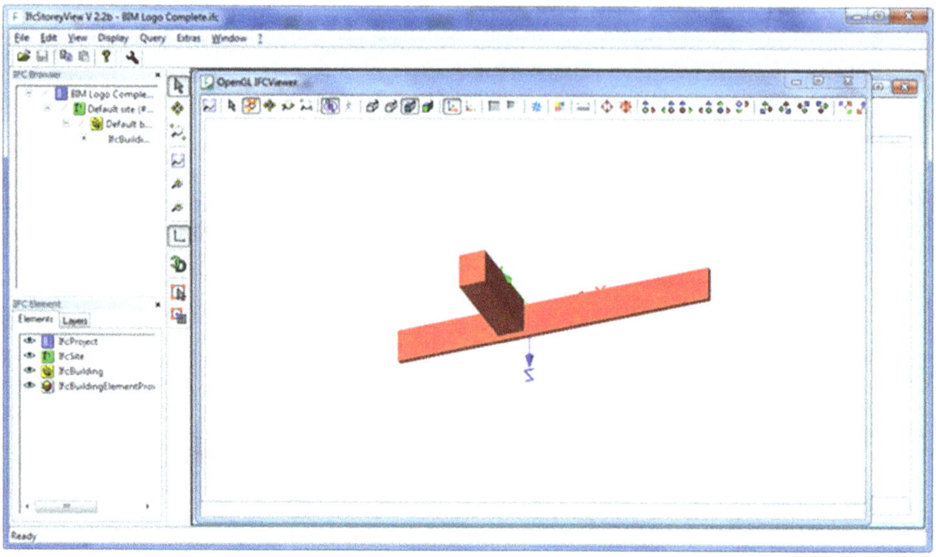

Figure 14. Revit 2012 (tested 2012)

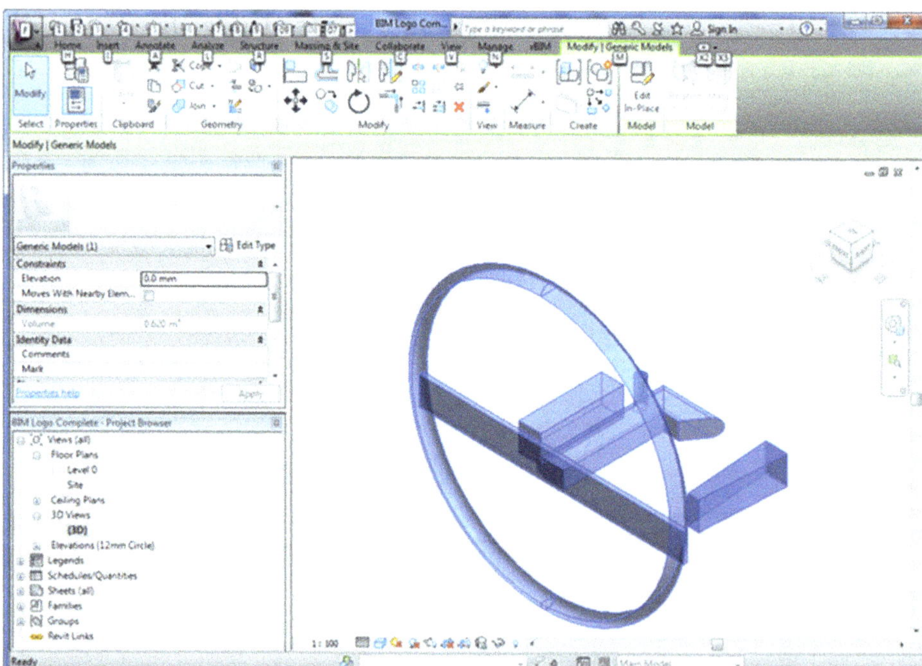

Figure 15. Revit 2017(tested 2017)

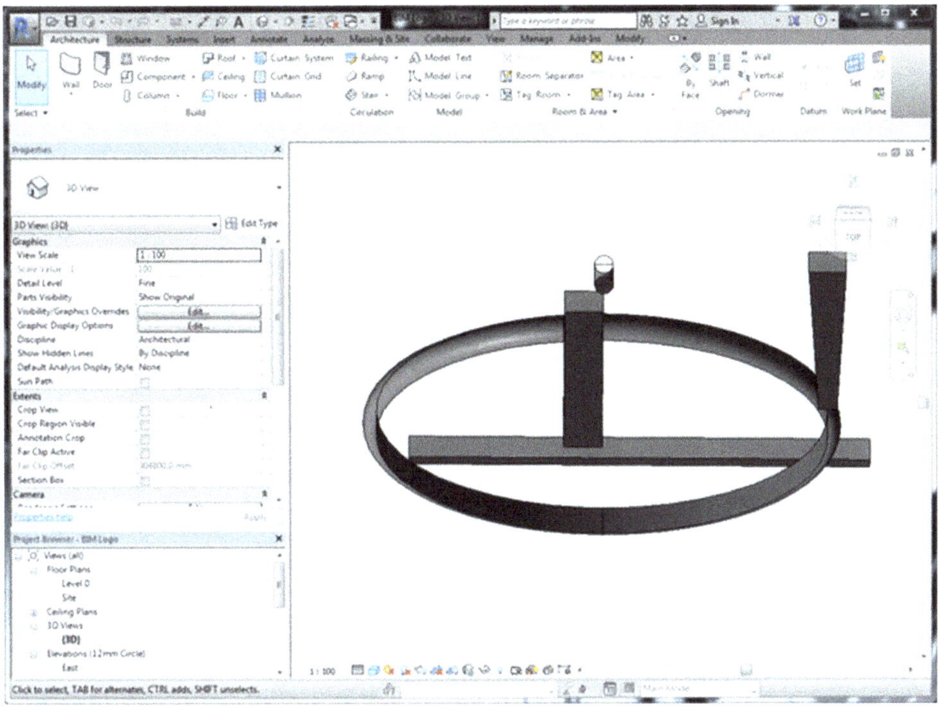

Figure 16. Solibri Model Viewer (tested 2012)

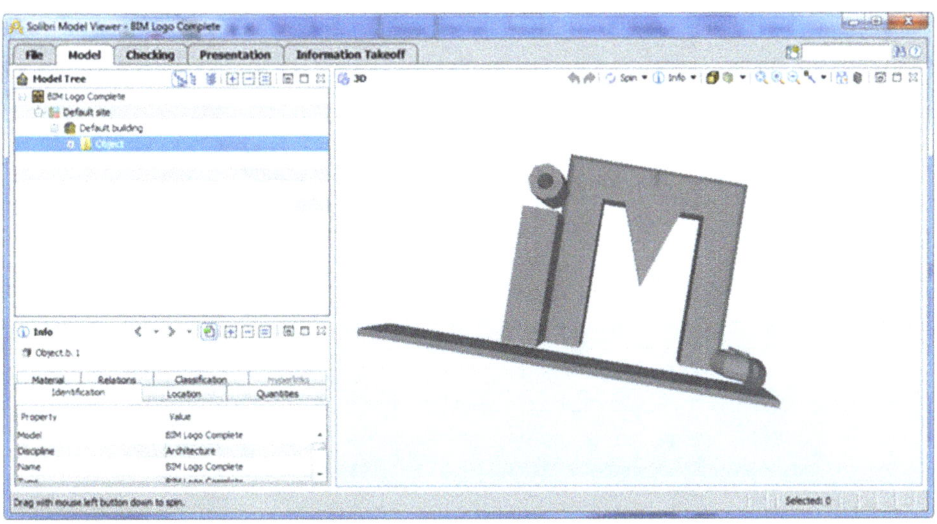

Figure 17. Solibri Model Viewer v. 9.7 (tested 2017)

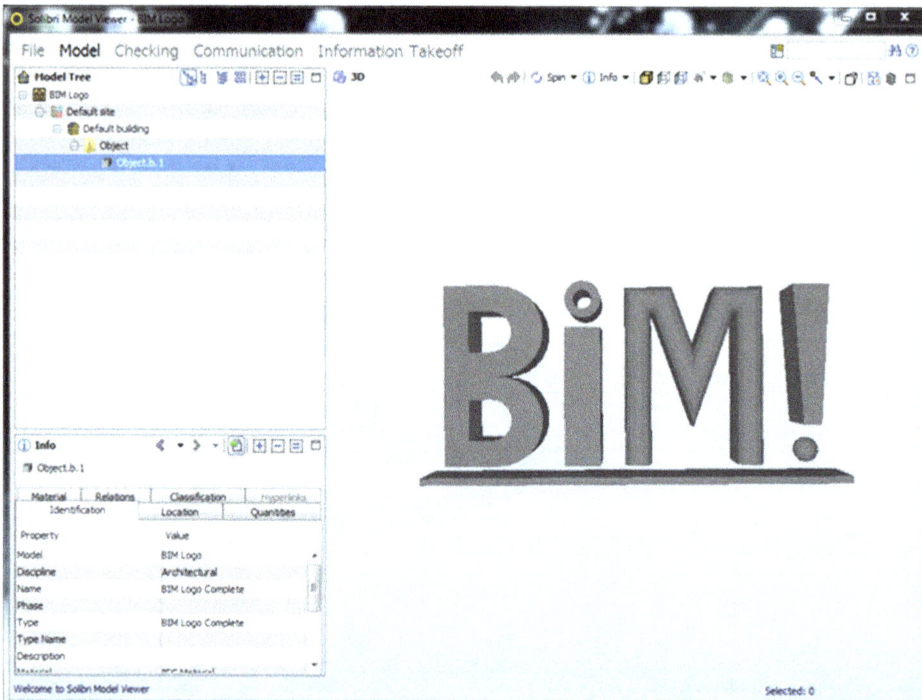

Figure 18. Tekla Structures v.18.1(tested 2012)

Figure 19. Tekla Structures v.2017sp2 (tested 2017)

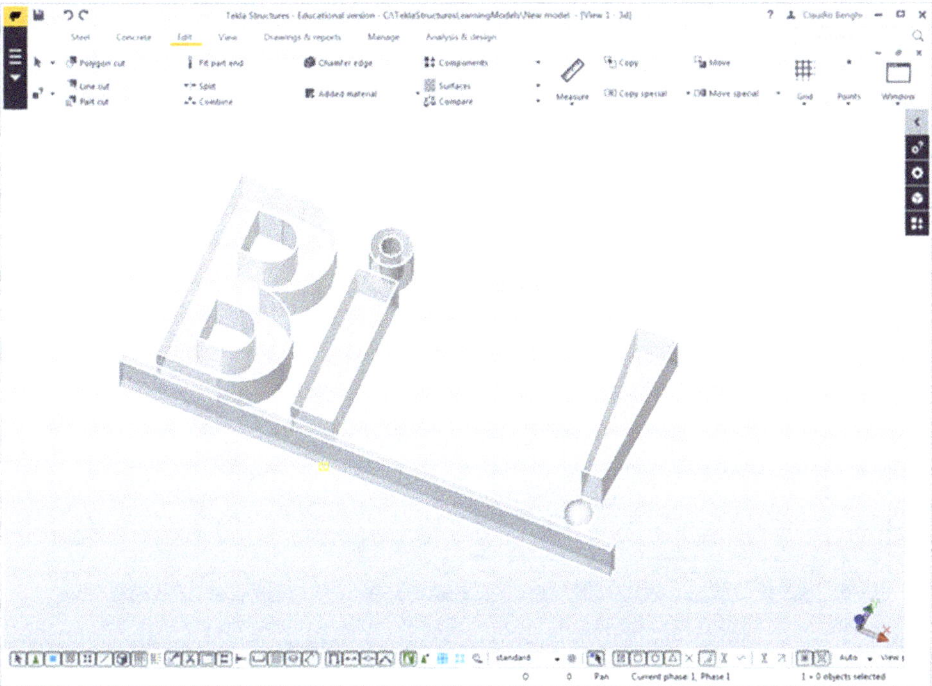

Figure 20. Tekla BIMsight (tested 2012)

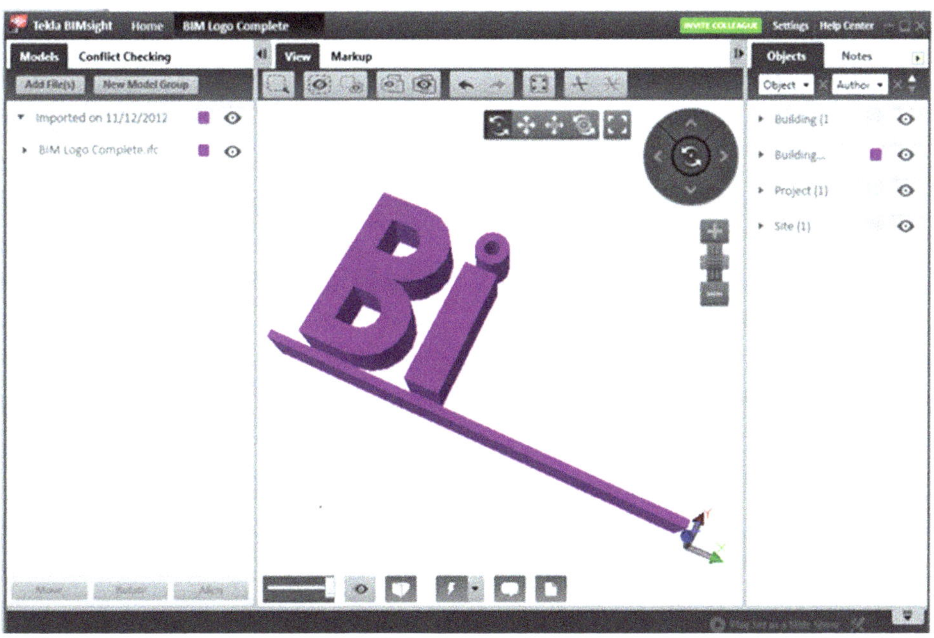

Figure 21. Tekla BIMsight (tested 2017)

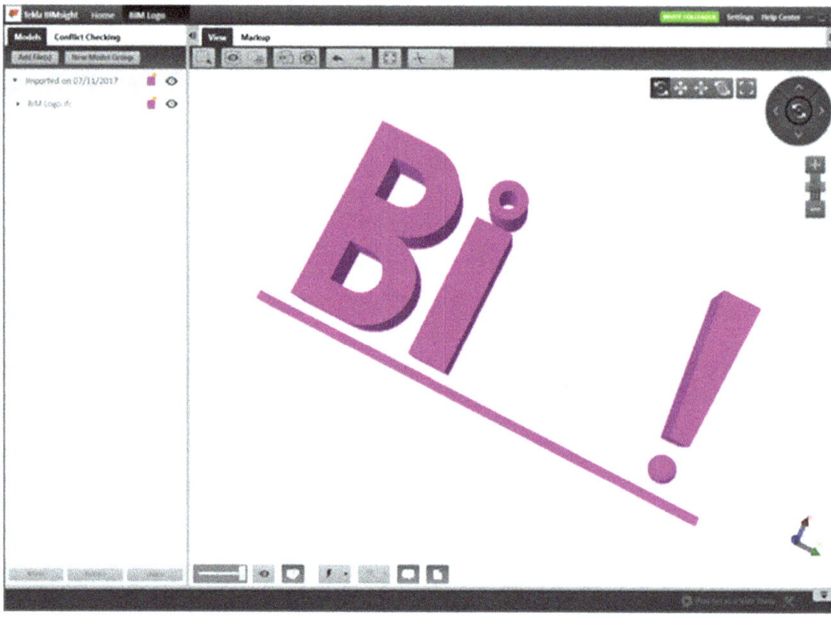

Figure 22. VectorWorks (tested 2012)

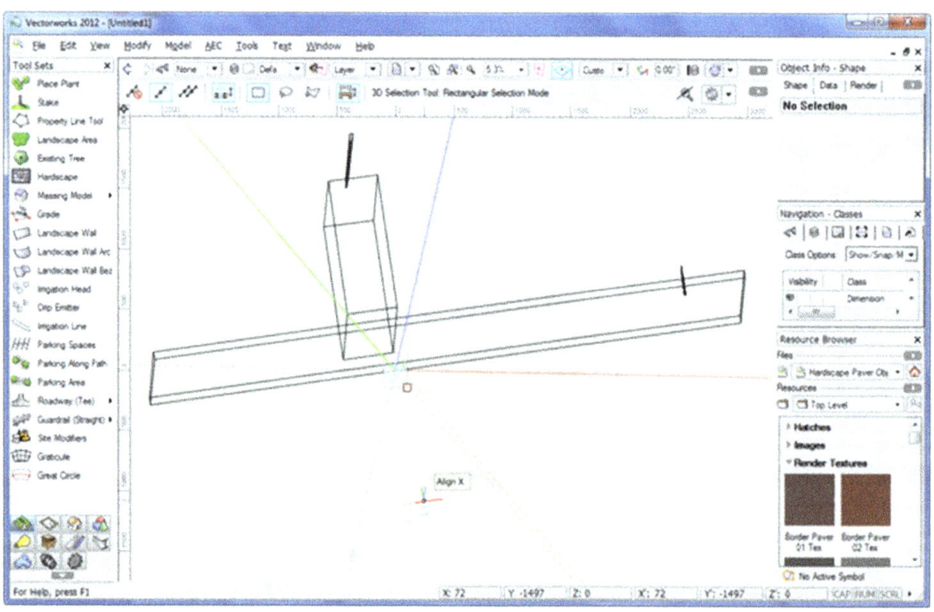

Figure 23. VectorWorks (tested 2017)

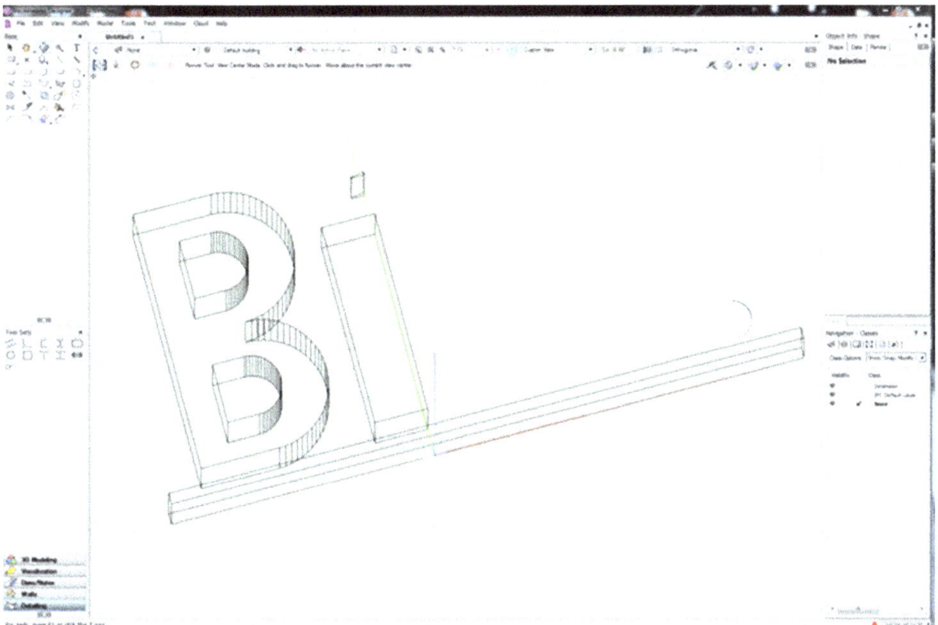

Figure 24. XbimXplorer, the items listed in the properties page display the geometric primitives used

Chapter 3
Topologically Consistent Space, Time, Version, and Scale Using Alexandrov Topologies

Patrick E. Bradley
Karlsruhe Institute of Technology, Germany

Norbert Paul
IEP Industrie Elektronik Paul GmbH, Germany

ABSTRACT

A novel approach to higher dimensional spatial database design is introduced by replacing the canonical solid–face–edge–vertex schema of topological data by a common type SpatialEntity, and the individual "bounded-by" relations between two consecutive classes by one separate binary relation BoundedBy on SpatialEntity defining an Alexandrov topology. This exposes mathematical principles of spatial data design. The first consequence is a mathematical definition of topological "dimension" for spatial data. Another is that every topology for spatial data is an Alexandrov topology. Also, version histories have a canonical Alexandrov topology, and generalizations can be consistently modeled by continuous foreign keys between LoDs. The result is a relational database schema for spatial data of dimension 6 and more, seamlessly integrating space-time, LoDs, and version history. Topological constructions enable queries across these different aspects. Giving points coordinates amounts can give rise to topological inconsistencies which can be measured with topological invariants.

DOI: 10.4018/978-1-5225-5625-1.ch003

INTRODUCTION

2D and 3D spatial models are well established for spatial data modeling, and there exist standards like CityGML in geo-spatial modeling and IFC for architectural models. These standards cover the representation of geometric objects – mostly using an incidence-based representation like, for example, (Rossignac & O'Connor, 1989; Hazelton et al., 1990). Currently there is active research on spatio-temporal queries (Sakr & Güting, 2009) as well as 4D spatio-temporal modeling (Anh, Vinh, & Duy, 2012), and also on considering other aspects like scale (v. Oosterom & Stoter, 2012) as additional dimensions of spatial data. Besides that, research on *n*D modeling provides generic spatial data models without a fixed dimension upper bound (Lienhardt, 1994) and often gives a formal definition of "topological dimension" of spatial data. Lienhardt proposes an *n*D data model called G-Maps where spatial entities and their topology are only implicitly given. Both (Bradley & Paul, 2014) and (Lienhardt, 1991) contain a broad discussion of the differences between G-Maps and incidence-based models as the one presented in this article. Topology has its own sub-discipline called "dimension theory" (Engelking, 1978) where the possible definitions of "topological dimension" are investigated. Among these, the Krull dimension (Hartshorne, 1993, p. 5) is particularly applicable for spatial data and is proposed in this article as a standard definition of spatial data dimension.

As a result of investigations if data of dimension beyond 4 have practical applications and if this could possibly lead to a "combinatorial explosion of complexity", and inspired by van Oosterom's ideas (v. Oosterom & Meijers, 2011), this article demonstrates the mathematical foundations of generic *n*D spatial modeling and its use for combining 3D spatial data, time, scale, and versioning into an integrated 6D+ model. In particular, it will be shown that sensible integration of scale increases the dimension of spatial data by more than one, hence the "+" in 6D+. Even more, it has turned out that further increasing the dimension can even *decrease* the complexity of the data model.

First a short introduction into the basic concepts of mathematical topology is given. This also covers a discussion of incidence graphs which turn out to be topological spaces. Then, as a small example application, an SQL representation of Egenhofer's (1991) nine-intersections model is presented. Surprisingly, there are SQL queries that prove the yet unverified mathematical statements in (Egenhofer, 1991) wrong. As the main intent is to expose *mathematical principles* of *n*D modeling, the relational model, formulated by Codd (1990), is used because with set theory it has the same mathematical foundation as topology (Bradley & Paul, 2010). The principles described here are also applicable to object oriented modeling. They also serve as a basis for a generic topological database model as an extension of the relational model: instead of tables, there are spaces on which queries operate, and then they return spaces as a result. Then, the relational database schema of the 6D+ model is developed step by step. By assigning coordinates to points, topological inconsistencies can occur. The concept of topological consistency is defined rigourously, and it is explained how to measure topological inconsistency. Finally, an implementation of concepts from this article is discussed.

TOPOLOGY

Topology is sort of a mathematical "brother" of set theory with its well-known relation $o \in M$ which denotes if an object o is element of a set M. In topology there is a weaker $o \in \text{cl}\, M$ which denotes that o is at least "close" to M and a stronger $o \in \text{int}\, M$ which means that o is even completely surrounded by M. "Weaker" and "stronger" means that $o \in \text{int}\, M \Rightarrow o \in M \Rightarrow o \in \text{cl}\, M$ always holds. These modified set memberships depend on each other and are specified by first fixing some "universal" set X and then providing an *additional* structure T_X for X:

Definition

Topology, topological space: when X is an arbitrary set and T_X is a set of subsets of X, then T_X is called a topology for X if it satisfies the following three axioms:

1. X and ∅ are elements of T_X.
2. For every two elements A, B of T_X the intersection $A \cap B$ is an element of X.
3. For every subset S of T_X the union $\cup_{A \in S} A$ is an element of T_X.

When T_X is a topology for X, then the pair $\underline{X} := (X, T_X)$ is called a topological space, an element of X is called a point in $\underline{X}$, and the elements of T_X are called the open sets of $\underline{X}$. Each complement $X \setminus N$ of an open set N is called closed in $\underline{X}$.

The topology is then used to define the above mentioned relationships between points and sets:

Definition

(Interior, boundary, closure) When $\underline{X} := (X, T_X)$ is a topological space, M a subset of X and p a point in $\underline{X}$, then p is said to be

- Close to M in $\underline{X}$, when every open set N_p in $\underline{X}$ which contains p as an element intersects M. We say that two sets intersect if their intersection is not empty.

The set of all points close to M is called the closure of M with respect to X and denoted by cl_X M.

- An interior point of M in X if there exists an open set N_p in X that contains p as an element and that is a subset of M. The set of all interior points of M in X is called the interior of M with respect to X and denoted by int_X M.
- A boundary point of M in X if p is close to M but not an interior point of M in X. The set of all boundary points of M in X is called the boundary of M with respect to X and denoted by bd_X M.

The closure of a set is its minimal closed superset and its interior is its maximal open subset. When the space *X* is known from the context we need not always mention it and may write "int *M*" instead of "int_X *M*". However, these notions are meaningless without a space—be it explicitly mentioned or implicitly given. Each of the three operators determines the topology and may serve as an alternative definition of a topological space. For closure (and interior) these properties that can serve as alternative axioms are:

Closure Axioms:

$$\mathrm{cl}\, \emptyset = \emptyset$$

$$\mathrm{cl}\, A \supseteq A$$

$$\mathrm{cl}\, (A \cup B) = \mathrm{cl}\, A \cup \mathrm{cl}\, B$$

$$\mathrm{cl}\, \mathrm{cl}\, A = \mathrm{cl}\, A$$

Interior Axioms:

$$\mathrm{int}\, X = X$$

$$\mathrm{int}\, A \subseteq A$$

$$\mathrm{int}\, (A \cap B) = \mathrm{int}\, A \cap \mathrm{int}\, B$$

$$\mathrm{int}\, \mathrm{int}\, A = \mathrm{int}\, A$$

From the properties also follow the monotonies that $A \subseteq B$ implies $\mathrm{int}\, A \subseteq \mathrm{int}\, B$ and $\mathrm{cl}\, A \subseteq \mathrm{cl}\, B$. Hint: $A \subseteq B$ is equivalent to $A \cup B = B$ and to $A \cap B = A$.

To give some examples we now denote closed intervals with boundaries *a* and *b* by [*a*, *b*], and open intervals by (*a*, *b*). Then the boundary of the real square [0,

1] $\times$ [0, 1] in the real plane $\mathbf{R}^2$ is the outer boundary line $\{0, 1\}\times[0, 1] \cup [0, 1]\times\{0, 1\}$ and its interior is the open unit square $(0, 1) \times (0, 1)$, whereas the boundary of the *rational* unit square $[0, 1]\times[0, 1] \cap \mathbf{Q}^2$ in $\mathbf{R}^2$ is the real unit square $[0, 1]\times[0, 1]$. Note that the latter boundary has a non-empty interior.

Examples of spaces:

- The n-dimensional Euclidean space $\mathbf{R}^n$ has the so-called *natural* topology T_d which is defined by the Euclidean distance d, a function that maps every pair a, b of points in $\mathbf{R}^n$ to their distance $d(a, b)$ —a non-negative real number. A set N is open in $\underline{\mathbf{R}}^n := (\mathbf{R}^n, T_d)$ iff (if and only if) for every point p in N there exists a distance $r_p > 0$ such that all points q in $\mathbf{R}^n$ satisfy $d(q, p) < r_p \Rightarrow q \in N$.
- When X is an arbitrary set and $R \subseteq X \times X$ is an *arbitrary* binary relation on X then the set

$$T(R) := \{ N \subseteq X \mid \forall (a, b) \in R: b \in N \Rightarrow a \in N \} \tag{1}$$

is the topology *generated by R*. These are the so-called *Alexandrov topologies*, and can be easily represented in a relational database by considering X an entity type and R a reflexive $n{:}m$-relationship type on X modeled in the classical way each database adept is confronted with in his first lectures.

The relational representation (X, R) of a topological space $(X, T_X) = (X, T(R))$ generalizes all topological data models which are based on topological associations like, for example, edge-vertex or face-edge associations modeled as an incidence graph. On the other hand *every* topology for a finite set can be modeled in this relational way which is also the most space efficient approach possible: *Every* data model for arbitrary topologies for a set X of n elements has a worst-case storage space complexity *lower* bound of $\Omega(n^2)$ for spaces of dimension 1 or more (Paul 2010). Hence, he who fears dimension-dependent "combinatorial explosions of complexity" fears spaces of dimension greater than zero without any reason. We will now prove that, first, $T(R)$ is a topology and, second, that every topology for a finite set is generated by a binary relation on that set:

Proof

To show that $T(R)$ is a topology we have to check the three axioms:

The first axiom is easy to see, because replacing N by either $\varnothing$ or X in Eqn. (1) – the definition of $T(R)$ – gives the implications $b \in \varnothing \Rightarrow a \in \varnothing$ and $b \in X \Rightarrow a \in X$ which are both trivially true.

The second axiom can be proven in a stronger version that the intersection of an *arbitrary* set of open sets is open: Let $S \subseteq T(R)$ be a set of open sets and let $\cap S$ be their intersection. We have to check $\forall\ (a, b) \in R$: $b \in \cap S \Rightarrow a \in \cap S$: Let $(a, b) \in R$ be arbitrarily chosen. Then from $b \in \cap S$ follows $a \in \cap S$ because from $b \in \cap S$ follows $b \in A$ for every $A \in S$. But every such A is in $T(R)$ and therefore satisfies $b \in A \Rightarrow a \in A$. Hence, every $A \in S$ contains a from which follows $a \in \cap S$. But when the intersection of every subset of $T(R)$ is an element of $T(R)$ then, of course, by $A \cap B = \cap\{A, B\}$ the intersection $A \cap B$ of two elements A, B from $T(R)$ is also in $T(R)$.

The third axiom can be shown similarly to the second by simply replacing "∩" by "∪", saying "there exists an *A*" instead of "for every *A*", and omitting the last sentence.

We now show that for every topology T_X for a finite set X there is a relation R^* with $T_X = T(R^*)$: When X is finite then T_X is finite, too. Then by induction on the size of any subset $S \subseteq T_X$ one easily sees $\cap S \in T_X$. But then for each point $p \in X$ the intersection N_p of all open sets U_p satisfying $p \in U_p$ is open. N_p is *the* smallest open set containing p, and is called the "minimal neighborhood of p". We define $R^* := \cup_{p \in X} N_p \times \{p\}$ and show $T(R^*) = T_X$: Assume $A \in T(R^*)$. When $p \in A$ is assumed then p has a minimal neighborhood $N_p \in T_X$. But by $N_p \times \{p\} \subseteq R^*$ every point $a \in N_p$ satisfies $(a, p) \in R^*$ and therefore a must be in A. Therefore $p \in N_p \subseteq A$ holds and hence we have $p \in \text{int}\, A$ with respect to T_X. So every point p in A is an interior point of A with respect to T_X which gives $A \in T_X$. On the other hand, when $A \in T_X$ holds, the minimal neighborhood N_b of every point $b \in A$ with respect to T_X is a subset of A. Now from both $(a, b) \in R^*$, and $b \in A$ and by $(a, b) \in N_b \times \{b\}$ and $N_b \subseteq A$ we have $a \in A$ and, hence, $A \in T(R^*)$. □

Note that R^* is transitive and reflexive whereas R is an arbitrary relation. Two relations R and S generate the same topology iff they have the same transitive and reflexive closure $R^* = S^*$. So we can remove tuples of shape (a, a) and tuples (a, c) if $a\ S\ b\ S\ c$ holds to save storage space without changing the topology. But this cannot improve the complexity bound of $O(n^2)$ for arbitrary topologies. However, when R is minimal with respect to $R^* = S^*$, hence when no such redundant tuple (a, c) exists, then we call (X, R) the *incidence graph* of the topological space $(X, T(R))$. On the other hand, when R is acyclic, then R^* is a partial order on X which yields the *poset representation* (X, R^*) of the topological space $(X, T(R^*)) = (X, T(R))$. Poset representations and incidence graphs are popular approaches in spatial data modeling (Rossignac & O'Connor 1989; Hazelton & Leahy & Williamson 1990). Here we have just seen that these do not only somehow "represent" topological associations. They are topological spaces on their own. As there is no mathematical notion of "inconsistent" topological spaces, the topology-defining relations are completely arbitrary. We will discuss topological consistency constraints later.

EXAMPLE APPLICATION

We will give a brief example application of a relational representation of a topological space. The following space results from partitioning the real plane $\mathbf{R}^2$ into 81 cells (Figure 1).

For convenience of the reader we have numbered the cells such that the dimension of a cell in X equals the number of odd digits of its identifier. Note that all cells of all dimensions are collected into one set. For example cell 11 is the lower left two-dimensional "tile". Cell 12 is the horizontal edge between tile 11 and tile 13 terminated by vertex 22, the common edge-point of tile 11 and tile 33. The table R represents the topology: Tuple R(11, **12**) denotes e.g. that **12** $\in$ bd{11}. The grid has no outer boundary just as $\mathbf{R}^2$ has no outer boundary. As $(X, T(R))$ represents a *quotient space* of $\mathbf{R}^2$ all topological properties of subsets of X in $(X, T(R))$ correspond to similar properties of corresponding subsets of $\mathbf{R}^2$. For example {11} is open in $(X, T(R))$ and so is the subset of real points in $\mathbf{R}^2$ the number 11 stands for.

Egenhofer (1991) introduced the so-called nine-intersections. Each set A of points in a topological space X defines three sets int A, bd A, and $X \setminus A$ which are disjoint iff A is closed. If we define ext $A := X \setminus \text{cl } A$, then two such sets A and B result in nine pairs of possibly intersecting sets which gives a Boolean matrix

$$I(A, B): \{\text{int}A, \text{bd}A, \text{ext}A\} \times \{\text{int}B, \text{bd}B, \text{ext}B\} \rightarrow \{\text{true}, \text{false}\} \qquad (3)$$

*Figure 1. The space (X, T(R)) that results from partitioning the real plane $\mathbf{R}^2$ into 81 cells; the horizontal edge between 11 and 13 is boundary cell **12**, itself bounded by cell 22, which is marked by a black bullet in the right picture.*

X
id
11
12
13
...
21
22
...
99

R	
ida	idb
11	**12**
11	21
13	**12**
...	
12	22
...	
98	88

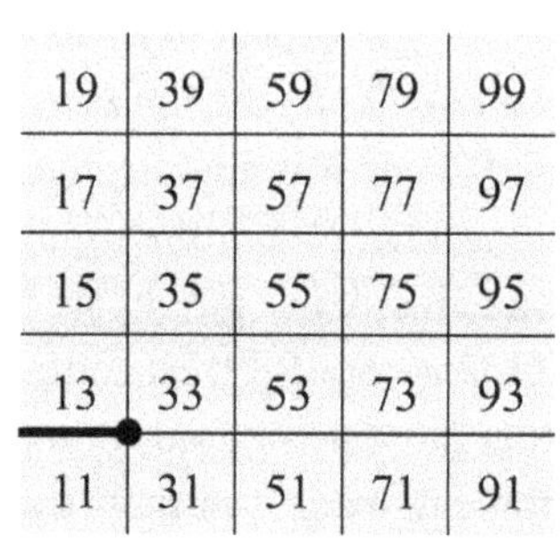

that tells for each pair *A*, *B* whether their corresponding sets intersect. With our relational database model (*X*, *R*) with schema *X(id)* and *R(ida*, *idb*) of the real plane we can compute nine-intersections in SQL and establish test cases for them. First we need the transitive and reflexive closure R^* of *R*:

```
create view poR(ida, idb) as
  select id as ida, id as idb from X  union
  select R.ida, R.idb from R          union
  select R.ida, R2.idb
  from R join R R2 on (R.idb = R2.ida);
```

As the dimension is 2 no recursive join computation is needed for R^*. We assume that each relational representation (*X*, *R*) of a topological space has such an associated view on R^* named poR where "po" stands for "preorder". Now for subsets *A* and *B* of *X* we define the following views:

```
create view intA(id) as
  select A.id  from A
  where not exists
    (select poR.ida  from poR
     where poR.idb = A.id
       and poR.ida not in (select id from A));
create view clA(id) as
  select distinct X.id  from X, poR, A
  where X.id = poR.idb and poR.ida = A.id;
create view bdA(id) as
  select clA.id  from clA
  where clA.id not in (select id from intA);
create view extA(id) as
  select id  from X
  where id not in (select id from clA);
```

We advocate that at least the operators int and cl should be built-in features of future RDBM systems. With the same views for *B* their nine-intersection matrix can be computed by

```
select
 exists(select * from intA ia, intB ib where ia.id=ib.id)
 as iaib,
 exists(select * from bdA da,  intB ib where da.id=ib.id)
```

```
as daib,
exists(select * from extA xa, intB ib where xa.id=ib.id)
as xaib,
exists(select * from intA ia,  bdB db where ia.id=db.id)
as iadb,
exists(select * from bdA da,   bdB db where da.id=db.id)
as dadb,
exists(select * from extA xa,  bdB db where xa.id=db.id)
as xadb,
exists(select * from intA ia, extB xb where ia.id=xb.id)
as iaxb,
exists(select * from bdA da,  extB xb where da.id=xb.id)
as daxb,
exists(select * from extA xa, extB xb where xa.id=xb.id)
as xaxb;
```

which returns a result tuple with nine Boolean values where each denotes if the corresponding sets intersect. For example iaxb ⇔ int $A \cap$ ext $B \neq \emptyset$.

Egenhofer (1991) restricts these nine-intersections to special subsets of a topological space. In order to decide if A and B meet the four preconditions established in (Egenhofer 1991) we defined the following SQL-predicates:

1. Each set A and B must be "full dimensional":

```
exists(select * from intA) and exists(select * from intB)
```

2. Each set A and B must be non-empty:

```
exists(select * from A) and exists(select * from B)
```

but this already follows from precondition 1.

3. All six sets involved must be connected: Connectivity can be tested by defining the zero-dimensional "connection space" $(X, R \cup R^T)$ associated to the space (X, R). R^T is the transposed of R, hence $a\, R^T\, b \Leftrightarrow b\, R\, a$. The closure of a point in $(X, R \cup R^T)$ is its connected component in (X, R). Then we can query connectedness of a space (X, R) in SQL: We first define the connection topology:

```
create view RuRt(ida, idb) as
select ida, idb from R   union          -- relation R
select idb as ida, ida as idb from R; -- R transposed
```

Note that this is a different topological space with *identical* point set! This is only possible when the topology is an additional structure similar to the approach in mathematics. A topology hard-coded into the entities of *X* would make this much more difficult. Then using the preorder view *poRuRT* of *RuRT* we have the SQL-predicate for connectedness of (*X*, *T*(*R*))

```
(select count(*) from X)^2 = (select count(*) from poRuRT)
```

because *poRuRT* contains all pairs of points in X that are connected by some path in (*X, T(R)*). To test connectedness of subsets of a space we need the so-called subspace topology. In order to topologically correctly restrict a relation to a subset, it must be transitive in general. The straight-forward definition of the subset topology of *M* is the restriction of *poR* to *M*:

```
create view poRM(ida, idb) as (
   select ida, idb  from poR
   where ida in (select id from M)
     and idb in(select id from M));
```

Then connectedness of *M* can be queried by

```
create view connectedM(connected) as (
with recursive CCM(ida, idb) as (
  select id as ida, id as idb from M   union
  select CCM.ida as ida, MB.id as idb
  from CCM, M MA, poRM, M MB
  where CCM.idb = MA.id and
            (MA.id = poRM.ida and MB.id = poRM.idb) or
            (MB.id = poRM.ida and MA.id = poRM.idb))
  select 'yes' as connected
  where (select count(*) from M)^2
        = (select count(*) from CCM));
```

This query returns a table with a column named “connected” and a tuple with value ‘yes’ if *M* is connected and an empty table if *M* is not connected. The SQL predicate then simply is

```
exists(select * from connectedM)
```

For all six views *intA*, *bdA*, *extA*, *intB*, *bdB*, and *extB* we have defined *connectedIntA*, etc. accordingly. An intersection (or natjoin) of them tells if all of them are connected.

4. Egenhofer's fourth requirement is that the sets A and B must be regular, i.e. cl int $A = A$ and cl int $B = B$: This could also be tested by queries in a manner similar to the above. A and B could also be regularized by using cl int A instead of A and cl int B instead of B. However, for our test case we have chosen a different approach where regularity need not be tested at all: We only take closures of open sets because these are the regular sets.

Before we present our test results we prove the above statement, that a set is regular if and only if it is the closure of an open set.

Proof

We first change our notation and write int(N) and cl(N) with parentheses to facilitate readability. As int(A) is open every regular set is the closure of an open set. Conversely, every closure of an open set is regular: Let N be open. Every set N satisfies $N \subseteq \mathrm{cl}(N)$. By monotony of interior and closure we have $\mathrm{cl}(\mathrm{int}(N)) \subseteq \mathrm{cl}(\mathrm{int}(\mathrm{cl}(N)))$. As N is open we have $\mathrm{int}(N) = N$ and therefore $\mathrm{cl}(N) \subseteq \mathrm{cl}(\mathrm{int}(\mathrm{cl}(N)))$. On the other hand $\mathrm{int}(\mathrm{cl}(N)) \subseteq \mathrm{cl}(N)$ holds from which $\mathrm{cl}(\mathrm{int}(\mathrm{cl}(N))) \subseteq \mathrm{cl}(\mathrm{cl}(N)) = \mathrm{cl}(N)$ follows. Both inclusions give $\mathrm{cl}(\mathrm{int}(\mathrm{cl}(N))) = \mathrm{cl}(N)$ which says that cl(N) is regular. □

Exploiting the above proven property we have created test cases by choosing only from the tiles as sample sets and then taking their closure. These closures of non-empty tile sets satisfy all preconditions except connectivity. Rectangles with unconnected boundary and unconnected exterior can be easily excluded a priori. Then by iterating all these rectangular subsets of our example grid, the above SQL statements surprisingly returned five more nine intersection patterns for input that satisfies the above preconditions but that have been claimed to be impossible in (Egenhofer 1991). Namely the patterns that showed up are where each tuple denotes a non-empty intersection of the two corresponding sets. These patterns for example occur with the following sets:

(a) $A = X$ $\quad B = X$

(b) $A = \mathrm{cl}\,\{11\}$ $\quad B = X \setminus \{11\}$

(c) $A = \mathrm{cl}\,\{11, 13, 31, 33\}$ $\quad B = X \setminus \{11\}$

Table 1.

A	B
int	int

A	B
bd	Bd
int	Ext
ext	Int

A	B
bd	int
int	bd
int	int
int	ext
ext	int

A	B
int	int
int	bd
int	ext

A	B
int	int
bd	int
ext	int

(d) $A = X$ $B = \mathrm{cl}\ \{11\}$

(e) $A = \mathrm{cl}\ \{11\}$ $B = X$

The sets satisfy all preconditions and have also been accepted by the queries that check them. Whereas the non-rectangular cases (b), and (c) have not been created automatically, they have equivalent automatically created counterparts. The corresponding subsets of $\mathbf{R}^2$ also turned out to meet these preconditions that have been claimed in (Egenhofer 1991) to exclude the above intersection patterns found. For example, test case (c) could represent

$$A = \{\ (x, y) \in \mathbf{R}^2 \mid x \leq 4, y \leq 4\ \}, B = \{\ (x, y) \in \mathbf{R}^2 \mid x \geq 2 \vee y \geq 2\ \}$$

in the real plane $\mathbf{R}^2$.

To wrap all up, a relational representation of topological spaces is possible, even facilitates the assessment of spatial reasoning approaches, and, for example, can help to create test cases for spatial reasoning concepts. It can even asses mathematical claims and it also challenges the claim that SQL is "insufficient for spatial data" (Egenhofer 1994). Whereas some queries involved may look rather complicated they realize standard topological operators – thereby demonstrate that these operators are feasible. Hence, we advocate to realize them as built-in operators of relational database management systems. Now after having seen a more theoretical application we now want to direct our attention to a topological model of more practical use.

DIMENSION

3D spatial data represent four kinds of entities: *Vertices* usually are zero-dimensional discrete points in $\mathbf{R}^3$. *Edges* are one-dimensional manifolds between their boundary vertices. *Faces* are two-dimensional manifolds enclosed by one or more loops of boundary edges. *Solids* are three-dimensional manifolds enclosed by boundary faces which constitute a cavity within which the solid resides. Such cavity is often called a *shell*. Most 3D models establish a "chain" of four classes with two consecutive classes connected by a "bounded-by" association (cf. (Zlatanova, Rahman & Shi, 2004) for an overview). Note that the chain length equals the model dimension.

According to 3D models, time can be modeled by the real line **R**. Each moment in time can be represented by a real number, thus resembling a vertex in $\mathbf{R}^3$. A time span is an open interval (t_1, t_2) bounded by starting point t_1 and ending point t_2, thus resembling an edge. This gives two classes of temporal entities: *Moment* and *Timespan* where each one-dimensional time span is bounded by two zero-dimensional moments. This "bounded by" association can be considered a special case of an association "chain" of length 1 which, again, corresponds with the dimension.

When data changes over time several versions of the data exist. In versioning software two consecutive versions along a version history are commonly connected by an edge. Versions can fork and merge and so a version history is a directed acyclic graph (DAG) with two classes, a *Version* and a *Transition* and each transition is bounded by an initial and a terminal version. So we could say that the "dimension" of a version history is one: there is only one association from *Transition* to *Version*. We will later take an alternative view on the version history with a different dimension.

Spatial data is often organized hierarchically at different levels of detail (LoD). For smooth transitions between LoDs it is often proposed to interpolate between consecutive LoDs by continuous morphing (v. Oosterom, & Meijers, 2011). Then between two consecutive LoDs one can assume an edge bounded by two LoDs giving a one-dimensional space.

Thus we have four types of spaces which we might call "elementary spaces": The "spatial" space which is the Euclidean 3D, the one-dimensional temporal space, a one-dimensional version space, and another one-dimensional LoD space which is essentially a linear graph $\cdots \to \bullet \to \bullet \to \cdots$.

A combination of these elementary spaces can give higher dimensional "combined spaces". In such a combined space a 3D *Solid* s and a 1D *Timespan ts* can be combined to a 4D pair (s, ts) which represents a 4D entity in space-time: the trajectory of s during ts. With a zero-dimensional *Moment* in time t, the pair (s, t) becomes a 3D space-time element representing solid s at moment t. When ts is bounded by t then the 4D-entity (s, ts) is bounded by a 3D-entity (s, t) and for each lower-dimensional nD element x in the bounded-by association of s there is a pair (x, t) in a corresponding association of (s, t). Therefore in the 4D model the length of the association chain has increased by 1. Within that model each pair (a, b) satisfies the dimension formula $\dim(a, b) = \dim a + \dim b$.

As database management systems (DBMS) only model finite sets we need a definition of "dimension" for finite spaces. As seen above, nD spaces consist of entities, possibly distributed over several classes, and a bounded-by relation of chain length n. Hence "dimension" of spatial data is the maximal length of a chain of entities such that each is bounded by a consecutive element. We call this the *combinatorial dimension* of spatial data. The note (Bradley & Paul, 2013) proves that this combinatorial dimension is equivalent to the topological Krull dimension (Hartshorne, 1993, p. 5). Note that even a simple sequence $(n, n - 1, ..., 0)$ can be considered a set $\{n, n - 1, ..., 0\}$ with "association" $a\ S\ b \Leftrightarrow a - 1 = b$ which then has dimension n and obviously no "combinatorial explosion of complexity" takes place.

Spatial Dimension and Consistency Rules

Usually, DBMS consistency rules are design tools, and it lies at the discretion of the user to make use of them. However, in spatial data modeling, "topological consistency" is often mentioned, but the rules to tell "consistent" from "inconsistent" spaces vary in the literature. We see two reasons for that: First, as mentioned above, the user should be entitled to decide which spatial data he considers "consistent" and which he does not. Second, the term "topological (in)consistency" is unknown in mathematics. Topology provides many well-defined topological *properties* which may or may not be used for a particular application as a consistency rule. But when consistency constraints are established for query input we see no advantage to enforce them also for query results. For example, in an Alexandrov space a *vertex* is an element that is not bounded by another element. The room connectivity graph of a building consists of rooms which are the 0D vertices, and the doors are their connecting 1D edges. Each door is then "bounded by" its two rooms. This possibility of "shifting down" the dimension and inverting it by "flipping" the boundary relation is a characteristic of Alexandrov spaces. That "flipping" is intimately related to the Poincaré duality. A topological database may provide a query which first selects doors and rooms thus removing all door frames and room boundary walls from the result. Then by querying the dual space we immediately get the room connectivity graph where the room volume objects are of dimension 0 and their connecting doors are of dimension 1.

One consistency rule is usually hard-coded into the classical sequential class schema: The only possible association between a face and one of its boundary vertices is indirectly via an intermediate edge. The schema inhibits direct face-vertex associations that "bypass" the edge class. However, a vertex within a face may represent a collapsed city within a region at a lower LoD. By specifying a class *SpatialEntity* as superclass of *Solid*, *Face*, *Edge*, and *Vertex* and replacing the associations between two consecutive classes by one association of *SpatialEntity* to itself, even that rule can be made optional. For all these reasons we discourage from partitioning spatial entities into a sequence of "dimension classes" and prefer collecting them into one topological point set such that the associated topology determines the dimension of the entities.

When the hard-coded dimension constraint becomes optional this immediately leads to a directed acyclic graph (DAG) of spatial entities. A DAG defines a partial ordering on its elements, and since 1937 it is well-known that partial orderings are essentially the same as the so-called T_0 *Alexandrov topologies* (Alexandrov, 1937), or, short, *spatial data models are topological spaces*. We will now provide an initial relational database schema for 3D spatial data based on this observation:

$$X(\underline{id}, attributes),\ R(\underline{ida}, \underline{idb}),\ \text{FK}:(ida) \to X,\ \text{FK}:(idb) \to X\ . \tag{4}$$

$$Vertex\ (\underline{vid}, x{:}\mathbf{R}, y{:}\mathbf{R}, z{:}\mathbf{R}),\ \text{FK}:(vid) \to X$$

Table X contains the spatial entities, and table R specifies the "bounded-by" relation. Primary key attributes are underlined. FK: $(a) \to T$ denotes a foreign key reference from the attribute(s) a to the primary key of the table T. Attribute *attributes* represents the set of "semantic", i.e. non-spatial, attributes. Although every relation $R \subseteq X \times X$ defines the Alexandrov space (X, T(R)) according to Eqn. (1) that space is T_0-separable iff R is acyclic (Adamson, 1995, Ex. 133). T_0–separability will be our only consistency rule enforced here, but, from the mathematical viewpoint, it is optional, too. Now for each topology-defining relation R on X we assume pre-order view $poR = R^*$ on the transitive and reflexive closure of R as presented above. As the dimension is somewhat bigger we use the SQL-query for arbitrary preorders which must use recursive join:

```
create view poR as
with recursive po(ida, idb) as (
   select id as ida, id as idb from X  union
   select po.ida, R.idb
   from po join R on (po.idb = R.ida))
select * from po ;
```

This schema allows spatial data of arbitrary dimension, but for the moment we assume that R chains have length ≤ 3, so the topological dimension matches the "geometric dimension". This is the number of the (floating point) coordinate attributes of *Vertex*. Also, the *id* of all vertices in X should occur in table *Vertex*. For obvious reasons, we call a pair (X, R) of a set X and a binary relation R on X a *topological data type*, or simply a *space*.

Figure 2. The process of a 1D house in Lineland constructed at time t_0, extended by a portion X at time t_1, and demolished at time t_2, is modeled as a 2D space-time complex in two steps. The middle complex repeats each element at a given point or period of time, and the lower complex collapses some elements, to get fewer entities to store.

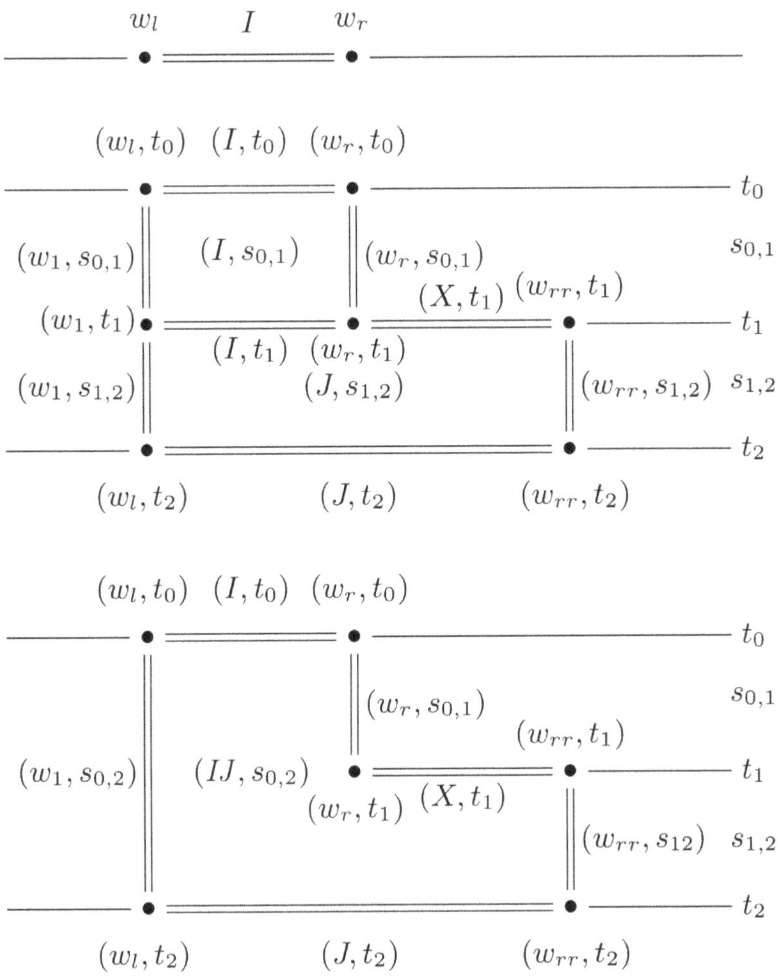

Temporal Dimension

Here, we establish the 4D *space-time* which models changes of a 3D space over time. For illustrative reasons, a 2D space-time of a 1D "building" example in Lineland (Abbott, 1884, Ch. 13) changing over 1D time will be discussed first. Imagine in 1D Lineland a "house" with interior *I* and boundary "walls" w_l to the left and w_r to the right as depicted on top of Figure 2. The boundary of *I* are the vertices w_l and w_r.

The house is erected at time t_0, modeled by "tagging" *I*, w_l and w_r with t_0 by taking the pairs (I, t_0), (w_l, t_0), and (w_r, t_0). The spatial "bounded-by" associations from *I* to w_l and from *I* to w_r are carried over as tagged space-time associations from (I, t_0) to (w_l, t_0), and from (I, t_0) to (w_r, t_0). The interior *I* during a time-span $s_{0,1}$ is also tagged to produce its trajectory $(I, s_{0,1})$. This element is bounded in the horizontal, or "spatial", direction by $(w_l, s_{0,1})$ because *I* is bounded by w_l . Here the "tag" $s_{0,1}$ does not change in this boundary association. The element $(I, s_{0,1})$ is also bounded in the vertical, or "temporal", direction by (I, t_0). Here the boundary association fixes element *I* and is taken from the boundary association between $s_{0,1}$ and t_0 . The space at time point t_1 models the before-after change derived by a 1D overlay of the spatial model of *I* before, and the spatial model of *J* after the change. Each overlay entity has a reference to its corresponding entity of the input spaces. These are two continuous partial functions *p* (like "past") and *f* (like "future") from the overlay space back to the two input spaces, also called *attaching maps* (Jänich, 1995, Ch. 3, §7). Now we paste the overlay onto the "past" entities by specifying that a tagged image element $(I, s_{0,1})$ is bounded by an element (x, t_1) if $p(x) = I$ holds. We also paste onto the "future" by specifying that $(w_l, s_{0,1})$ is bounded by (w_l, t_1) because of $p(w_l) = w_l$. Interestingly (w_r, t_l) is a boundary element of the past wall trajectory $(w_r, s_{0,1})$ because of $p(w_r) = w_r$, but it is a boundary element of the future interior trajectory $(J, s_{1,2})$ because of $f(w_r) = J$. This intermediate formal step would create a lot of redundant data when implemented explicitly. So tagged space-time entities not changing over time should be automatically collapsed into one. Additionally, the user may specify further entities that may change over time but are considered "identical" before and after that change. In our example automatic identification should be carried out on $(w_l, s_{0,1})$, (w_l, t_1), and $(w_l, s_{1,2})$, whereas $(I, s_{0,1})$, (I, t_1), and $(J, s_{1,2})$ are manually "collapsed" into $(IJ, s_{0,2})$. This then gives a topological "quotient space" (Jänich, 1995, Ch. 3) depicted in the lower part of Figure 1.

The same construction applied to 3D spatial models gives a topological 4D space-time. But when the 3D model is represented by a chain of three associations, and when the temporal model also has its classical layout, we get a grid of eight different classes of spatio-temporal entities and ten different space-time-boundary associations as depicted in Figure 3. This complexity could be somewhat reduced by an approach like (Hazelton & Leahy & Williamson 1990) where, for example, TimeSolid and IntervalFace are unified into a POLYHED with a Boolean attribute "WORLD" that tells whether a POLYHED in question is "time-like" (IntervalFace) or "space-like" (TimeSolid). This approach would then only give five spatial entity classes. However, with only two classes SpatialEntity and TemporalEntity, each with a relation BoundedBy there would be only one spatio-temporal class where each entity consists of a pair (spatial, temporal), and only one boundary relation associating each pair (spatial, temporal) with all (*ds*, temporal) for every boundary

Figure 3. Directly combining the classical sequential space model with the same classical approach for a temporal model creates eight classes for time-space entities and ten different incidence relations

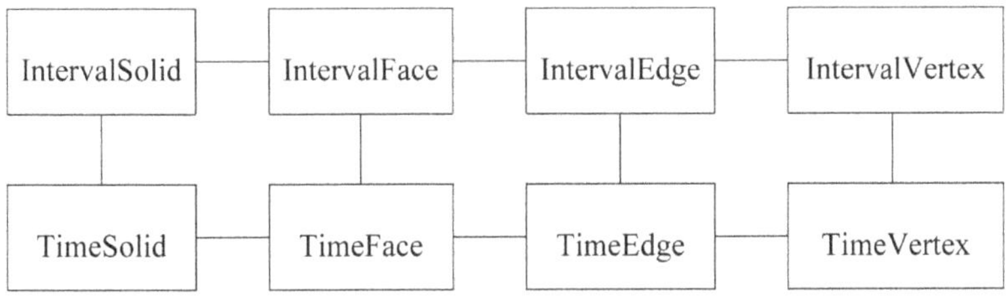

element *ds* of spatial and, dually, with the pairs (spatial, *dt*) for every boundary time point *dt* of temporal. This yields the topology of the so-called product space (Jänich, 1995, Ch. I, §3). So the class SpaceTimeEntity with a relation BoundedBy (cf. Figure 4) allows to model arbitrary space-time configurations.

Considering *id* a surrogate key for space-time-entity pairs, the relational schema for 3D spatial data can be easily extended into 4D: *Vertex*(<u>*vid*</u>, *x*:**R**, *y*:**R**, *z*:**R**, *t*:**R**). Note the simplicity. Like (Sakr & Güting, 2009), it allows moving objects between two time points that may be geometrically interpolated by a query. A simple topological SQL query for the space at a given time point *t* with no interpolation is:

```
create view Xt as
with mmX(id, tmin, tmax) as (
  select X.id, min(V.t) as tmin, max(V.t) as tmax
  from (X join poR on (X.id = poR.ida))
       join Vertex V on (poR.idb = V.vid)
  group by X.id)
select X.*
from X join mmX on (X.id = mmX.id)
```

Figure 4. The one space-time class and bounded-by association obtained by combining the spatial and temporal models in a non-classical way

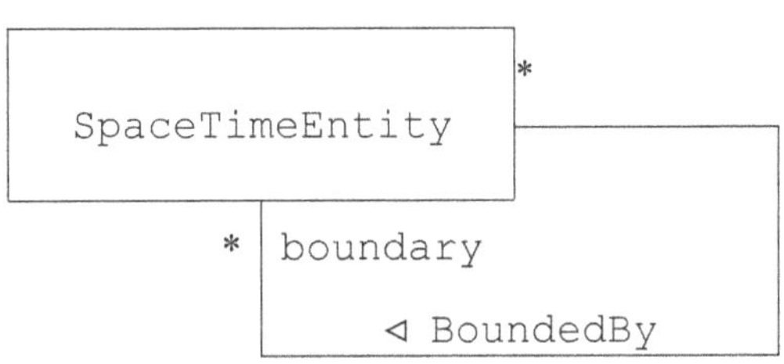

```
where (mmX.tmin < V.t and V.t < mmX.tmax)
   or (mmX.tmin = V.t and V.t = mmX.tmax);
```

The sub-query *mmX* first associates each element *X.id* with its topological closure cl{*X.id*} by joining it to *poR*. The join with *Vertex* selects all vertices of cl{*X.id*}. The group-by clause then computes the time interval for *X.id*. The query then selects a subset *Xt* of *X* where the interval contains *t*. As the pair (*X*, *R*) defines a topological space (*X*, *T*(*R*)), there is a relation *Rt* that generates the subspace topology $T(R)|_{Xt}$. Simply restricting *R* to *Xt* is generally wrong (Bradley & Paul, 2010). Naively restricting *poR* to *Xt* as in our initial example is correct but expensive. An optimal *Rt* is achieved by passing that restriction to Codd's OPEN operator which returns a minimal relation with the same transitive closure as the input relation (Codd, 1979, p. 427). Hence

$$Rt := \text{OPEN}(\{(a, b) \in poR \mid a \neq b, a \in Xt, b \in Xt\}) .$$

Just as the set *Xt* is a Θ-selection of *X*, the space (*Xt*, *Rt*) can be considered a topological Θ-selection of (*X*, *R*), hence a "topologized" basic relational query operator. The topology *T*(*Rt*) is the minimal topology for which the inclusion function i: $Xt \rightarrow X$ is continuous (Adamson, 1995, Ex. 1). Bradley & Paul (2010) exploit this observation to define a relationally complete topological database query language. Note that an edge in *X*, representing the trajectory of a vertex *v* from t_0 to t_1 with $t_0 < t < t_1$, topologically becomes a vertex in the result space *Xt* which does not contain the edge's space-time-vertices.

VERSIONING

The version graph is a DAG whose vertices resp. edges correspond to the different versions resp. modifications of a space. An edge (*i*, *j*) indicates a modification of version *i* into version *j*. Clearly, not all versions need to be stored explicitly. By storing the initial version, any other version *v* can be reconstructed by applying all modifications on the paths to *v* in the version graph. But it is advisable to redundantly store more versions e.g. the current version but we will not delve into the matter of balancing redundancy avoidance against robustness and speed. Note that we only consider topological changes here.

The possible topological modifications of a pair (*X*, *R*) are: First, a point can be added, and a point $x \in X$ can be removed. The latter forces relation *R* to be modified by removing all pairs (*y*, *x*) and (*x*, *z*) with $y, z \in X$. But if *y*, *z* are different from *x* with $(y, x) \in R$ and $(x, z) \in R$, then (*y*, *z*) must exist in the transitive closure R^+ of *R* and must, hence, eventually be added to R. The reason is that the modified set $X \setminus \{x\}$ should be a subspace of (X, R), meaning that the (possibly indirect) bounded-by association $y\, R^+ z$ must be retained in $X \setminus \{x\}$ by the induced bounded-by relation. Further, a pair of points can be added to or removed from R. These are the elementary modifications and a general modification is a sequence of elementary modifications. The following relation schema:

X(*id*, *version*, *atts*), FK: (*version*) → *VX*,

R(*ida*, *idb*), FK: (*ida*) → *X*, FK: (*idb*) → *X*

stores the versions of a space. The *version*-attribute gives the version in which an object or bounded-by association appears for the first time, and *VX* is the object table of the *version space* described below. This yields a single space containing all objects appearing in some version. The following schema stores in which version an object or bounded-by association is deleted:

DelX(*id*, *version*), FK: (*id*) → *X*,

DelR(*ida*, *idb*, *version*), FK: (*ida*, *idb*) → *R* .

The version numbers and transitions are stored as:

VX(*version*), *VR*(*fromv*, *tov*), FK: (*fromv*) → *VX*, FK: (*tov*) → *VX*

which is the *version space* $V=(VX, VR)$, a space of arbitrary combinatorial dimension. Since the version graph is a directed acyclic graph, the consistency rule for the version space is that of a T_0-space.

Figure 5. A topological merge of texts violating the consistency rule "linear DAG"

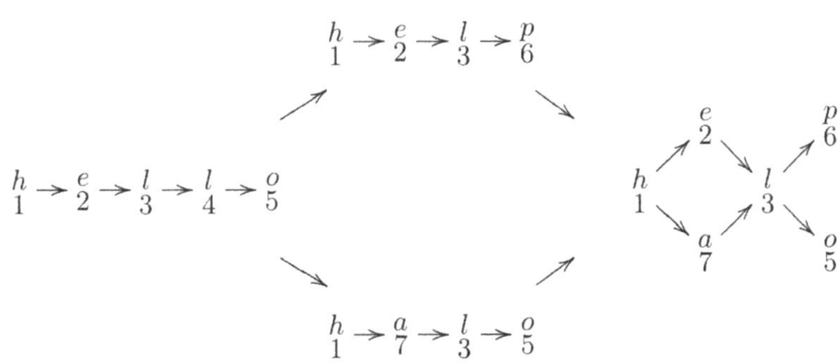

The *topological merge* of versions (X_1, R_1), (X_2, R_2) which may be modifications of a common space (X, R) is a space (Y, S) fitting into the diagram (Figure 6).

The space (Y, S) is simply the union space $(Y, S) = (X_1, R_1) \sqcup (X_2, R_2) := (X_1 \cup X_2, R_1 \cup R_2)$. A conflict can occur if the resulting space (Y, S) violates some consistency rule which we then call a *consistency conflict*. For example, when an object with some identifier *id* is modified in X_1 but left unchanged in X_2, then the union would violate the primary key consistency rule. We call a consistency conflict with primary key violation an *inherent conflict*. In the case of a conflict, a warning statement should prompt the user to resolve the matter.

Example. The merge of texts can also be viewed as a topological merge by considering a string a linear DAG $\bullet \rightarrow \bullet \rightarrow \cdots \rightarrow \bullet$ whose semantic attribute takes values in an alphabet. If the resulting space is again a linear DAG without inherent conflict, then the topological merge of texts is valid. However, even without inherent conflict the merge may not represent text. The topological merge of texts in Figure 5 illustrates such a consistency conflict.

As backtracking is the only way to produce existing versions, it follows that every version space is a T_0-space. We allow any finite T_0-space as a possible version space. In particular, a version history need not have a unique starting point, or even be connected. This allows to begin with different parts of some spatial model. Each are successively modified, modifications are merged, and in the end one unique realization is obtained through a final merge. Every step produces valid topological data, and at each merge occasional consistency conflicts are resolved whenever they occur.

To store a DAG as a topological data type comes natural. An alternative would be a one-dimensional simplicial complex. However, this increases the size by the relation which associates edges with boundary vertices, plus the additional orientation information of edges. Hence, increasing the dimension does in fact lead to a decrease in complexity!

Figure 6. Topological merge

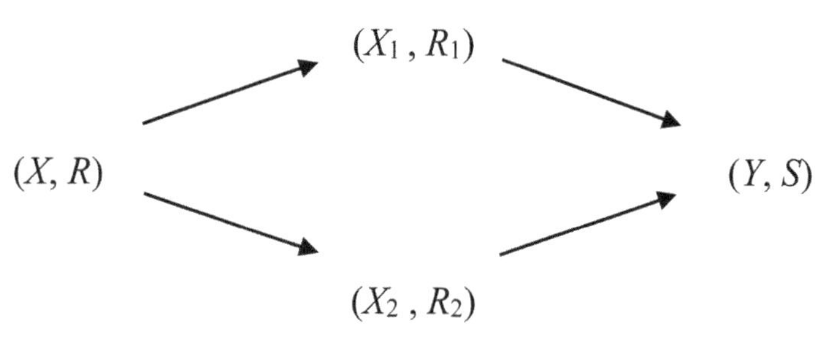

Figure 7. The family of surjective and continuous GeneralizesTo-functions and the BoundedBy-relation on objects

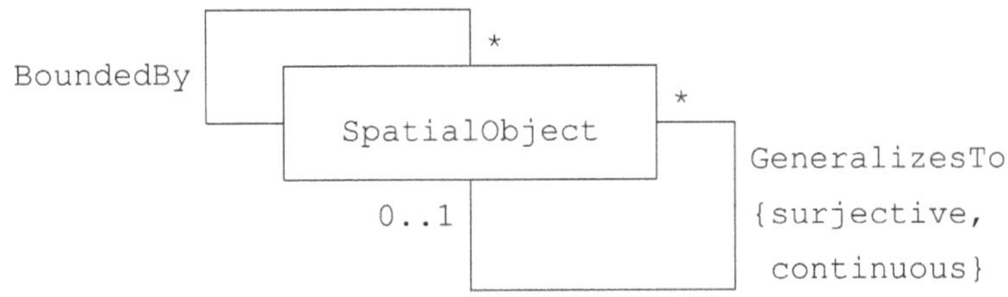

Levels of Detail

In order to enable queries across different levels of detail it is necessary to link several objects in one LoD with their aggregate object into which they collapse in the next LoD. As all objects in the finer LoD have a unique counterpart in the coarser LoD, we have a *generalization function* $g: F \to C$ from the fine space F to the coarse space C. This function is not injective as it establishes a one-to-many relationship. One important consistency rule is that objects that are "close" to each other must generalize to objects which are also "close" to each other. Respecting "closeness" is nothing but to require that g be *continuous* (Alexandrov, 1937, p. 508). More precisely, g is a *continuous function* between spaces (F, R) and (C, S) if every bounded-by association for F is mapped to a (possibly indirect) bounded-by association in C: either $g(x1) = g(x2)$ or

$$(x_1, x_2) \in R \Rightarrow \exists\, y_1, \ldots, y_m \in C\colon g(x_1)\, S\, y_1 S, \ldots, S\, y_m\, S\, g(x_2)\,.$$

This rule is equivalent to the usual definition of continuous function from topology (McCord, 1966, §4): namely that the pre-image of an open set be open (Jänich, 1995, Ch. I, §5). For example, the subspace topology from the last paragraph of Sec. 4 is defined in such a way that the inclusion function is continuous. Another consistency rule is that g be surjective. Then every object in the coarse space is indeed the generalization of an object in the fine space. For the classical model

$$Solid \to Face \to Edge \to Vertex \qquad (5)$$

a continuous function g for generalization purposes implies the explicit modeling of up to $16 = 4^2$ possible types of association pairs, because g can map every class to any other class, and there is no reason to forbid any mapping of one class to some other class. Extending this to spaces of arbitrary dimension n (e.g. space-time etc.) gives $(n + 1)^2$ different LoD associations to be modeled explicitly in order to form one single generalization function. So, the classical model (5) considerably increases the complexity for describing functions. But if LoD associations are modeled on one common class of primitive objects, then the complexity of the class model decreases substantially. In the UML diagram of Figure 7, this class is called SpatialObject. Instead of several dimension-dependent bounded-by associations there is simply one generic such BoundedBy association between arbitrary objects. Functions are incorporated as a GeneralizesTo association respecting the consistency rule: "continuous functions that are surjective between two consecutive LoDs". Of course, further consistency rules can be imposed if necessary.

Example

Assume the generalization of a region subdivided into polygons *A*, *B*, *C*, as depicted in Figure 8. There is a generalization function to *A'*, *B'*, C'. Now polygon *C* generalizes to vertex *C'* in the interior of polygon *B'*. However, the classical model (5) does not allow vertex *C'* to be in a bounded-by association with a face because there is no intermediate edge. Consequently, the topological information that there is a path from anywhere in *A'* to *C'* inside the generalized region cannot be inferred from the classical model without resorting to the underlying geometry. But then the geometry and topology of the generalized region are by force inconsistent: geometry says that vertex *C'* lies inside face *B'*, but the classical model forces *C'* to be topologically disconnected from *B'*. Consequently, the generalization function is not continuous! As the statement "*B'* is bounded by *C'* " cannot be explicitly modeled, it has to be inferred from the position of *C'*. Also a possible position error of *C'* causing this vertex to be outside *B'* cannot be corrected by the topological model.

This is remedied by topological data types. A direct bounded-by relation *R'* associates *B'* with the four lines adjacent to *B'* which are themselves, through *R'*, bounded by the four corners of *B'* which are not bounded by other objects. Hence,

Figure 8. A generalization to a non-classical space

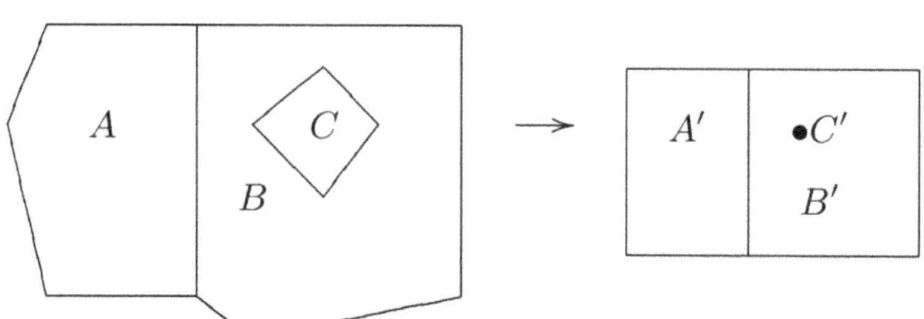

the latter are vertices, implying that *B'* must be a face. And *R'* directly associates *B'* with *C'* which in turn is not bounded by an object. Hence, *C'* is a vertex at the boundary of object *B'*. Now the generalization function is continuous.

We now demonstrate interpolation between different LoDs. Assume a chain $X_1 \to \cdots \to X_m$ of generalization functions, and each space X_i be embedded in some real vector space $\mathbf{R}^n$. We assign to each X_i an extra level coordinate $i \in \mathbf{R}$, leading to an embedding into $\mathbf{R}^{n+1} = \mathbf{R}^n \times \mathbf{R}$, and each LoD i sits inside some slice $\mathbf{R}^n \times \{i\}$. The generalization function g_i associates each $x_i \in X_i$ with some $g_i(x_i) \in X_{i+1}$. If x_i and $g_i(x_i)$ are vertices, then they have, by our model, coordinates in $\mathbf{R}^{n+1}$, and an interpolating function between x_i and $g_i(x_i)$ can be defined. By assigning to *every* element x_i of every X_i coordinates (e.g. by taking a representative in the interior of the geometric realization of x_i), it becomes possible to define interpolation functions between any object x_i and $g_i(x_i)$. The interpolation function can be given as a family of continuous functions $f_{x,i}$: $[0, 1] \to \mathbf{R}^{n+1}$ with $f_{x,i}(0) = x_i$ and $f_x(1) = g_i(x_i)$, and as long as $t \in [0, 1)$ the objects $f_{x,i}(t)$ are considered a placeholder for x_i inside X_i, but when $t = 1$ they become generalized to $f_{x,i}(1) = g_i(x_i)$. So the topology of X_i stays unchanged while $t \in [0, 1)$ and becomes that of X_{i+1} as soon as $t = 1$. The geometry gradually shrinks from one LoD to the next. By giving coordinates to all elements, and not only the vertices, it becomes possible in a geometric realization of a continuous zoom, for each vertex to have a unique trajectory, and so a unique position inside each slice between two LoDs. This includes also those vertices generalizing to elements which are not vertices. In fact, unique trajectories become possible for all elements through the positions of their geometric representatives (i.e. coordinates) in $\mathbf{R}^{n+1}$.

From above, we derive the following relational schema for generalizations which allow interpolations between LoDs. Notice that the table *Vertex* from Sec. "Spatial Dimension and Consistency" is now called *Point*, as it may contain elements which are not vertices.

X(<u>id</u>, <u>lod</u>, gid, glod, atts), CFK(*R*): *(gid, glod)* → *X*

R(<u>ida</u>, <u>idb</u>, <u>lod</u>), FK: *(ida, lod)* → *X*, FK: *(idb, lod)* → *X*

Point(<u>pid</u>, <u>lod</u>, x, y, z, t), FK: *(pid, lod)* → *X* .

Here, CFK(R): *(gid, glod)* → *X* denotes a *continuous foreign key*, i.e. the references from the tuples in *X* having *(gid, glod)* ≠ (NULL, NULL) to *X* are a continuous function with respect to the topology *T(R)*. The attribute *R.lod* in both foreign keys from *R* to *X* gives a disjoint union of a family of spaces indexed by that attribute. This models each LoD in its entirety, whereas (v. Oosterom, & Meijers, 2011) propagates the interesting alternative idea that objects in one LoD not collapsing with other objects in the next LoD need not be repeated in the model. It would be interesting to compare both approaches.

Integrating the Different Spaces

Here, we put together the individual schemas for space-time, version and scale to one single schema. The different data models are integrated by collecting all tables, incorporating the attributes, and providing the foreign keys. As the LoD-attribute is part of the primary key for the point set, all elements (not only vertices) can be given space-time coordinates. This leads to the following tables:

$X(\underline{id}, \underline{lod}, gid, glod, version, atts)$, $R(\underline{ida}, \underline{idb}, \underline{lod})$, $Point(\underline{pid}, \underline{lod}, x, y, z, t)$

$DelX(\underline{id}, \underline{lod}, version)$, $DelR(\underline{ida}, \underline{idb}, \underline{lod}, version)$, $VX(\underline{version})$, $VR(\underline{fromv}, \underline{tov})$

And the corresponding foreign keys are:

FK: $X.version \rightarrow VX$, CFK(R): $(X.gid, X.glod) \rightarrow X$,

FK: $(R.ida, R.lod) \rightarrow X$, FK: $(R.idb, R.lod) \rightarrow X$,

FK: $(Point.pid, Point.lod) \rightarrow X$, FK: $(DelX.id, DelX.lod) \rightarrow X$,

FK: $(DelR.ida, DelR.lod) \rightarrow X$, FK: $(DelR.idb, DelR.lod) \rightarrow X$,

FK: $VR.fromv \rightarrow VX$, FK: $VR.tov \rightarrow VX$

All spaces made of polytopes in Euclidean space-time $\mathbf{R}^4$ can be consistently modeled together with their versions and generalizations according to this schema. Although the combinatorial dimension can be arbitrary, the element coordinates are x, y, z, t. Introducing more coordinates increases the dimension of the embedding space $\mathbf{R}^n$. Other semantic data can be linked to the model by extending the database schema.

The Integrated 6D+ Space

So far, we have a system of models for what we called "elementary" spaces. To show how these can be combined into one space, we first extract the LoD *space* (*LoDX*,

LoDR) by two queries that project *Xv* and *Rv*, a version *v* of the stored spaces, onto its two *lod* attributes:

```
create view LoDX as
  select lod from Xv union select glod as lod from Xv;
create view LoDR as select lod, glod from Xv;
```

Another query can convert this into the 1D edge graph (*V*, *RV*), *V* being the "union" of *LoDX* with *LoDR* . This can be achieved by duplicating the identifiers *x* in LoDX into pairs (*lod*, *glod*):= (*x*, *x*) to make it union compatible with *LoDR*. The relation *RV* contains ((*a*, *b*), (*a*, *a*)) and ((*a*, *b*), (*b*, *b*)) for every pair (*a*, *b*) in *LoDR*. Adding the graph *Gv* of the generalization function in *Xv* to *Rv*, after making them union compatible, yields a new topology $T(Gv \cup Rv)$. The equi-join of spaces (*Xv*, $Gv \cup Rv$) and (*V*, *RV*) on *Xv.lod* and *V.lod* gives an *equi-join space*, or a topological *pullback* (Hatcher, 2002, p. 406), of dimension ≥ 5. This is the space presented in (v. Oosterom, & Meijers, 2011) and it is also a combinatorial variant of the *mapping telescope* (Hatcher, 2002, p. 312). However, for each LoD *i*, except the coarsest, this space contains two redundant copies of one input space: the space at LoD vertex (*x*, *x*) and a homeomorphic copy at LoD edge (*x*, *g*(*x*)). So the integrated 5D+ space has redundant information that usually occur at table joins.

One task of database design is normalization by a lossless factoring of tables with redundancies into smaller tables, hence, take the opposite way we just went. It would be possible to also integrate the whole versioning history into a huge 6D+ space, but this would have even more redundancies. For the above schema this means that a query can integrate all spaces into one huge space but this will then have anomalies. So it may serve as a derived database view for integrated space-time-version queries but its anomalies make it unsuitable for explicit storage. In short: We propose the future development of a topological relational database design theory extending its relational counterpart that has proved so successful in recent years.

Topological Consistency

Although mathematical topology does not know the notion of "topological consistency", different more or less vague definitions of this notion can be found in various places in the literature on topological models for spatial data. We will give a precise definition based on the idea that assigning coordinates to points of an abstract model produces another model which is related to the abstract model.

A topological model of some object usually consists of an abstract model together with a mapping to the surrounding space $\mathbf{R}^n$ (where n = 2,3 or 4) by assigning coordinates to its points. Ideally, the coordinization yields an *embedding*, meaning that the abstract model is homeomorphic to the embedded model. However, if the mapping is not an embedding, then topological inconsistencies arise. An example is given in Figure 9. The image in the left represents an abstract model, and the image on the right occurs for a mapping to the surrounding plane which is not an embedding.

The following topological inconsistency arises: In the topology of the abstract model *X* (defined by relation *R*), the vertex represented by a circle is in the boundary of the slanted edges only. However, in the geometric realization given by *m*(*X*), the circle is in the boundary also of the horizontal edge. Hence that edge has three boundary vertices!

The approach for capturing topological inconsistencies is as follows: For a mapping *m* of an abstract model *X* into the surrounding space, one defines the *overlay space* *o*(*X*), given by *m*(*X*) and all non-empty intersections $i = m(a) \cap m(b)$ for *a*, *b* in *X*, by defining a relation *S* on *o*(*X*). Namely,

1. $(a, b) \in S$, if $(a, b) \in R$
2. $(i, a) \in S$, if *i* does not equal *m*(*a*)

We say that the topological model for *X* is *topologically consistent,* if *X* and *o*(*X*) are homeomorphic. In this case, the mapping *m* is an embedding. This notion of topological consistency was introduced in (Bradley, 2015) in a slightly different formulation.

In order to verify topological consistency in the above sense, one needs to check that *X* and *o*(*X*) are homeomorphic using the mapping *m*. If the cardinalities of *X* and *o*(*X*) are different, then one concludes that the topological model for *X* is not topologically consistent. In case these two sets have the same cardinality, then one

Figure 9. Left: an abstract model X. Right: X is mapped to the surrounding plane with image m(X), but m is not an embedding

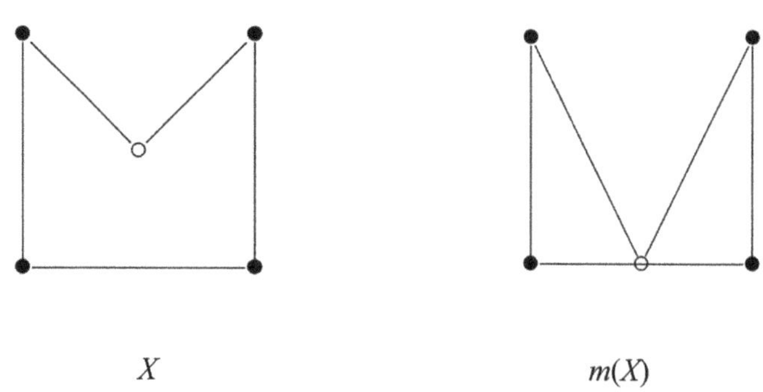

needs to check whether the mapping *m* is one-to-one, in which case the topological model for *X* is topologically consistent. If in this case *m* is not one-to-one, then *m* does not define an interesting topological model for *X*.

In order to quantify topological inconsistencies, topological invariants can be used. A *topological invariant* is any property which does not change under homeomorphisms. A measure of topological inconsistency is then given by comparing topological invariants of *X* and *o*(*X*). We propose to use the *Betti numbers* for this task. The *zeroth Betti number* is the number of connected components, and the higher Betti numbers can be interpreted as the number of holes of a dimension: the *first Betti number* as the number of loops, the *second Betti number* as the number of voids, and so on. If the *n*-th Betti number of *X* differs from that of *o*(*X*), then the topological model for *X* is topologically inconsistent, and their difference says how topologically large the inconsistency (in dimension *n*) is.

Implementation

In Sec. "Temporal Dimension", the subset *Xt* at a time point *t* of a set *X* in space-time was obtained by a relational *selection* which was then converted into a topological subspace. We have also seen topological variants of relational *union*, *Cartesian product*, and *join*. In fact, all query operators of Relational Algebra (Codd, 1990) can be turned into corresponding topological constructions which operate on spaces and return spaces (Bradley & Paul, 2010). Here, we shortly describe a first prototype of an experimental implementation of this topological query language: There exist two classes of topological constructions (Adamson, 1995, Ch. 3): the initial spaces and the final spaces. A relational query operator on some input *spaces* first operates on their point sets in the conventional manner and returns a result *set X* – a database table containing the query result tuples. Now the result tuples are always linked with the entities of the input spaces by functions: either from the result *X* back to the input (intersection, selection, or join), giving an *initial space*, or the function maps input entities to *X* (like union, projection), giving a *final space*.

The prototype is programmed in Common Lisp and has its own simplified Relational Algebra in Lisp syntax. A space can be defined by the space constructor who takes a set, an incidence relation, and which returns a space. Each basic operator, like select, natjoin, or project, has its spatial counterpart that acts on the point sets, constructs the corresponding topology for the result set and returns both as a space. The operators can be arbitrarily nested. The Common Lisp Object System (CLOS) dispatches the operators to their corresponding methods at runtime. For example, on a relation obj the projection (project (atts) obj) returns the projected relation, but when obj is a space the projection returns the corresponding projected space. In particular, many of the above presented complex SQL queries are provided as built-in features or carried out in the background when needed. The experiences gained shall help to produce a topological relational DBMS to carry out the above topological data modeling by built-in features. It would also provide topological consistency rules, and could start the discussion of topological data modeling rules and end in a spatial data modeling theory that extends the current relational modeling theory.

CONCLUSION

It has been shown how the relational model can be used for topological data in a very natural way, and, in particular, it has been proven that the relational model does not impose any limitations for finite topological data modeling. A relational database schema based on Alexandrov topologies, that seamlessly integrates 4D space-time data, version histories, and different levels of detail (LoD), is presented. Such a topology can always be represented by a directed acyclic graph, and it imposes fewer restrictions than the canonical Solid – Face – Edge –Vertex models. The gained flexibility and simplicity alleviates more sophisticated spatial data modeling and exposes mathematical topology in spatial data. This has practical consequences because topology is a fundamental concept in mathematics. So it is likely to have more to offer for spatial data modeling than is momentarily used. A first contribution is a precise definition of "spatial data dimension" and a method for measuring topological consistency. A topological version space allows the recovery of different versions of a spatial model by using queries based on topological constructions. Among the new consistency rules, "continuity" of foreign keys allows consistent modeling in different LoDs.

ACKNOWLEDGMENT

This work was partially funded by the Deutsche Forschungsgemeinschaft with grants BR3513/12-1 and BR2128/18-1.

REFERENCES

Abbott, E. A. (1884). *Flatland: A romance of many dimensions*. London: Seeley.

Adamson, I. T. (1995). *A general topology workbook*. Birkhäuser.

Alexandrov, P. (1937). Diskrete Räume. *Matematicheskii Sbornik*, *44*(2), 501–518.

Anh, N., Vinh, P. T., & Duy, H. (2012). A study on 4d gis spatio-temporal data model. *Proceedings of the fourth int. conf. on knowledge and systems engineering (KSE) 2012*, 34-38. doi: 10.1109/KSE.2012.1

Bradley, P. (2015). Supporting data analytics for smart cities: An overview of data models and topology. In A. Gammerman, V. Vovk, & H. Papadopoulos (Eds.), *Statistical Learning and Data Sciences SLDS 2015, LNCS 9047* (pp. 406–413). Springer. doi:10.1007/978-3-319-17091-6_35

Bradley, P. E., & Paul, N. (2010). Using the relational model to capture topological information of spaces. *The Computer Journal*, *53*(1), 69–89. doi:10.1093/comjnl/bxn054

Bradley, P. E., & Paul, N. (2013). *Dimension of Alexandrov topologies*. Retrieved August 25, 2017, from http://arxiv.org/abs/1305.1815

Bradley, P. E., & Paul, N. (2014). Comparing G-maps with other topological data structures. *GeoInformatica*, *18*(3), 595–620. doi:10.100710707-013-0191-1

Codd, E. F. (1979). Extending the database relational model to capture more meaning. *ACM Transactions on Database Systems*, *4*(4), 397–434. doi:10.1145/320107.320109

Codd, E. F. (1990). *The relational model for database management: version 2*. Boston: AddisonWesley.

Egenhofer, M. J. (1991, January). Reasoning about binary topological relations. In *Advances in spatial databases* (pp. 141–160). Springer Berlin Heidelberg. doi:10.1007/3-540-54414-3_36

Egenhofer, M. J. (1994). Spatial SQL: A query and presentation language. *Knowledge and data engineering. IEEE Transactions on*, *6*(1), 86–95.

Engelking, R. (1978). *Dimension theory*. Amsterdam: North-Holland.

Hartshorne, R. (1993). *Algebraic geometry*. New York: Springer.

Hatcher, A. (2002). *Algebraic topology*. Cambridge University Press. Retrieved August 12, 2013, from http://www.math.cornell.edu/~hatcher/AT/ATpage.html

Hazelton, N., Leahy, F., & Williamson, I. (1990). On the design of temporally-referenced, 3-D geographical information systems: development of four-dimensional GIS. *Proceedings of GIS/LIS'90*, 357-372.

Jänich, K. (1995). *Topology*. Springer.

Kuratowski, K. (1968). *Topology* (Vol. 2). Cambridge, UK: Academic Press.

Lienhardt, P. (1991). Topological models for boundary representation: A comparison with ndimensional generalized maps. *Computer Aided Design, 23*(1), 59–82. doi:10.1016/0010-4485(91)90082-8

Lienhardt, P. (1994). N-dimensional generalized combinatorial maps and cellular quasi-manifolds. *International Journal of Computational Geometry & Applications, 4*(03), 275–324. doi:10.1142/S0218195994000173

McCord, M. C. (1966). Singular homology and homotopy groups of finite topological spaces. *Duke Mathematical Journal, 33*(3), 465–474. doi:10.1215/S0012-7094-66-03352-7

Paul, N. (2010). Basic topological notions and their relation to BIM. In J. Underwood & U. Isikdag (Eds.), *Handbook of research on building information modeling and construction informatics: Concepts and technologies*. IGI Global. doi:10.4018/978-1-60566-928-1.ch019

Rossignac, J., & O'Connor, M. (1989). SGC: A dimension-independent model for pointsets with internal structures and incomplete boundaries. *Proceedings of the IFIP Workshop on CAD/CAM*, 145-180.

Sakr, M. A., & Güting, R. H. (2009). Spatiotemporal pattern queries in secondo. In SSTD 2009: Advances in spatial and temporal databases (pp. 422-426). Springer.

Van Oosterom, P., & Stoter, J. (2012). *Principles of 5d modeling, full integration of 3d space, time and scale*. Geospatial World Weekly.

Zlatanova, S., Rahman, A. A., & Shi, W. (2004). Topological models and frameworks for 3D spatial objects. *Computers & Geosciences, 30*(4), 419–428. doi:10.1016/j.cageo.2003.06.004

Chapter 4
BIM FM:
An International Call for Action

Ricardo Codinhoto
The University of Bath, UK

Vincenzo Donato
Politecnico di Torino, Italy

Julie Comlay
The University of Bath, UK

Kemi Adeyeye
The University of Bath, UK

Arto Kiviniemi
The University of Liverpool, UK

ABSTRACT

Despite significant progress for the adoption of BIM in AEC, currently its adoption for FM has been sparse, scarce, and extraneous. There are few cases in the world where robust adoption has taken place that are able to demonstrate success and are willing to disseminate the positive impact of BIM FM on sustainability, operational efficiency, and cost reduction. To date, there is no approach, motivation, or support in place to enable the extensive adoption of BIM for FM worldwide. In the UK, for instance, the UK BIM initiative, mandate, and the Digital Built Britain cannot count on the participation of FM stakeholders; the government has only started promoting initiatives that could trigger an extensive BIM approach, generating benefits for organizations and more importantly, society as a whole. In this chapter, data from authors' various research projects has been put together to generate an agenda for BIM FM implementation. The findings reveal that unless an intervention, such as a mandate for FM services suppliers, is put in place, very little will happen with regards to BIM FM.

DOI: 10.4018/978-1-5225-5625-1.ch004

A SLIGHTLY DIFFERENT VIEWPOINT

The year is 2018 and much of what is proclaimed in the academic press, popular media and all other means of communications is that humanity has disturbed the balance of life in our planet to an extent that might be irreversible. Global warming and climate change constantly make headlines as the explanation for floods, landslides, draughts, storms, the melting of the polar caps and rising sea levels, as well as irregular weather patterns. Amongst the causes of these is the continuous accumulation of pollutant gases, such as CO2 and methane, in our atmosphere. Many of these gases come from the collective impact of the way we live our lives individually, and that has to change.

Building use is at the core of the problem. The study by Kleips et al. (2001) revealed that, on average, American and Canadian adults spend approximately 87% of their time in enclosed buildings. Give or take, a person dying at the age of 83 (UK life expectancy) would have spent 72 years of their life inside a building. While individually that might not be significant, collectively it is a critical problem that requires attention as buildings consume approximately 40% of all energy production and generate circa 36% of global CO2 and other greenhouse gases (European Commission, 2017).

Building occupancy is the subject explored within this article, with a focus on the use of Building Information Modelling (BIM) to support facilities management (FM). BIM for FM had its first significant application at the Sydney Opera House (Ballesty et al., 2006; Schevers et al., 2007) and has seen a number of applications worldwide since. This paper will not significantly address the matter of applications as discussed by Becerik-Gerber et al., (2012), Arayici et al., (2012) or the compilation of case applications presented by Teicholz (2013) and Volk et al., (2014). Instead, it will refer to the authors own earlier publications about various cases (e.g. Codinhoto et al., 2013a; Kiviniemi and Codinhoto, 2014; Comlay 2015 and Comlay and Codinhoto, 2017), to form the argument as to why the extensive adoption of BIM for facilities management has not happened, and as such is failing to deliver critical environmental and economic benefits.

The authors have drawn evidence from various independent pieces of research related to BIM FM carried out by them since 2011. Whilst evidence from systematic data collection forms the basis of the arguments presented in this article, anecdotal evidence from the participation of the authors in various activities related to the advancement of BIM FM and experience from conducting research in this field is also used. The aim of this article is not to contribute to the scientific advancement of BIM FM, rather it aims to present state-of-the-art of BIM FM and contribute to expanding its practical adoption.

SOURCES OF EVIDENCE, KNOWLEDGE AND BELIEFS

The sources of evidence, knowledge and belief that underpin the discussions presented in this article were derived from various longitudinal research projects undertaken at different organisations. A synopsis of the aims and methods for these projects is presented in the following paragraphs and links to additional information are provided. In addition, the authors are referring to their experience and participation in research and educational programmes and professional groups such as Sennatti Properties, BuildingSmart, BIM TaskGroup and UK BIM Alliance.

Manchester Town Hall Complex Project (MTHCP)

This project started in 2010, was completed in 2014 and is used as a case study. The project consisted of a major construction redevelopment (12500sqm) of a public library and the town hall built in the 1870s, with further work in the 1930s, in Manchester UK. In this £40m project, the client established that BIM had to be used in the design, construction and operational phases. Data from this project was gathered within three different stages.

Stage 1: Design and Construction (2011)

The aim of this research was to explore issues related to BIM implementation in design and construction and handover. In this project, BIM was implemented in a bottom-up approach and as such the research team were able to identify implementation issues as they emerged and discuss them with the project team at the point of occurrence. Several techniques were used for data collection including non-participant observations of design development; interviews with 11 members of the multidisciplinary team; archival documental analysis and BIM capability and maturity assessment using the NBIMS CMM. Data analysis focused on assessing the capability maturity level of the design and construction team; mapping formal and informal contractual relationships amongst stakeholders; listing the contracted and completed scope of works; listing BIM deliverables for design and construction; identifying the Information Exchange Standards utilised; registering the BIM implementation process utilised and eliciting expected BIM FM benefits. The team comprised of five researchers collecting data over six months. Details related to data collection and results are presented in Codinhoto et al. (2011).

Stage 2: FM Implementation and Handover Stage (2013)

Based on the same project, the focus of the analysis was on the identification of the FM processes in place and those offering potential to be supported by BIM; the BIM capabilities used to support FM and its maturity level; and the identification of facilitators and barriers to the use of BIM FM. In addition, the level of BIM maturity of the design and construction team was reassessed 2 years after project inception and the NBIMS CMM was used for FM purposes. The tools and techniques used for data collection included 18 semi-structured interviews with facilities managers; a 2-hour validation workshop; modelling of FM services using the swimming lanes process model technique; archival analysis of the Facilities Management Output Specification for Statutory Servicing and Reactive Maintenance. Details related to data collection and results are presented in Codinhoto et al. (2013b).

Stage 3: Post Occupancy Evaluation (2016)

A post occupancy evaluation was carried out two years after project completion in 2014. The focus of the research was to identify the existence of performance gaps (if any) and the level of satisfaction of building users. Data was collected through survey questionnaire (n=120 respondents including visitors and staff); 40 hours of building use observation; 10 in-depth interviews with staff and visitors; on-site Environmental Performance Measurement using 12 Raspberry PI sensors (temperature, humidity, lighting and motion, and CO2 concentration and 30 thermal spot check measurements with additional thermal comfort questionnaire survey. The team also accessed the energy model produced in 2010 using IES VE software. Details related to data collection and results are presented in Shen and Codinhoto (2017).

Salford Royal Foundation Trust (SRFT)

The case for investigation was a major hospital in the North of the UK. The research started in 2008 investigating a £200 million project for the redevelopment of a large hospital complex in Salford. The project was 80% funded through Private Finance Initiative (PFI) and 20% through public capital and included the redevelopment of various buildings with four different age profiles (1850-1899; 1900-1949; 1950-1975 and 1976-1999). The project involved the redesign of services and existing facilities (refurbishment) as well as the design of new facilities. Data from this project was gathered within two different stages.

Stage 1: Service and Building Design Integration

The focus of the research was to identify the impact of service design on building design and building design on service delivery. Data was collected through archival analysis of the (re)design of the services and facilities. In addition, seven in depth semi-structured interviews were carried out with project directors and service and building design coordinators to map the process of designing services and facilities and identify, according to interviewee's perspectives, facilitators and barriers to the integration of service and building design. Additional evidence was gathered through documents such as service descriptions and building plans. Research findings were validated through a workshop. Details related to data collection and results are presented in Codinhoto et al. (2008).

Stage 2: Enablers, barriers, maturity and challenges for BIM FM adoption

The research took place in 2014, two years after project completion and the results were presented as a dissertation for the award of master in Building Information Modelling (Comlay, 2015). The method involved work shadowing to support the observation of FM practices, working processes and procedures and its evaluation with regard to: types of tasks undertaken, time line for resolution, management of current processes, allocation of work, use of facilities, impact on service delivery i.e. time, availability of space, space utilisation, workflows and task interdependencies. Supporting evidence included, FM documentation, archival records, interviews, and direct observation. A total of 38 hours of digitally recorded data was compiled, transcribed and a qualitative analysis undertaken. In addition, two interviews with 'Thought Leaders' were undertaken for validation purposes. Details related to data collection and results are presented in Comlay (2015).

The University of Bath Campus (UoB)

This research project was ongoing at the time this article was written and based on waste management modelling for the University of Bath campus. With 18,000 students and staff, the campus can be considered as a small-scale town composed of buildings of different sizes and uses (education, accommodation, catering, banking, healthcare, etc.) and its own infrastructure systems such as water provision, transport hubs, energy. The aim of the research was to investigate how information modelling can support the reduction of operational costs, environmental impacts and generation of waste. This research consisted of two interrelated projects.

- **Data Requirements for BIM FM:** This research was conducted with the aim of investigating the design and construction bias within Employers Information Requirements protocols. The research method included an extensive literature review of EIR's, BEP's, BIM for facilities management, information management, standards, guidelines, processes, Big Data and data analytics. EIR and BEP exemplars were sourced from large owner/client organisations based in the UK. Each organisational EIR was compared to the BIMSmart (2013) exemplar EIR template and BEP's were compared to the Cpix BEP (2013) template to determine a range of adaptations. Elements of comparative analysis were used to identify emerging trends within categories, identifying variables and emphasis of use (Cragun et al., 2015). Investigating EIR's against the BIMSmart template and the efficacy of BEP's to respond to EIR's to determine the use profile of the BIMSmart template. This was done by using the BIMSmart template as a baseline and identifying the total number of clauses within each EIR and how many clauses have been added comparatively. Details related to data collection and results are presented in Comlay and Codinhoto (2017).
- **OPEX Waste Management Modelling:** The Department of Estates and Facilities granted access to existing building documentation (in paper or digital – CAD format) to support LOD200 information modelling that consisted of: A 'topographical model' using REVIT and AutoCAD created to capture physical environment properties in question; A 'bin zone model' with the approximate catchment area of each waste collection point (bins) using Voroni diagrams; A 'directional space model' using the Dijkstra algorithm to indicate the optimum path with respect to distance and waste volume, during waste collection; a depth max model was used to describe the visibility of each waste collection point and its relationship with waste generation; finally, Smart Move was used to model the flow of people (or 'crowding') within the space. The outputs of the model included the optimum path for bin collection and a relationship between bin arrangement with respect to people and waste generation. The final information model also supports the generation of campaigns for waste generation reduction and service efficiency.

The plethora of research projects presented here is not exhaustive, but it reflects the level of expertise of the authors in relation to longitudinal research across design, construction and operation phases of building projects. It is this experience (built on evidence) that underpins the statements and claims presented in this article.

BIM IS YET TO CONTRIBUTE TO ORGANISATIONS AND CITIES

Much work still remains to be done with regards to the full adoption of BIM for FM purposes. Worldwide, the adoption of BIM for design and construction progressed rapidly between 2011 and 2017, but while a huge amount of information generated in design and construction is handed over to facilities managers, very little is utilised (RIBA, 2017). Here, Lillrank's (1995) theory of transfer of complex systems is used to speculate why that is. As represented in Figure 1, what is happening is that the new ideas and practice of information modelling as applied to design and construction have been abstracted and 'packaged' in various levels to be useful. At the FM end, application requires the ideas to be 'unpacked' and tailored to FM processes. To some extent, the abstraction can be at a very low level for CAPEX (which has many similarities with design and construction processes). This is a possible explanation for the larger number of reported cases of the use of BIM for refurbishment/maintenance of existing buildings (Volk et al., 2014). However, for OPEX (where arguably the most relevant benefits of information modelling can be obtained) higher levels of abstraction are required, thus modelling has to be done afresh without previous knowledge as it is dissimilar to design and construction, thus creating a knowledge gap. Because FM teams are experiencing huge budget reductions (Naylor, 2017), the capacity for carrying out BIM FM implementation (modelling) is simply non-existent. In addition, while BIM knowledge evolved within the building sector (and more recently infrastructure), there is an acknowledged lack of BIM expertise within FM preventing its extensive adoption (Cracknell, 2012).

Figure 1. Transfer of Complex Information through abstraction and application (adapted from Lillrank, 1995).

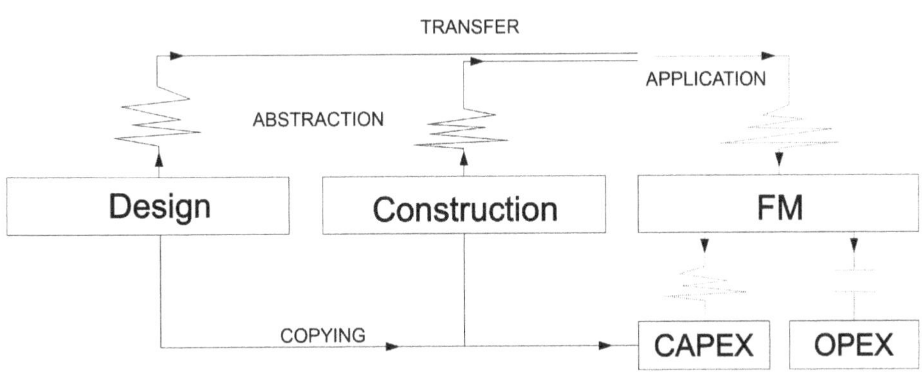

Indeed, multidisciplinary literacy is a reason why BIM has not been used more in FM and in particular OPEX. The difficulties of transferring BIM from design and construction to FM are, to some extent, related to a general lack of understanding regarding what FM entails. In part, this is caused by the lack of general knowledge relating to the impact that service and building have on each other (Codinhoto et al., 2008; Tzortzopoulos et al, 2009) but also from a lack of general guidance for how FM representatives can contribute during the development of design and construction projects (Codinhoto et al., 2013a).

Academically, the confusion arises from (several) definitions of FM with contrasting meanings (e.g. British Standards Institute (BSI), 2007; Atkin and Brooks, 2009; Bungar, 2012 and Alexander, 2013). Thus the literacy issue is on both sides, i.e. FM teams lacking BIM expertise and design and construction teams lacking specific FM knowledge. Part of the problem is related to the fact that FM definitions do not have a clear scope. In this respect, early discussions by Thomson (1990) and Tay and Ooi (2001) have emphasised how FM definitions place particular emphasis on maintenance and cleaning, sometimes extending the definition to incorporate the provision of support services such as porterage and reception. In a more inclusive manner, Alexander (2013) states that FM concerns the management of quality, value and risk associated with the occupancy of buildings and the delivery of customer services which links asset management and the provision of support services. This view is also shared by Atkin and Brooks (2009) and the British Institute of Facilities Management (BIFM) which sees FM as a "*critical professional and strategic business discipline*". For Atkin and Brooks (2009) the current view of FM covers a wider range of activities such as financial management, change management, health and safety, contract management and ICT and therefore its processes are related to the strategic, tactical and operational levels of an organisation. Similarly the BSI (2007) defines FM as "*the integration of processes within an organisation to maintain and develop the agreed services which support and improve the effectiveness of its primary activities*" (BSI, 2007). Also part of the spectrum of definitions is the idea of asset management. The BSI (2014) defines asset management as "*involves the balancing of costs, opportunities and risks against the desired performance of assets, to achieve the organizational objectives. The balancing might need to be considered over different timeframes.*" In other words, the asset is not the building but the artefacts within buildings. It is worth mentioning that in certain countries such as the USA, the term Asset Management refers to FM. For clarification purposes, in this article FM refers to the support services and operations that are essential for the successful delivery of core services and operations of a business. As shown in Figure 2, that involves services for maintaining facilities (buildings and surrounding areas) and assets (e.g. furniture and equipment) in an appropriate

Figure 2. Conceptual scope of Facilities management

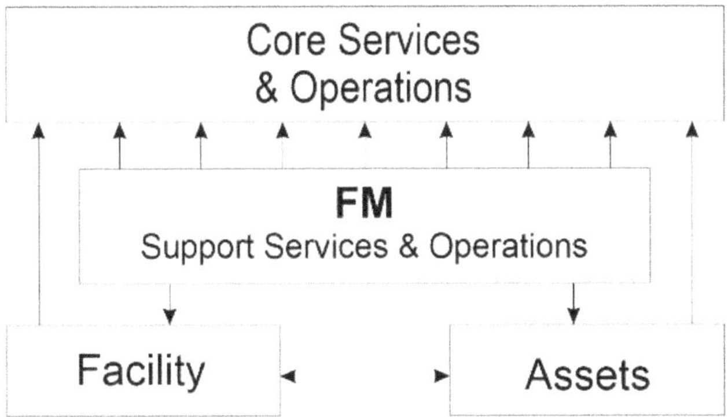

offering condition (i.e. quantity, quality and location) as well as providing support services (e.g. porterage, security) and operations (e.g. waste collection) that are essential for the core business to succeed.

Supporting services refers to the several FM services (Table 1) that are, in general, organised in key priority areas, each having its specific subdivision to facilitate the monitoring of performance indicators and compliance to regulatory standards. In the table below, IT is considered a unit of its own and even though related to FM, it is separated from it and delivered by a specialist team with its own budget and own

Table 1. Facilities management services (source: Codinhoto et al., 2013b)

Facilities Management

Health & Safety	Fire & Safety	Security	Maintenance Systems	Stat. test & inspect.	Operational	IT
• Occupational safety and health, including compliance with local, state, and national laws and agencies, such as the Occupational Safety and Health Administration (OSHA), the Environmental Protection Agency (EPA), and EN 54 • Industrial hygiene, including indoor air quality • Display Screen Regulations • Safety Rules for Contractors • Permit to Work System • Risk Assessments • Control of Substances Hazardous to Health	• Fire protection and safety • Fire Risk Assessments • Fire Doors • Fire stopping • Wet risers • Fire extinguishers • Sprinkler systems • Fire alarm systems • Smoke/heat detectors	• Security • Access control • Security guarding • Intruder alert • CCTV	• Heating, ventilating, air conditioning and refrigeration • preventive and predictive maintenance • Corrective maintenance/ Reactive repairs • Building automation systems • Building fabric and decorative • Grounds maintenance and horticulture • Computerized maintenance management system	• Lifting equipment • Work equipment • Legionella • Pressure systems • Asbestos • Man-safe systems (window cleaning, roof access, etc.) • Electrical portable appliances and fixed wiring • Lightning conductors • Emergency lighting • Fire protection systems • Compliance audits	• Help desk • Asset management • Disabled Access • Cleaning • Waste management • Environmental Issues • Reception • Meeting room management • Mail Room • Photocopying • Vending • Office space planning, layout, and furniture placement • Car Parking • Specifying, tendering and contracts' negotiation • Pest control • Daily inspection of escape routes and fire exits	• Hardware Maintenance • Software maintenance • System management • Upgrading • License control

service level agreements (which in general is a cause of intra-departmental conflict within the organisation – Codinhoto et al., 2013b).

Another issue causing conflict is the fact that FM is often not seen as directly contributing to the core business. As argued by Thomson (1990), facilities management is more than construction, real estate, building operations, maintenance, cleaning and reception. For Thompson, FM is related to business planning - where building design is linked to service design according to business objectives and together they influence organisational strategy development. Planning is at the core of FM (Thomson, 1990; Barrett and Baldry, 2009; Alexander, 2013) and it is its integration into company management strategy that determines whether FM is or is not perceived as adding value to the core business. However, planning and value management are quite often neglected by CEO's as key FM functions. In general, what is seen is the adoption of piecemeal approaches that are not process orientated but focused instead on small short-term gains rather than long-term rewards. Consequently, facilities managers have difficulties in systematically identifying areas where value can be added. There is little or no integration between FM and corporate thinking (Cairns, 2003; Alexander, 2013).

To date, only piecemeal small data is used but there is substantial unexplored potential for generating Big Data that can support core business activities. As argued by Comlay and Codinhoto (2017) BIM can generate opportunities for organisations to leverage information for service delivery, improving productivity, reliability and the multiple use of profile information. This transformational aspect of digital information modelling implementation for FM operates at department and organisational level, it also has the capacity to be disruptive (Reinhardt & Gurtner, 2015). As digital expertise and the sophistication of data capture and management develops within an organisation, what also develops is the feasibility for the integration of data at various levels within and beyond the organisation. From an organisation's perspective, that means that they can better identify and manage inefficiencies. That information can also be used for city and regional planning purposes, as well as for supporting the achievement of country and global targets. However, frameworks for the holistic integration of information still do not exist. Standards for Smart Cities are in development (i.e. PAS 180 series) and its adoption is very limited. Even though much progress has been achieved, the limited use of small data settings locally with some exchange across areas is what prevails as depicted in Figure 3. Big Data, where individual information is integrated to form a whole, is to date only a concept.

Indeed, as argued by Codinhoto and Kiviniemi (2014) and Volk et al., (2014) the literature shows that there is very little that has been implemented extensively in FM and even less has been measured in terms of improvements made due to BIM. In addition, for those who have sought to adopted BIM for FM, the level of

Figure 3. Fragmented Small Data usage

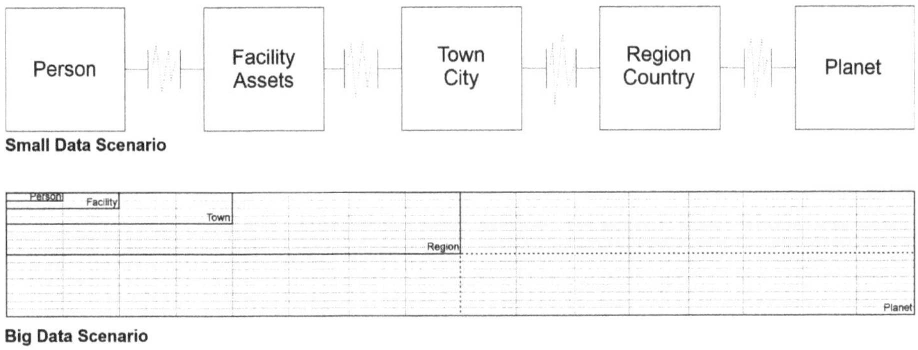

maturity is low, even though it is higher when compared with those with no BIM FM programme (Figure 4). BIM FM related articles reporting on case studies shows that the application of BIM for FM has been focused on Hard FM (CAPEX) to a great extent and in particular on the accuracy of 3D as-built models (LOD 500) and the link to digital statutory/maintenance/supplier information available through online services. In an inverted fashion, the emphasis is placed on BIM (a solution) rather than on the organisational problems (that BIM can resolve). In other words, it's a solution looking for a problem, rather than the other way round. Still, there are cases where the correct approach have been taken. One of the best publicly documented examples of successful implementation of BIM in FM is Sydney Opera House (CRC 2007, Linning 2015). BIM did not come from the design and construction process, which regarding the age of the building would of course have been impossible, however the building has subsequently been modelled specifically for FM purposes. This project demonstrates excellently the possibilities of BIM in FM and basically similar data delivery could be implemented in real construction process, although it would require significant changes in the current work processes. Some more recent examples of successful implementation of BIM for FM are presented in Teicholz's book BIM for Facility Managers (2013). Among those are Texas A&M Health Science Center and the School of Cinematic Arts in the University of Southern California. In this respect, several indicators of BIM for FM successful implementation exist. For instance, financial indicators related to estimating costs associated with: operations and maintenance, energy, building functions, real estate, plant, etc; indicators associated with the need for preventive maintenance of fabric, systems, and components; functional indicators related schedule maintenance of building space for operational use (e.g. cleaning schedule), etc. To a large extent, most of the indicators reported in the literature were obtained through the testimony of managers rather than from an information model (Codinhoto and Kiviniemi,

Figure 4. BIM FM capability maturity level at the Manchester Central Library. Comparative view with the wider sector (Source: Codinhoto et al., 2013a)

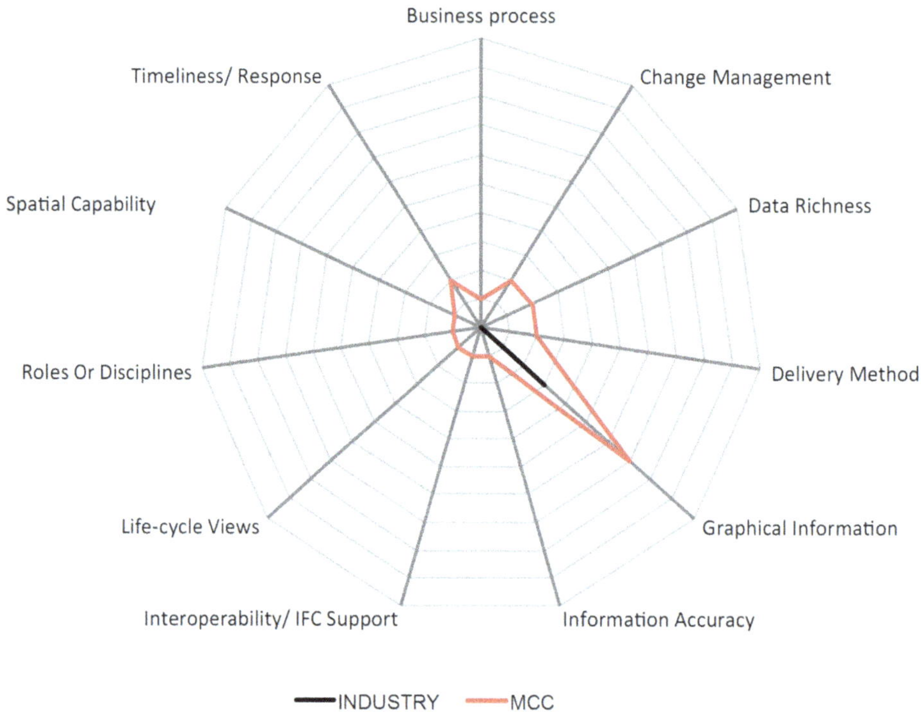

2014). Among the many projects where the owner and project team have developed the design and construction processes considering the content and value of the FM information, most were clustered in silos focused on CAPEX (Volk et al., 2014).

With regards to information silos in FM, little movement has been seen towards FM integrated data. Currently, a significant proportion of data used by FM teams is contained in information silos which limits the use profile of data for FM service delivery through a lack of integrated data (Codinhoto & Kiviniemi, 2014; Bilal et al., 2016). For instance, in Table 2 details the data silos both digital and paper that were identified in SRFT. Digital data for the CSD is stored on a dedicated server in an organised file structure. Alchemy is the e-document management system which is not a networked licence and is problematic in terms of installation, reliability and accessibility. These underlying issues affect the use of the digital data available to the team members within the different departments within SRFT. As discussed in Comlay and Codinhoto (2017), more recent approaches that deliver a sustainable model continue to value data that is delivered through the multiple reuse of data for different purposes, regardless of the primary motivation for the collection of

Table 2. Information assets and silos (Source: Comlay, 2015)

Team	Data Type	Software	Process	Digital	Paper Copy
O & M	Permits	X	Manual	X	✓
	BMS		Automated	✓	
	Helpdesk – Reactive Maintenance Jobs	Catalyst	Automated	✓	✓
	PPM	X	Manual		✓
	Digital Door locks		Digital control panel	✓	
O & M Capital	Payment Certificates	Excel	Manual Input		✓
O & M Capital	Orders	Integra	Manual Input	✓	✓
O & M Capital	Drawings	Autocad	Digital Archive	✓	
Capital	Meter Readings	X	Manual Collection	X	✓
Capital	Energy Data	Excel	Automated calculation	✓	
Finance	Orders	Integra Pro daCapo	Manual input	✓	✓
Performance	Balanced Score Card	Pro daCapo	Manual Input	✓	

the data (Batra, 2014). The search for more integration suggests organisations will become increasingly agile and responsive to market trends that are relevant to the organisation. For instance, debates around circular economy (e.g. Cruz et al., 2015; ARUP, 2016) indicate that information requirements for physical assets are becoming even more significant and increasing in volume due to organisations endeavouring to demonstrate they have met global strategies for carbon reduction, budget constraints, economic turmoil and instability. In addition, governmental bodies and taxpayers are more likely to require organisational accountability for environmental performance of OPEX activities (Crown, 2013). Despite BIM standards and demonstration studies being available, change is slow for BIM FM implementation and while there is engagement in the debate by FM teams, the status quo continues with isolated pots of data (Bilal et al., 2016). In the UK, the 2011 BIM Mandate did not count with the participation of FM providers and while Government Soft Landings (GSL) and COBie addressed the issues of data exchange between BIM for design and construction and an asset information model (AIM), there is no government (client) push. The improvement of FM service delivery through the integration of data silos and a platform for integrated digital data is currently expected to only be delivered if a business need is identified.

Although BIM promises to generate benefits and overcome problems in the management of buildings that can affect the core business (Becerik-Gerber et al., 2012; Arayici et al., (2012); Liu & Issa, 2015; Nicał & Wodyński, 2016), the obstacles to its adoption are significant and challenging (BIS, 2011). Authors such as Bernstein and Pittman (2004), Kiviniemi et al. (2008), Forns-Samso et al. (2011) and Wang et al., (2013) have discussed a series of key obstacles to BIM implementation that require further attention. These include legal issues, business-related issues, people and technical issues, and developments to resolve these issues have not evolved since early publications.

For Kiviniemi et al. (2008), the legal issues of using data, including the lack of adoption of e-procurement routes (Grilo and Jardim-Goncalves, 2011) and undefined responsibilities of data content in the models and the legal status of these models compared to other documents, has yet to be addressed in FM contexts. Very little exists for design and construction, but even less is available for FM that would give facilities managers confidence when implementing BIM. This issue is reflected when considering the lack of FM information needs that are generally available at early stages of design of new buildings (Liu & Issa, 2013) where stakeholder engagement would unlock significant potential for reduced lifecycle costs (Wang et al., 2013). Moreover, the use of BIM for FM is not sufficiently integrated within existing FM IT systems. According to Ammari & Hammad (2014), Yalcinkaya & Singh, (2014) and Motamedi, Hammad, & Asen (2014) this is related to the current focus of existing FM information systems on work orders and asset inventory. Data entry necessary for FM purposes is presently undertaken manually, rather than through sensors or other automated data collection methods.

With regard to business, decisions that resolve business challenges such as the allocation of roles, responsibilities and rewards that apply to the different stakeholders are still a barrier. The great majority of FM providers have not experienced using BIM and there is a lack of clarity regarding the impact BIM would have on current roles and responsibilities. Thus, there is a need for well-defined transactional business process models to ease the flow of information and connect processes. Changes in business processes and business relationships are necessary for benefiting from BIM (Bernstein and Pittman, 2004; Konukcu, & Koseoglu, 2012; Liu et al., 2013; Deshpande, Azhar, & Amireddy, 2014, Zou, Jones, & Kiviniemi, 2015). These changes are effected within an organisation, but also beyond that, achieved through the integration of information from the supply chain as exemplified by Jalaei & Jrade (2014). The benefits of using data analytics, as demonstrated by Barton and Court (2012) and Liu (2015) can deliver productivity gains of approximately 5% and budget savings of 6%.

Additionally, there are issues relating to how people work in the BIM implementation environment. For instance, issues related to people's fear and resistance to change and to potential changes to roles within the organisation. In the context of increasingly constrained budgets, change is perceived as "more work to do", even though BIM may create opportunities for doing the same work in a more efficient manner, i.e. "to do less work", but that is not seen by facilities managers. Yan and Damian (2008), Forns-Samso et al. (2011) and Kassem, et al., (2015) found that an unwillingness to change processes; and the allocation of time and human resources to the training process, are major obstacles to the adoption of BIM.

Finally, there are technical problems that relate to software, particularly in terms of data exchange and interoperability; which are still problematic (Bernstein and Pittman, 2004; Yan and Damian, 2008). Technical issues are, by far, the most explored subject in BIM literature and a common theme in design, construction and FM. Authors such as Steel, Drogemuller and Toth (2012) and Stapleton et al., (2014) agree that interoperability issues must be addressed prior to digitalisation. Robust solutions to the problem of data exchange do not exist, Hallberg & Tarandi, (2011) and many others suggest that the standardisation of IFC will aid the adoption of a maintenance strategy within FM process by controlling the build-up of information, making it more efficient than a traditional database. In this respect, some progress has been made and reported through the use of cloud solutions (Forns-Samso et al., 2011; Redmond, Hore, Alshawi, & West, 2012; Juan & Zheng, 2014). Similarly, in the UK, the government recommendation has been the use of COBie (Cabinet Office, 2012).[1].

The implementation of BIM is a complex process that can end in financial losses if not managed properly. The change from traditional non-intelligent data to BIM-based intelligent information systems impacts in many areas within an organisation. Early works such as Jung and Gibson (1999) and Jung and Joo (2011) are still relevant in discussing key areas impacted by this change. For these authors, there are 3 main areas of concern: BIM technology (property, relation, standards and utilisation), BIM perspective (project, organisation and industry) and Construction Business Function (e.g. planning, design, estimating, scheduling, etc.). Knowledge management, thus, becomes essential for managing corporate knowledge, and failing to capture knowledge from BIM models results in significant costs and risks, therefore more focus must be placed on knowledge modelling (Konukcu, & Koseoglu, 2012; Motawa & Almarshad, 2013; Liu et al., 2013; Deshpande, Azhar, & Amireddy, 2014, Zou, Jones, & Kiviniemi, 2015 and Mignard & Nicolle, 2015; Donato, 2017).

In this respect, as in any business organisation, the clear identification of the need and the resulting impact of every implementation process should be measured to promote learning. Thus, the first step in the implementation of BIM is the identification of the current problems that the organisation that may be improved with the use of BIM and how much improvement can be achieved (e.g. reduction of maintenance delays, reduced time to assess information, etc.) through the implementation (Codinhoto et al., 2011). The decision to implement should not be based simply on a short-term cost-benefit analysis as, from a business perspective, investments made by an organisation can result in improvements that are necessary for maintaining the organisation's competitiveness. However, as argued by Comlay and Codinhoto (2017) data digitalisation can only be justified if the benefits from it are robust (Bilal et al., 2016). Isolated digitalisation is not the solution, but an integrated approach to improve process and workflows, increasing transparency in the data, and accessibility to the data, deliver a platform to use analytics as a predictor of the future (Bilal et al., 2016). Data analytics can support the development of strategies to future-proof organisations (Wamba et al., 2017). However, Giel et al. (2015) argues that owners are overwhelmed by the change management required.

Adding to the problem is the fact that clients' information needs are not known by contractors who often ask in the preparation of EIRs "what information do you need?", but also by clients who respond "What information can you give me?" The lack of definition by clients of the information, documentation and deliverables required throughout design and construction and handover are problematic (Beck, 2012; Cotts, Roper, & Payant, 2010). This is as a result of the multiplicity of choices, the complexity of processes, BIM guides and BIM standards together with the data use profiles of multiple users (Giel, et al., 2015), and the lack of specificity for FM within all these resources. Giel et al. (2015) outline a pragmatic model for transitioning to a digital landscape through the connection of digital information, not necessarily requiring highly developed geometric modelling expertise from the outset. An additional constraint concerns construction supply chain, which are not very experienced in delivering digital information modelling for FM but are regarded as the primary change agents to develop the expertise of clients accessing the construction market (CIC, 2014). Giel et al. (2015) with a limited survey indicates that 47% of client respondents do not have their BIM information requirements in place for FM. There are good examples of developing FM practice, in Manchester Town Hall the BIM FM implementation was driven by the client. The FM team as client at the Manchester Library Project, sought BIM education and experienced the difficulties of changing the status quo at a time when the benefits of BIM for FM were unknown and unproven. This demonstrates the importance of internal BIM champions able to drive change and validate the benefits of implementation (Comlay and Codinhoto, 2017).

Performance of information management within FM can be enhanced by taking a 'lean' perspective. For instance, Jylhä and Suvanto (2015) identified that improvements to the quality of information for FM service delivery minimises the additional time required for data validation and reliability checks. Missing data as a result of data atrophy and omitted data, is reduced while productivity is increased. Problem solving is achieved in a shorter timescale because data is transparent, visible and easily available, as demonstrated in 'Lessons-learned' within the BIM task group website. Ventilation motor replacement was reduced from 4 weeks and 14 work hours to 1 day and 3 work hours, in turn reducing costs by £286 and disruption by 27 days. This gains observed in the MTHCP can be achieved even in a situation where the level of BIM maturity of the FM team is not high. Figure 5 shows the level of capability maturity of the FM team whose achieved the presented financial gains above and it is evidence that implementation of digital information modelling and geometric modelling within an FM environment will lead to value creation, as also stated by Jylhä & Suvanto, (2015).

Yes, there are benefits, there are gains, but where do we draw a line for what information should and shouldn't be modelled? As argued by Lucas (1999), initiatives such as Cloud technology and Internet of things (IoT) have unlocked a colossal unprecedented potential for data connectivity, thus making the data stream in FM extremely large and complex. Figure 6 shows an example from MTHCP where all documentation from facility and assets where stored in a cloud repository and

Figure 5. Capability maturity assessment of facilities managers at MTHCP (Source: Codinhoto et al., 2013a)

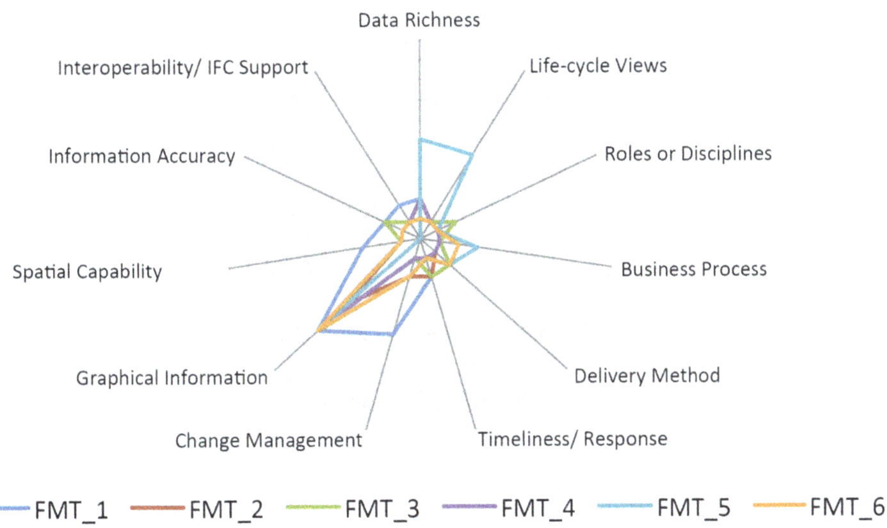

Figure 6. Graphical information display linking 3D model and relevant building documentation store in cloud (Source: Codinhoto et al., 2011)

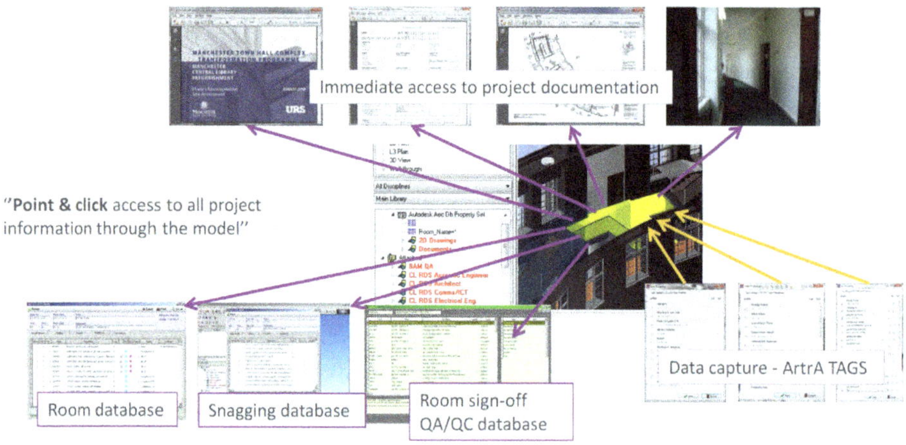

linked to the 3D model through a standardised nomenclature. Consequently, the transition to digitised data for FM adds to the intricacy of the working environment and the potential for data overload (Irizarry, Gheisari, Williams, & Walker, 2013). Considering that the digitalisation of building complexes is happening in stages as building stocks are renovated, that means that FM teams are managing two information platforms in parallel, a set of historic text and 2D drawings together with a digital based information platform received from recent CAPEX BIM projects. In general, this is an inefficient process (Irizarry et al., 2013) as isolated database queries and spreadsheets are challenging tools to use for delivering insight and identifying patterns in collected data (Berinato, 2016).

While the problem is understood, the solution for integration of legacy estate FM information within a digital information modelling platform is under developed worldwide, despite the application of digital information modelling processes, work flows, information sharing and technology for existing built assets, having the potential to create added value for organisations (Crown, 2013). In Figure 7, for instance, a model of waste movement was created for the UoB to support waste management. The model was built from a 3D model but uses a different approach (graph theory) to map flows. As suggested by Giel et al. (2015), the key benefits of digitalised information are in the potential for: asset management, building systems analysis, ease of transfer of data to computerized maintenance management system [CMMS], space management and control, scheduling of maintenance, GPS/GIS integration, disaster planning etc. This view is confirmed by the Ministry of Justice

Figure 7. UoB library: graph theory model for waste movement

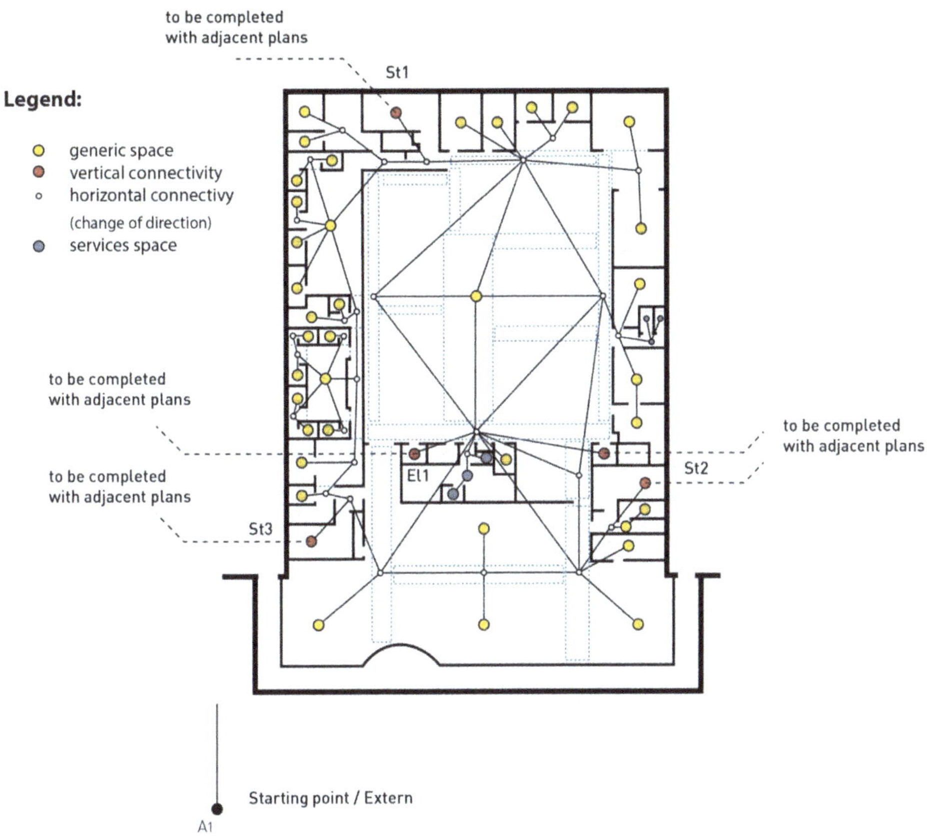

(UK) and Sydney Opera House which continue to develop and publicise strategies to transition and operate digitally modelled information for FM with reported gains of accessibility to information, resource utilisation and increased productivity (Linning, 2015).

Paradoxically, the management of information requirements and their data demands, adds a layer of complexity to O&M service delivery that is generally unquantifiable. However, it is known that successful implementation which is focussed on how to capture and use data; and is appropriate to the organisation, assists with improved decision-making and achieves an increasingly flexible team, able to respond to changing organisational requirements (Barton & Court, 2012). Refining the resolution of the granularity of information required, to map data requirements for digitalised information requirements of organisations, necessitates a data audit of the existing situation to understand the inefficiencies and identify the development needs.

The data collection needs identified should be sufficient to deliver FM, not just a catchall of JIC [just in case] information, also known as 'data bloat'. High information granularity is as bad as low granularity. In instances of lack of knowledge around digital data there is a tendency to request everything, a 'low granularity' of data, which presents significant storage issues and incurred costs and does not address the issues facing FM (Bilal et al., 2016). Individually designed professional BIM models represent on average 100Mb of data, a single construction project with the potential range in models, visualisations, analytical models, AIM model and data, has significant data storage and maintenance needs. The granularity of data, resolved as the quantification of how much fine data is required (low granularity) and strategic level data (high granularity) presents significant cost implications (Comlay and Codinhoto, 2017).

Furthermore, to be effective, exchangeable and useful, data formats have to be standardised. Worldwide, various classification systems exist that respond to the need for a structured way to organise information so it can be seamlessly exchanged. For instance, local standardisation initiatives have developed in the United Kingdom (e.g. COBie and Uniclass 2015), the United States (e.g. Masterformat and Omniclass), Finland (e.g. Talo 2000), Denmark (e.g. Cuneco), Japan (JCCS), Brazil (e.g. NBR 15965-1), Australia (e.g. Natspec), Portugal (e.g. ProNIC – Salvado et al., 2016), etc.. In fact, many of these relatively successful approaches have only achieved limited adoption (Biscaya, 2013; RIBA, 2017) and as IT systems become more sophisticated and complex, the need to find more robust and effective ways of exchanging information is constant.

Several 'new' improved systems have been developed within the European context (such as the ISO 12006-3:2007; ISO 16739:2013 and ISO 29481-2:2012). Amongst prominent solutions already implemented or in early stages of its implementation are OmniClass (Ominiclass, 2011), Uniclass 2015 (RIBA, 2017) and ISO 12006-2: 2015 which maps into other classification systems. These systems comprehensively identify classes for the organization of information, indicate their relationship and enable cross-referencing with other systems. Information types include: geometrical, functional, technical and cost data as well as maintenance data. These systems have contributed considerably to the improvement of information structuring and exchange in recent past (Mêda et a., 2017) and the challenge is to keep them effective in the face of increased complexity.

As dicussed by Comlay and Codinhoto (2017) BIM Standards such as ISO/TS 12911:2012, ISO 29481-1:2010, BSi PAS1192:3, etc., aim to achieve the quality, validity and reliability of data and information; this approach to standardisation reduces variability, atrophy, replication and redundancy in the data (Corrocher, 2013). Standards reduce technical uncertainty and complexity and result from innovation and knowledge sharing between an extensive range of experts and specialists (Corrocher,

2013). Compliance with BIM standards has seen the commensurate development of the IFC standard to deliver interoperability for BIM tools. However, the IFC standard is under constant revision to meet the needs of the industry, which has affected the information requirements of clients where all file types are required, due, in part to a lack of confidence in IFC data (Giel et al., 2015). The use of standards in the digital construction platform ensures consistent information is issued and released in appropriate formats, achieving good levels of efficiency for information exchange and compatibility (Giel et al., 2015). Roles and responsibilities require clear definition together with key deliverables and processes for managing the quality and timing of data exchanges (Biddle et al., 2012).

BIM standards and guidelines currently available apply in the main to new build CAPEX projects. PAS 1192-3: 2016 and BS 8536-1-2015 are specifically for facilities management, however, legacy estate is not explicit within these standards. BS EN 19650-2 and BS EN 19650- 2, are international standards under development, due Spring 2018 that will also be relevant to facilities management (BSI, 2017). These proposed ISO standards engage with BIM for the whole life cycle of a built asset and may be applied irrespective of procurement strategy, organisation typology and size, however, the standards are aimed at CAPEX projects i.e. refurbishment and do not address digital data for an existing built environment. Currently the standards are acting as an enabler and as an '... engine of innovation', primarily for design, construction and refurbishment. The current standards are unbalanced at present, as they do not engage with legacy estate (Corrocher, 2013). Unfortunately there remains no framework for 'how' large client/owners can migrate their legacy estate to a BIM platform.

In addition, a series of standards for smart cities (PAS 180, PAS 181, PAS 182, PAS 183, PAS 184, PAS 185, PD 8100:2015 and PD 8101:2014) is emerging in parallel. In this respect, it is expected that smart cities will have a certain amount of data that is the result of a compilation of data from individuals and buildings. In other words, there are overlaps between building and city data sets and the current lack of adoption of standards at building level will create difficulties for the creation of smart cities platforms.

For clients to achieve the level of information required by the FM team to operate and maintain their buildings, the information requirements of the client should be sufficiently detailed to receive useful information from the supply chain that meets the current state and the projected future state and meet BS 8536:2015 - FM briefing for design & construction. Overall, FM teams are engaging with the highly complex paradigm of digital information modelling, with all its intricacies without the support of fully developed guidance, standards and templates that will assist in an easier transition to a digitalised landscape.

Information structuring and exchange in construction has been a constant challenge. Despite much progress being made, a unified classification system still does not exist, due several reasons including inherent cultural differences informing the generation of (unrelated) taxonomies. While an agreed classification does not exist, software providers use their own approaches for the classification of information, thus causing problems of integration (Laakso and Kiviniemi, 2012; Monteiro et al., 2014). This means that client organisations face the paradox of receiving information from best of breed solutions used by designers and contractors when standardised solutions are better for data use. Best of breed software performs specialized functions better than an integrated system, however each is limited by its specialty area, with cross connectivity and integration challenges and unique data structures.

KEY ISSUES TO BE ADDRESSED FOR EXPANDING BIM FM ADOPTION

In light of the discussion presented here, various issues are highlighted below that require addressing if BIM FM is to be adopted at a level where it makes a significant environmental and economic contribution to organisations, cities and society.

Firstly, the issue of literacy must be addressed. There is evidence of FM literacy issues by designers and contractors. In this respect, the early involvement of FM representatives in the design process helps minimise issues, however, FM services design is never clear throughout the design process as focus is placed on core services offered by the client. In addition, the analysis of EIRs and BEPs shows that the templates for capturing data requirements have been developed by designers and contractors, so while useful during the design and construction phases they do not capture the necessary information for hard and soft FM purposes. What aggravates the problem of communication between parties is the fact that there are also BIM literacy issues amongst facilities managers. In this respect, to say that there are BIM literacy issues amongst facilities managers is not to say that Facilities Management has not been digitalised. The adoption of BIM FM has been increasing since 2011, but the pace is too slow and with limited emphasis to CAPEX. While literacy remains an issue, facilities managers will still be requesting COBie from a 'just in case' perspective without benefiting from using the information given through it.

Secondly, there is an assumption amongst FM providers that their current level of BIM FM capability and maturity is very low. In all assessments of maturity carried out by the authors, the self-evaluation done by facilities managers proved worse than that carried out by the independent research team. Partially this problem is related to the strong association that FM providers make between BIM and 3D models. To some extent, that association is pertinent as CAPEX in particular can be facilitated by information that is built in digital 3D objects. However, that is not the only source of information, as data for FM purposes can come from various sources such as spreadsheets, documents and databases. Thus, even though much of the effort related to BIM implementation has been placed on hard FM (maintenance of building fabric, statutory, planned and reactive maintenance alongside environmental control) and much less has been done for soft FM services such as waste management, security, building resilience, etc. The potential for its full adoption is latent and depends on skills for information modelling, i.e. connecting disconnected pieces of information sets using the native format they are generated in.

Thirdly, because the potential for BIM FM has not been exploited, the utilisation of meaningful POE for organisations' benefit is patchy, at best. The mentality that POEs are a burden still remains. Managers perceive that conducting POEs can create expectations amongst building users that existing problems will be resolved. In general that is never the case, as interventions require funding that is usually not available. In addition, it also reveals that POEs are perceived as a one-off event rather than a continuous monitoring activity. As a consequence, opportunities for capturing detailed service design processes are missed and consequent communication with designers and contractors in the advent of new CAPEX projects suffers. In addition, while BIM FM of new or refurbishment projects have experienced some level of BIM implementation, there has been no push for modelling the existing building stock. There is a lack of strategy for how digital information from new buildings will be managed with existing analogue information of existing buildings in multi-site complexes such as hospitals, universities, industrial complexes, etc.

Fourth, is the issue of guidance, which has been developed through the creation of standards. However, these have been created with a focus on different levels of data analysis and further integration is needed. As an example from the UK, PAS 1192:2 and PAS 1192:3 the level of data analysis is "building" whereas the PAS 180 series focus on cities. Studies looking at Big Data are, in general, embryonic and the lack of an agreed framework and strategic thinking for content and data format means that more work should be expected. For that reason, some may be put off and prefer to wait until a data structure is more definitive, hindering the adoption of standards.

Fifth, is the lack of BIM FM research buy-in. BIM FM research, to a great extent, falls within applied research. As such, it requires the participation of organisations through the provision of access to data. However, there is an inertia in the sector to buy into the idea of BIM FM research. Huge financial pressures placed on

FM means that teams have been reduced to bare minimums and the operation of facilities is done at full capacity, on a fire-fighting approach. FM teams are reactive to managing their service delivery model and are managing significant levels of backlog maintenance that requires additional funding outside of the FM budget. In addition, 3rd party FM contractors are in high demand due to limited availability of FM speciality firms and are choosing to ignore implementing BIM for FM. The large demand from clients and shortage of contractors empowers 3rd parties. Because demand is high, many prefer to release the contract rather than upskill their own organisation. This therefore, makes it difficult for clients to specify BIM for FM as FM contractors are refusing to comply.

Finally, and more importantly, there is a need for a BIM FM mandate. Mirroring the UK 2011 mandate, a general mobilisation of BIM FM efforts has to happen. Governments must act as client and regulator as no links or umbrella organisations exists for FM and there are no means by which buildings with different functions such as health, education, retail, leisure, housing, etc., from public or private organisations can be reached, if not by means of contracts and tax incentives and penalties. Some may feel discouraged at the enormity of the task before them. It will demand a strategy for upskilling the FM task force. It will require coordination so that a framework for data generation at various levels is agreed. It will require collaboration amongst researchers, practitioners and users so that change can take place and standards are fully adopted. It will take time for adaptation and adjustment but the environment and society cannot continue to pay for avoidable built in inefficiencies in the way we utilise buildings.

REFERENCES

Alexander, K. (Ed.). (2013). *Facilities management: theory and practice*. Routledge.

Ammari, K. E., & Hammad, A. (2014). Collaborative BIM-based markerless mixed reality framework for facilities maintenance. *Computing in Civil and Building Engineering - Proceedings of the 2014 International Conference on Computing in Civil and Building Engineering*, 657–664. 10.1061/9780784413616.082

Arayici, Y., Onyenobi, T. C., & Egbu, C. O. (2012). Building information modelling (BIM) for facilities management (FM): The MediaCity case study approach. *International Journal of 3D Information Modelling, 1*(1), 55-73.

ARUP. (2016). *The Circular Economy in the Built Environment*. Retrieved from http://publications.arup.com/publications/c/circular_economy_in_the_built_environment

Atkin, B., & Brooks, A. (2009). *Total Facilities Management, Facilities* (3rd ed.; Vol. 12). Oxford, UK: Wiley-Blackwell.

Ballesty, S., Morris, J., Ding, L., Drogemuller, R., Mitchell, J., Schevers, H., . . . Spink, G. (2006). An Integrated Collaborative Approach for FM – Sydney Opera House FM Exemplar, Clients Driving Innovation: Moving Ideas into Practice. Gold Coast, Australia: Academic Press.

Barrett, P., & Baldry, D. (2009). *Facilities management: Towards best practice*. John Wiley & Sons.

Barton, D., & Court, D. (2012). Making Advanced Analytics Work For You. *Harvard Business Review*, *90*(10), 78–83. PMID:23074867

Batra, S. (2014). Big data analytics and its reflections on DIKW hierarchy. *Review of Management*, *4*(1), 5–17.

Becerik-Gerber, B., Jazizadeh, F., Li, N., & Calis, G. (2012). Application Areas and Data Requirements for BIM-Enabled Facilities Management. *Journal of Construction Engineering and Management*, *138*(May), 431–442. doi:10.1061/(ASCE)CO.1943-7862.0000433

Beck, K. (2012, Spring). The state of Wisconsin: BIM – digital FM handover pilot projects. *Journal of Building Information Modeling*, 22-23.

Berinato, S. (2016). Visualizations that really work. *Harvard Business Review*.

Bernstein, P. G., & Pittman, J. H. (2004). *Barriers to the adoption of building information modeling in the building industry*. Autodesk building solutions.

Biddle, B., Curci, F. X., Haslach, T. F., Marchant, G. E., Askland, A., & Gaudet, L. (2012). The expanding role and importance of standards in the information and communications technology industry. *Jurimetrics*, 177–208.

Bilal, M., Oyedele, L. O., Qadir, J., Munir, K., Ajayi, S. O., Akinade, O. O., & Pasha, M. (2016). Big Data in the construction industry: A review of present status, opportunities, and future trends. *Advanced Engineering Informatics*, *30*(3), 500–521. doi:10.1016/j.aei.2016.07.001

BIS. (2011). *A report for the Government Construction Client Group Building Information Modelling (BIM)*. Working Party Strategy Paper. Retrieved from: http://www.bimtaskgroup.org/wp-content/uploads/2012/03/BIS-BIM-strategy-Report.pdf

Biscaya, S. (2013). *Coordination and Management of Information for Construction Design Projects – A Framework for Portugal*. Manchester, UK: University of Salford. Retrieved from http://usir.salford.ac.uk/29281/

BSI. (2007). *Facility Management - Part 1: Terms and definitions: BS EN 15221-1:2006* (Vol. 44). London: BSI Standards Publication.

BSI. (2014). Asset management. Overview, principles and terminology. *BS ISO, 55000*, 2014.

Bungar, S. (2012). *Facilities Management-UK Overview*. London: Facilities Management.

Cabinet Office. (2012). *The Government Soft Landings Policy*. Retrieved from BIM task group website: http://www.bimtaskgroup.org/gsl-policy-2

Cairns, G. (2003). Seeking a facilities management philosophy for the changing workplace. *Facilities*, *21*(5/6), 95–105. doi:10.1108/02632770310476705

CIC - Construction Industry Council. (2014). *Built Environment 2050: A Report on our Digital Future*. BIM 2050. Available at: https://www.google.co.uk/url?sa=t&rct=j&q=&esrc=s&source=web&cd=1&cad=rja&uact=8&ved=0ahUKEwjbyszd8pDYAhWpLcAKHckiBJgQFggpMAA&url=http%3A%2F%2Fcic.org.uk%2Fdownload.php%3Ff%3Dbe2050-cic-bim2050-2014-1.pdf&usg=AOvVaw3FocdkM1OqsWG9LLuznY2M

Codinhoto, R., Kiviniemi, A., Kemmer, S., & Da Rocha, C. G. (2011). *BIM Implementation: Manchester Town Hall Complex*. SCRI Report Series 6. Available at: http://live.scri.salford.ac.uk/wp-content/uploads/2011/12/MCC_Final_Research_Report.pdf

Codinhoto, R., Kiviniemi, A., Kemmer, S., & da Rocha, C. G. (2013a). BIM-FM implementation: an exploratory invesigation. *International Journal of 3-D Information Modeling, 2*(2), 1-15.

Codinhoto, R., Kiviniemi, A., Kemmer, S., Martin, E., Donato, V., & Tonso, L. (2013b). *BIM FM at the Manchester Town Hall Complex: A Case Study Investigation. Research Report*. The University of Salford.

Codinhoto, R., Tzortzopoulos, P., Rooke, J., Kagioglou, M., & Koskela, L. (2008). *Facilitators and barriers to the integration of healthcare service and building design*. Academic Press.

Comlay, J. (2015). *Facilities Management (FM) Organisational Transformation: Building Information Modelling (BIM) is the Proposition* (MSc dissertation). The University of Salford.

Comlay, J., & Codinhoto, R. (2017). Facilities Management: Granularity of Information for a Digital Information Platform. *Proceedings of the International Conference on Sustainable Futures (ICSF) Bahrain 2017.*

Corrocher, N. (2013). The Development of Short Message Services: Standard Organizations as Engines of Innovations. *Revue Economique*, *64*(1), 149–163. doi:10.3917/reco.641.0149

Cotts, D., Roper, K., & Payant, R. (2010). *The facility management handbook* (3rd ed.). New York: American Management Association.

Cracknell, D. (2012). *Sector BIM strategy: Improving BIM education & training*. The Higher Education Academy. Available at http://www.heacademy.ac.uk/assets/documents/disciplines/built_environment/BIM/BIM_Skills_CIC_David_Cracknell.pdf

Cragun, D., Pal, T., Vadaparampil, S. T., Baldwin, J., Hampel, H., & DeBate, R. D. (2015). Qualitative Comparative Analysis. *Journal of Mixed Methods Research*, *10*(3), 251–272. doi:10.1177/1558689815572023 PMID:27429602

CRC Construction Innovation. (2007). *Adopting BIM for Facilities Management: Solutions for Managing the Sydney Opera House*. Brisbane, Australia: Cooperative Research Center for Construction Innovation.

Crown. (2013). *Construction 2025. (URN BIS/13/955).* London: HMSO.

Cruz, F., Chong, W. K., & Grau, D. (2015). Design for Disassembly and Deconstruction - Challenges and Opportunities. *Procedia Engineering*, *118*, 1296–1304. doi:10.1016/j.proeng.2015.08.485

Deshpande, A., Azhar, S., & Amireddy, S. (2014). A framework for a BIM-based knowledge management system. *Procedia Engineering*, *85*, 113–122. doi:10.1016/j.proeng.2014.10.535

Donato, V. (2017). Towards design process validation integrating graph theory into BIM. *Architectural Engineering and Design Management*, *13*(1), 22–38. doi:10.1080/17452007.2016.1208602

European Commission. (2017). *Energy Efficiency: Buildings*. Available at https://ec.europa.eu/energy/en/topics/energy-efficiency/buildings

Forns-Samso, F., Bogus, S. M., & Migliaccio, G. C. (2011). *Use of Building Information Modeling (BIM) in Facilities Management*. Paper presented at the 3rd International/9th Construction Speciality Conference, Ottawa, Ontario, Canada.

Giel, B. K., Mayo, G., & Issa, R. R. A. (2015). BIM Use and Requirements Among Building Owners. In R. R. A. Issa & S. Olbina (Eds.), Building Information Modeling - Applications and Practices. American Society of Civil Engineers (ASCE). doi:10.1061/9780784413982.ch10

Grilo, A., & Jardim-Goncalves, R. (2011). Challenging electronic procurement in the AEC sector: A BIM-based integrated perspective. *Automation in Construction*, *20*(2), 107–114. doi:10.1016/j.autcon.2010.09.008

Hallberg, D., & Tarandi, V. (2011). On the use of open bim and 4d visualisation in a predictive life cycle management system for construction works. *Journal of Information Technology in Construction*, *16*, 445–466. Retrieved from http://www.diva-portal.org/smash/record.jsf?pid=diva2%3A478341&dswid=-8750

Irizarry, J., Gheisari, M., Williams, G., & Walker, B. N. (2013). InfoSPOT: A mobile Augmented Reality method for accessing building information through a situation awareness approach. *Automation in Construction*, *33*, 11–23. doi:10.1016/j.autcon.2012.09.002

Jalaei, F., & Jrade, A. (2014). Developing a spatial data framework for facility management supply chain. In *Construction Research Congress 2014* (pp. 140–149). Academic Press. 10.1061/9780784413517.176

Juan, D., & Zheng, Q. (2014). Cloud and Open BIM-Based Building Information Interoperability Research. *Journal of Service Science and Management*, *7*(April), 47–56. doi:10.4236/jssm.2014.72005

Jung, Y., & Gibson, G. E. (1999). Planning for computer integrated construction. *Journal of Computing in Civil Engineering*, *13*(4), 217–225. doi:10.1061/(ASCE)0887-3801(1999)13:4(217)

Jung, Y., & Joo, M. (2011). Building Information Modeling (BIM): Framework for Practical Implementation. *Automation in Construction*, *20*(2), 126–133. doi:10.1016/j.autcon.2010.09.010

Jylhä, T., & Suvanto, M. E. (2015). Impacts of poor quality of information in the facility management field. *Facilities*, *33*(5/6), 302–319. doi:10.1108/F-07-2013-0057

Kassem, M., Kelly, G., Dawood, N., Serginson, M., & Lockley, S. (2015). BIM in facilities management applications: A case study of a large university complex. *Built Environment Project and Asset Management*, *5*(3), 261–277. doi:10.1108/BEPAM-02-2014-0011

Kiviniemi, A., & Codinhoto, R. (2014). Challenges in the Implementation of BIM for FM—Case Manchester Town Hall Complex. *Computing in Civil and Building Engineering*, 665–672. Retrieved from http://ascelibrary.org/doi/abs/10.1061/9780784413616.083

Kiviniemi, A., Tarandi, V., Karlshøj, J., Bell, H., & Karud, O. J. (2008). *Review of the development and implementation of IFC compatible BIM*. Erabuild Funding Organizations, Europe.

Klepeis, N. E., Nelson, W. C., Ott, W. R., Robinson, J. P., Tsang, A. M., Switzer, P., ... Engelmann, W. H. (2001). The National Human Activity Pattern Survey (NHAPS): A resource for assessing exposure to environmental pollutants. *Journal of Exposure Science & Environmental Epidemiology*, *11*(3), 231–252. doi:10.1038j.jea.7500165 PMID:11477521

Konukcu, S., & Koseoglu, O. (2012). Knowledge Management Through Bim in Construction Supply Chains. In *CIB W78 2012: 29th International Conference* (pp. 17–19). Beirut: Academic Press.

Laakso, M., & Kiviniemi, A. O. (2012). The IFC standard: A review of history, development, and standardization, information technology. *ITcon*, *17*(9), 134–161.

Lillrank, P. (1995). The transfer of management innovations from Japan. *Organization Studies*, *16*(6), 971–989. doi:10.1177/017084069501600603

Linning, C. (Producer). (2015). *BIM for Facility Management* [PowerPoint slides]. Retrieved from https://bips.dk/files/article_files/keynote_bim_i_fm_paa_sydney_opera_house.pdf

Liu, R., & Issa, R. R. A. (2013). Issues in BIM for Facility Management from Industry Practitioners' Perspectives. *Computing in Civil Engineering*, 411–418.

Liu, Y. (2015). Big data and predictive business analytics. *The Journal of Business Forecasting*, *33*(4), 40–42.

Lucas, P. (1999). *The trillion-node network*. MAYA Technical Report #MTR-00001.

Mêda, P., Hjelseth, E., & Sousa, H. (2017). Construction information framework—the role of classification systems. In *eWork and eBusiness in Architecture, Engineering and Construction: ECPPM 2016: Proceedings of the 11th European Conference on Product and Process Modelling (ECPPM 2016)* (p. 447). CRC Press.

Mignard, C., & Nicolle, C. (2015). A Building Information Management (BIM) Framework and Supporting Case Study for Existing Building Operations, Maintenance and Sustainability. *Procedia Engineering*, *118*, 1104–1111. doi:10.1016/j.proeng.2015.08.450

Monteiro, A., Mêda, P., & Martins, J. P. (2014). Framework for the coordinated application of two different integrated project delivery platforms. *Automation in Construction*, *38*, 87–99. doi:10.1016/j.autcon.2013.11.003

Motamedi, A., Hammad, A., & Asen, Y. (2014). Knowledge-assisted BIM-based visual analytics for failure root cause detection in facilities management. *Automation in Construction*, *43*, 73–83. doi:10.1016/j.autcon.2014.03.012

Motawa, I., & Almarshad, A. (2013). A knowledge-based BIM system for building maintenance. *Automation in Construction*, *29*, 173–182. doi:10.1016/j.autcon.2012.09.008

Naylor, S. R. (2017). *NHS Property and Estates: Why the estate matters for patients*. London, UK: Crown.

Nicał, A. K., & Wodyński, W. A. (2016). Enhancing Facility Management through BIM 6D. In *Creative Construction Conference* (pp. 656–661). Budapest: Academic Press.

Redmond, A., Hore, A., Alshawi, M., & West, R. (2012). Exploring how information exchanges can be enhanced through Cloud BIM. *Automation in Construction*, *24*, 175–183. doi:10.1016/j.autcon.2012.02.003

Reinhardt, R., & Gurtner, S. (2015). Differences between early adopters of disruptive and sustaining innovations. *Journal of Business Research*, *68*(1), 137–145. doi:10.1016/j.jbusres.2014.04.007

RIBA. (2017). *NBS National BIM Report 2017*. London: RIBA Enterprises Ltd.

Schevers, H., Mitchell, J., Akhurst, P., Marchant, D., Bull, S., McDonald, K., ... Linning, C. (2007). Towards digital facility modelling for Sydney opera house using IFC and semantic web technology. *Journal of Information Technology in Construction*, *12*(24), 347–362.

Shen, S., & Codinhoto, R. (2017). *Post-Occupancy Evaluation Report: Manchester Central Library Case Study* (Master Dissertation). The University of Bath.

Stapleton, K. A. J., Gledson, B. J., & Alwan, Z. (2014). Understanding technological interoperability through observations of data leakage in Building Information Modelling (BIM) based transactions. *Proceedings of the 32nd eCAADe Conference*, *2*, 515–524. 10.13140/RG.2.1.3882.5849

Steel, J., Drogemuller, R., & Toth, B. (2012). Model interoperability in building information modelling. *Software & Systems Modeling*, *11*(1), 99–109. doi:10.100710270-010-0178-4

Tay, L., & Ooi, J. T. L. (2001). Facilities management: A 'Jack of all trades'? *Facilities*, *19*(10), 357–363. doi:10.1108/EUM0000000005534

Teicholz, P. (Ed.). (2013). *BIM for facility managers*. John Wiley & Sons.

Thomson, T. (1990). The essence of facilities management. *Facilities*, *8*(8), 8–12. doi:10.1108/EUM0000000002119

Tzortzopoulos, P., Codinhoto, R., Kagioglou, M., Rooke, J. A., & Koskela, L. J. (2009). The gaps between healthcare service and building design: A state of the art review. *Ambiente Construído*, *9*(2), 47–55.

Volk, R., Stengel, J., & Schultmann, F. (2014). Building Information Modeling (BIM) for existing buildings — Literature review and future needs. *Automation in Construction*, *38*, 109–127. doi:10.1016/j.autcon.2013.10.023

Wamba, S. F., Gunasekaran, A., Akter, S., Ren, S., Dubey, R., & Childe, S. J. (2017). Big data analytics and firm performance: Effects of dynamic capabilities. *Journal of Business Research*, *70*, 356–365. doi:10.1016/j.jbusres.2016.08.009

Wang, Y., Wang, X., Wang, J., Yung, P., & Jun, G. (2013). Engagement of facilities management in design stage through BIM: Framework and a case study. *Advances in Civil Engineering, 2013*(30836). 10.1155/2013/189105

Yalcinkaya, M., & Singh, V. (2014, July). Building information modeling (BIM) for facilities management–literature review and future needs. In *IFIP International Conference on Product Lifecycle Management* (pp. 1-10). Springer. 10.1007/978-3-662-45937-9_1

Yan, H., & Demian, P. (2008). *Benefits and barriers of building information modelling*. Academic Press.

Zou, Y., Jones, S. W., & Kiviniemi, A. (2015). BIM and Knowledge Based Risk Management System A Conceptual Model. *Proceedings of CitA BIM Gathering 2015*, 63–68.

ENDNOTE

1 Although COBie is often seen as an Excel file, it is actually a subset (and partly extension) of IFC and can also be delivered in IFC format.

Chapter 5
Three-Dimensional Information Retrieval (3DIR):
A Graph Theoretic Formulation for Exploiting 3D Geometry and Model Topology in Information Retrieval

Peter Demian
Loughborough University, UK

Kirti Ruikar
Loughborough University, UK

Anne Morris
Loughborough University, UK

ABSTRACT

The 3DIR project investigated the use of 3D visualization to formulate queries, compute the relevance of information items, and visualize search results. Workshops identified the user needs. Based on these, a graph theoretic formulation was created to inform the emerging system architecture. A prototype was developed. This enabled relationships between 3D objects to be used to widen a search. An evaluation of the prototype demonstrated that a tight coupling between text-based retrieval and 3D models could enhance information retrieval but add an extra layer of complexity.

DOI: 10.4018/978-1-5225-5625-1.ch005

INTRODUCTION

In building modelling environments, information is increasingly being crammed into 2D/3D building and product models. This is particularly true given the rise of Building Information Modelling (BIM). The Three-Dimensional Information Retrieval (3DIR) project investigated information retrieval from these environments, where information or documents are linked to a 3D building model. In these situations, the 3D visualisation or 3D geometry of a building can be exploited when formulating information retrieval queries, computing the relevance of information items to the query, or visualizing search results. Managing such building information repositories in this way would take advantage of human strengths in vision, spatial cognition and visual memory (Lansdale and Edmonds, 1992; Robertson et al., 1998).

Information retrieval is associated with documents, and a critic might argue that documents are relics from the pre-BIM age that are no longer relevant in the era of BIM. However, the challenge of information retrieval is pertinent whether we are dealing with documents which are coarse grains of information or building object parameters/attributes as finer grains of information. Demian and Fruchter (2005) demonstrated that traditional retrieval computations can be applied with good results to 3D building models where textual or symbolic data are treated as very short documents. In this sense, it is almost a question of semantics whether the information being retrieved comes from object properties embedded in the BIM, or from external documents linked to the BIM. The challenge remains of retrieving non-geometric or textual information.

This chapter describes developments of the 3DIR project whose aim was to improve information retrieval when retrieving information or documents linked to a 3D artefact, or retrieving non-geometric information embedded in the model of the artefact. It proposes a formulation based on graph theory as a useful theoretical lens for research and software development for information retrieval from 3D models. The central objective was to develop an information retrieval toolset for documents/information linked to 3D building models which exploits 3D geometry and linked information. Such a toolset is essentially a search engine for retrieving information within a BIM platform.

RELATED WORK

Building design, construction and operation are information intensive activities. For example, even over a decade ago in the UK construction industry, on average, one computer-aided design (CAD) document was produced for every 9 m^2 of building floor space (Gray and Hughes 2001). Several researchers (Leslie, 1996; Veeramani and Russell, 2000; Ugwu, 2005) have reported the problem of "information overload" in the construction sector.

BIMs are following this general trend and becoming more information-rich. Regarding volumes of information specifically in BIMs, Demian and Walters (2014) identified BIM platforms as a particularly favourable communication medium in construction, compared to extranets, email and Enterprise Resource Planning systems. Charalambous et al. (2013) reported the advantages of BIM over documents and extranets. Although no absolute measures of the quantities of information were found, the implication from studies such as those is that BIMs are increasingly information-rich.

Information retrieval techniques have been used in construction to retrieve reusable designs (Demian and Fruchter 2005). Beyond text, Brilakis and Soibelman (2008) automatically identify particular features in construction site photographs with a view subsequently to using information retrieval techniques to manage collections of photographs. Bridging textual and geometric content, Caldas et al. (2002) propose techniques for automatically classifying construction documents based on project CAD components. Lin and Soibelman (2009) augment standard information retrieval techniques with formal representations of domain knowledge to improve the performance of a search engine for online product information. Rezgui (2006) similarly uses domain knowledge to formulate an ontology that informs the indexing and retrieval of construction content. These studies demonstrate how standard retrieval computations can be complemented when applied to building design and construction.

None of the studies encountered in the literature specifically exploit 3D data and 3D visualisations for information retrieval. This approach lies at the intersection of three academic fields: (1) BIM and CAD, (2) information retrieval and (3) information visualisation.

BIM and CAD

The state of the art in digital content management in building design and construction projects is being transformed by the emergence of Building Information Modelling (Eastman et al. 2011). Whereas CAD models classically attempted to model the geometry of buildings or building components in two or three dimensions (e.g. Eastman 1999, Emmitt and Ruikar 2013), Building Information Models include non-geometric content as well (Figure 1). This content includes the non-geometric attributes of physical building components (such as the cost of a component) as well as non-geometric entities. For example, Building Information Models can include entities to model the processes of design (Austin et al. 2000) and construction (Koo and Fischer 2000) and the organizations (i.e., teams and individuals) that execute

Figure 1. Contrast between a CAD model, a representation of building geometry, and a Building Information Model, a representation of context-sensitive objects (from Ruikar, 2014)

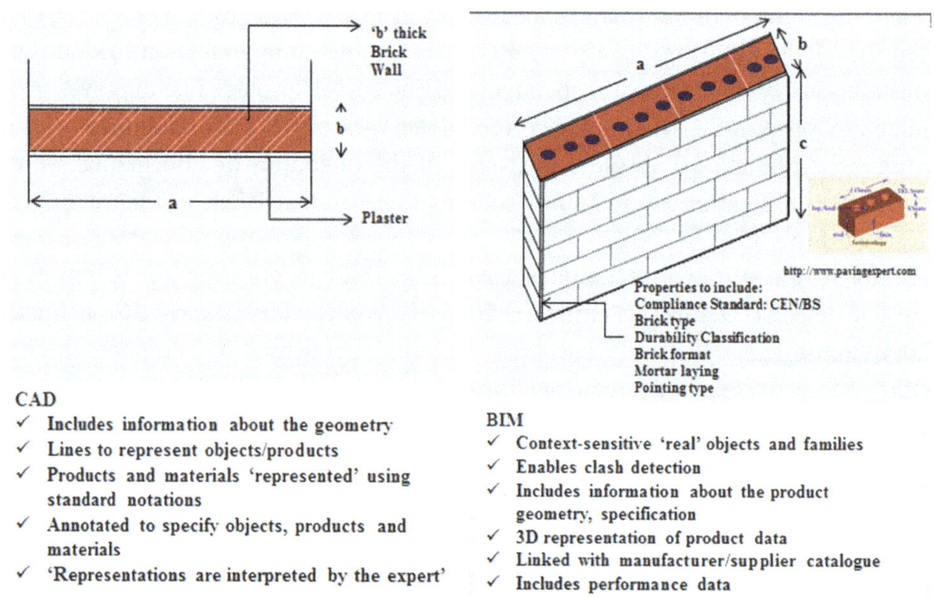

those processes (Kunz et al. 1998). In addition, BIM is not limited to the design and construction phases but can be extended to cover the entire life cycle of constructed facilities, from briefing/programming, through design, to facilities management and even disposal.

In the context of the 3DIR project, it is noteworthy that, although as noted above, CAD and BIMs nowadays include both geometric and non-geometric information, the geometric 3D model of the building is central, and is often expected to serve as a visual index that leads to the additional non-geometric content (Figure 1). This approach often fails, because such systems do not exploit human abilities in spatial cognition and visual memory. Non-geometric content does not leave enough *information scent* (Pirolli and Card 1999) in the geometric CAD model that enables the information forager to find it. This concept served as an important point of departure for the 3DIR project.

Information Retrieval

Information retrieval (IR) is concerned with systems that help users to fulfil their information needs. In particular, IR computations can quantify the relevance of information items based on user queries (Dominich 2008). Demian and Fruchter (2005) demonstrated that traditional IR techniques could be applied to retrieve information from BIMs and product models; the semantic information attached to 3D objects could be treated as very short documents and standard text document computations employed, giving reasonable retrieval results. As noted in the introduction to this section and under "BIM and CAD" above, information retrieval has recently been applied in managing the vast volume of information accumulated in building design, construction and operation.

Information Visualisation

The 3DIR project bridges the domains of information visualisation and scientific (or 3D) visualisation. Visualisation has been defined as "the use of visual representations to amplify cognition" (Section 1 of Card et al. 1999). Information visualisation (IV), in particular, refers to the visualisation of abstract data, unrelated to physical space. Such data (e.g. financial data, abstract conceptions, hierarchical and network data structures) have no obvious spatial mapping. One important branch of IV is the visualisation of collections of documents (Section 6 of Card et al. 1999). This aspect informed the visualisation of search results in 3DIR. Efforts to visualise document collections range in scale from visualising the whole internet, through visualising smaller document collections in information workspaces, to visualising an individual document (Card et al. 1999).

In construction, Wu and Hsieh (2012) identified the lack of a single interface which combined and visualised the information from the disparate project sources as an important cause of work breakdowns. They go on to propose the Project Information Integration Management Framework (PIIM Framework), operationalised in a software prototype in which data is presented in conjunction with the 3D model. Shaaban et al. (2001) propose different approaches for the application of IV in architecture. They present the task-driven approach as the most effective, which places the user's task and information needs at the centre, and considers his/her visual and cognitive processing. In support specifically of health and safety analysis, Zhang et al. (2015) superimpose information about health hazards and safety equipment on the 3D building visualisation. Gerrish et al (2017) demonstrate techniques for visualising building energy performance data, both in abstract representations as well as superimposed on floorplans.

This body of literature demonstrates the value of integrating 3D visualisation with IV. When visualising documents specifically, a range of document properties have been mapped to visual cues such as shape, colour and size.

Topology

As explained under "Results", the exploitation of topological relationships between objects in a 3D model emerged as a promising avenue for improving information retrieval from BIMs. In general language, topology is the "study of the way in which constituent parts are interrelated or arranged" (OED Online 2017). In mathematics, topology is the study of a collection of open sets, making a given set a *topological space* (Gemignani 1972). In spatial modelling, topology is concerned with the notions of "interior", "boundary", or "exterior". Paul (2009) examined how these notions could be captured by the Industry Foundation Classes (IFC), as buildings were modelled in 3D Euclidean space. Borrmann and Rank (2009a) present algorithms for the standard topological operators in 3D space: *within*, *contain*, *touch*, *overlap*, *disjoint* and *equal*; their prototype uses IFC-VRML files.

In this research, *topological relationships* are taken to include any relationships between 3D building elements in a model which may enhance information retrieval. These relationships might be strictly topological and concerned with interior/boundary/exterior of 3D components as noted above, more general spatial/directional relationships (Borrmann and Rank, 2009b), or even relationships as they occur in a very general semantic sense, albeit linked to spatial topology (Lin, 2013). In this sense, any two objects in a model sharing the same attribute (for example: two components supplied by the same manufacturer) can be said to be related. If a user searching for information is interested in the first object, but not the second object related to it, information from the second object can still be retrieved but ranked as less relevant.

The literature reviewed highlights the promise of systematically applying IR and IV techniques in BIM environments to exploit human cognitive strengths and facilitate more effective information management.

A GRAPH THEORETIC FORMULATION OF INFORMATION LINKED TO 3D MODELS

Graph theory provides a useful theoretical lens for studying information-rich 3D models and retrieving information in these environments. This theoretical lens can inform software research and development in this area. A *graph* in this context is a series of vertices connected by edges. Each edge joins exactly two vertices. Any graph X can be modelled mathematically by listing its set of vertices V(X) and set of edges E(X) (Aldous and Wilson (2003) give an introduction to graph theory). In the case of an information-rich 3D model, it is possible to distinguish between the set of 3D vertices V_{3D}, which are the 3D objects in the model, and information vertices V_i, which are linked information items, whether properties of the 3D objects treated as short documents or linked full text documents.

Similarly, for the edges in the graph theoretic formulation of a 3D model, it is possible to distinguish between two types of edges. The first more obvious type of edge is the edge joining a 3D object to one of its properties, i.e. an edge between a vertex in set V_{3D} and a vertex in set V_i. The set of this natural type of edge can be called E_n. It arises simply from the fact that 3D objects and their properties (or linked documents) are modelled as separate (but linked) objects. The second, more subtle, type of edge is that edge which encodes some topological relationship between two 3D objects, as discussed under "Topology" above and under "Exploitation of Model Topology in 3DIR" below, i.e. an edge joining two related V_{3D} vertices. The set of this type of topological edge can be called E_t. Such edges and the topological relations they model are one of the focal points of this current development in the 3DIR project. (If it were not for the edges in set E_t, the emerging graph would consist of two disjoint sets of 3D and information vertices, with each natural edge in E_n connecting an item in one set to an item in the other, i.e. a *bipartite graph*). Figure 2 gives an example of this formulation for a simple 3D model.

This formulation is useful because of its distinction between 3D and non-3D information objects, its classification of the different types of links between information items, and its allowance for topological relationships between 3D objects. These sets of vertices and edges can be used to develop an architecture for information retrieval from 3D modelling environments.

Figure 2. An example graph theoretic formulation of a simple 3D model consisting of a Roof, a Door and a Wall (all three objects having properties, and the Wall being linked to an external document)

V_{3D}	
1	Roof
2	Door
3	Wall

V_i	
4	Name: Red roof Type: Roof
5	Name: West door Type: Door Material: Glass
6	Name: South wall Type: External Wall Material: Concrete
7	This document is a reinforcement schedule for the South wall (external).

E_n	
a	1,4
b	2,5
c	3,6
d	3,7

E_t		
e	1,2	Touching
f	1,3	Intersecting
g	2,3	Hosting

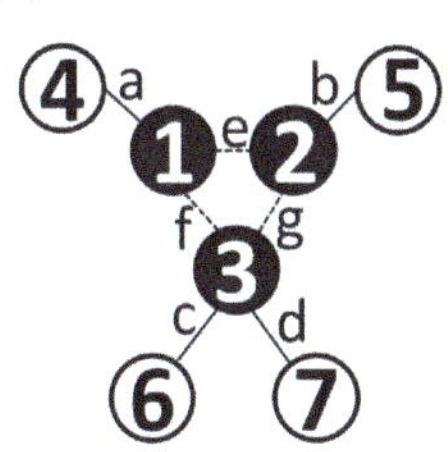

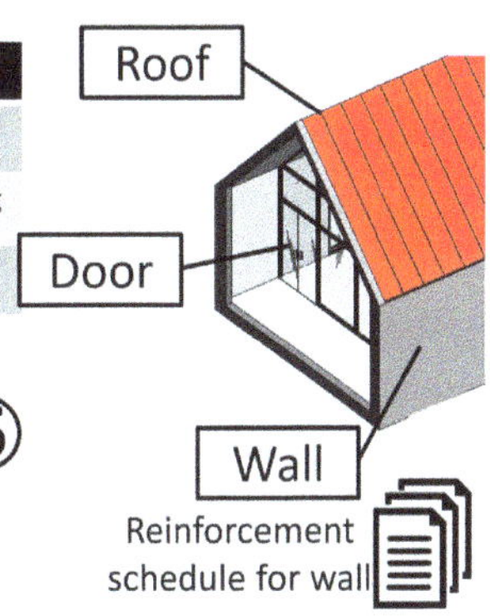

METHOD

Following reviews of literature as presented above, it appears that information management in BIM/CAD systems remains a challenge, and innovative information retrieval techniques are needed to address this. To inform the design of such a system incorporating novel information retrieval functionality, workshops were convened at "Contractors", a large multinational contractor, and "Architects", a renowned architectural practice in London. Armed with the needs identified at these workshops, and informed by the graph theoretic formulation in Figure 2, a software prototype was designed and developed as an add-in under the Autodesk Revit BIM platform. Although the ultimate aspiration is for any software development to remain platform independent and avoid favouring any particular commercial BIM environment, it was found that the Autodesk Revit Application Programming Interface (API) provided excellent opportunities for development and research prototyping. A platform-neutral format such as IFC would have been preferable for wider dissemination, but IFC authoring and viewing software did not provide the necessary API.

This *original* version of 3DIR was evaluated with the help of professionals from "Contractors" and "Architects". From this preliminary evaluation, the notion emerged of exploiting topological relationships in the model for information retrieval. In terms of the graph theoretic formulation, the idea was to exploit edges of type E_t to improve information retrieval and exploration. This notion was tested using three particular relationships or edges E_t as a proof of concept:

1. **The *Hosted by* Relationship:** This is built in to the Revit information architecture; for example, a particular window may be hosted by a particular wall.
2. **The *Intersecting* Relationship:** For example, if 3D volumes are used to model spaces, the volumes for two crossing corridors would intersect one another.
3. **The *Touching* Relationship:** For example, two adjacent walls may touch one another.

The *intersecting* and *touching* relationships (and the corresponding edges E_t) were able to be inferred using simple geometrical computations through the Revit API, whereas the *hosted by* relationship is explicitly encoded by Revit. Those three relationships were chosen as a proof of concept, as a set of fundamental relationships between 3D objects. (A more thorough exploration of topological relationships in 3D can be found in Ellul & Haklay 2007 and Ellul & Haklay 2009.) This *modified* version of 3DIR was re-evaluated with the help of a cohort of postgraduate students at Loughborough University.

RESULTS

The results of the initial workshops are presented under "Needs Analysis". The interface and system architecture of the original 3DIR system are described under "3DIR Interface and Prototype Development". The implementation of a basic set of topological relationships is described under "Exploitation of Model Topology in 3DIR". The results of the evaluation of the *original* (3DIR) and *modified* (3DIR+Topology) versions of 3DIR are presented under "Results of 3DIR Evaluation".

Needs Analysis

The ICT Director of "Architects", an international architecture practice based in London, was interviewed. The ICT Director spoke strategically about the shift from CAD to BIM (and, indeed, from hand drafting to CAD), the cost of software, the potential productivity gains and measuring the Return on Investment of these new tools. "Architects" had just adopted a new commercial BIM platform and, despite the software being perceived as extremely expensive, the productivity gains were evident: "it's taken a fraction of the time to produce the information we would normally produce with 2D drawings".

Following the interview with the ICT Director of "Architects", two architects joined the conversation and a focus group discussion was held. The architects echoed the productivity gains enabled by the new BIM platform. They noted that the BIM information architecture enabled much more information to be included almost effortlessly in the model. Expressed in terms of the graph theory formulation, models could now include vertices of the type in set V_i, which added great value to the V_{3D} vertices. The information bearing capacity of the new 3D models was used as an important communication medium for collaboration. This contrasted with the pre-BIM days described by the ICT Director: "in those days, we could still put some data into our models, but nobody did it because we had nobody to share it with." Beyond including data *within* models, all focus group participants acknowledged the difficulty of linking external documents to BIMs. If those links were in place, however, the focus group participants could clearly see the possibilities for improved information retrieval.

In addition to the interview and focus group at "Architects", a focus group was convened at "Contractors", a major UK contractor. The participants from "Contactor" did not feel that "documents" were going to remain relevant in the new era of BIM. They did acknowledge that models were becoming more and more information-rich, but not with traditional documents. *Interoperability* was an urgent issue which emerged repeatedly in that focus group. "Contractors" teams extensively used extranets on their projects. The participants identified the disconnection between 3D models and other project documents in extranets as a major obstacle to retrieving information, either from the documents or from the 3D models. Models in the extranet rarely contained links to other external information.

3DIR Interface and Prototype Development

The needs analysis exposed the complexity of information management in 3D environments. The assumption made at the outset of the 3DIR project was that links existed in standard practice between 3D components and textual information or documents. The needs analysis demonstrated that this assumption was questionable, although the situation rapidly changed as the research progressed. Even though links between documents and 3D components remain rare, as noted from the literature: the textual, numerical or symbolic parameters given to 3D components in most BIM information architectures can be considered as non-3D information linked to 3D components. In other words, 3D models included V_i information items (graph vertices) and implicitly included E_n links (graph edges) as links between these 3D objects and their properties. Isolating the challenge of exploiting such links for information retrieval, the following salient requirements (R) were distilled from the needs analysis:

R1. When formulating queries, users need the ability to search by keyword, 3D volume, by selecting a set of components from the model, or by any combination of these.

R2. When selecting a component or set of components from the model within which the user wishes to search, users would like the option of searching beyond this selection, based on relationships between components, i.e. topological relationships encoded as E_t edges.

R3. When visualising search results, users need to retain the standard text-based listing, but would also like search results somehow superimposed on the 3D model.

These requirements emerged when the findings from the needs workshops were used to formulate fictional archetypal *problem scenarios* (Rosson and Carroll 2001). Those in turn were developed into *activity scenarios, information scenarios* and *interaction scenarios*, culminating in a *usability specification*. Through this process, requirements R1, R2 and R3 were translated into the following corresponding system usability specification items (S):

S1. Multiple search modes are needed:

a. Clicking on a single component or collection of components should display all the textual information (whether parameters or external documents) linked to that/those component(s). In terms of the graph theoretic formulation, by clicking on a 3D object from set V_{3D}, the system should follow all edges in the set E_n which include this 3D object and retrieve linked V_i vertices.

b. The system should allow the user to filter the search by keyword, by selecting desired 3D components or by specifying a 3D volume.

S2. "Hops function": with a single component or collection of components selected, the system should give the user the option of searching or "hopping" outside this selection to related 3D components up to a specified maximum number of hops away. In terms of the graph theoretic formulation, E_t edges should be followed to make "hops" outside a select set of 3D objects, and expand the pool of 3D objects from which search results are retrieved. Search results can be ranked by "hops".

S3. Text search results listing is needed, together with as many visual representations of search results as the API allows:

a. Text listing should be available

b. Retrieved 3D components should be "selected"

c. Retrieved 3D components should be isolated, i.e. all other components in the model being temporarily rendered invisible

d. Retrieved 3D components should be highlighted by insetting a phantom coloured shape above them. The colour of the shape could then be used to denote the type of information retrieved (i.e. format of file or type of parameter) and the size of the shape could be used to denote the relevance according to the text retrieval computation.

A prototype was developed under the Autodesk Revit platform. Revit is a common commercial BIM platform and, as noted above, upon reviewing common BIM platforms, was found to have a powerful API. The source code was written in C#. The Apache Lucene open source library was used for the text indexing and search functions.

3DIR appears in the add-ins ribbon of the standard Revit interface. The first step when searching a building model using 3DIR for the first time is to "index" the model using that icon on the 3DIR toolbar. This will create an index of all text terms from the 3D object parameters or linked text documents. Once an index has been created, the "Search" tool can be used which brings up the dialogue box shown in Figure 3.

As the user enters keywords in the text box, search results are listed in real time in the dialogue box. In the example shown in Figure 3, the keyword "roof" is

Figure 3. The 3DIR interface. Left: Screenshots showing the software functions satisfying the specifications (S) which emerged from the user workshops. Right: Various presentations of search results.

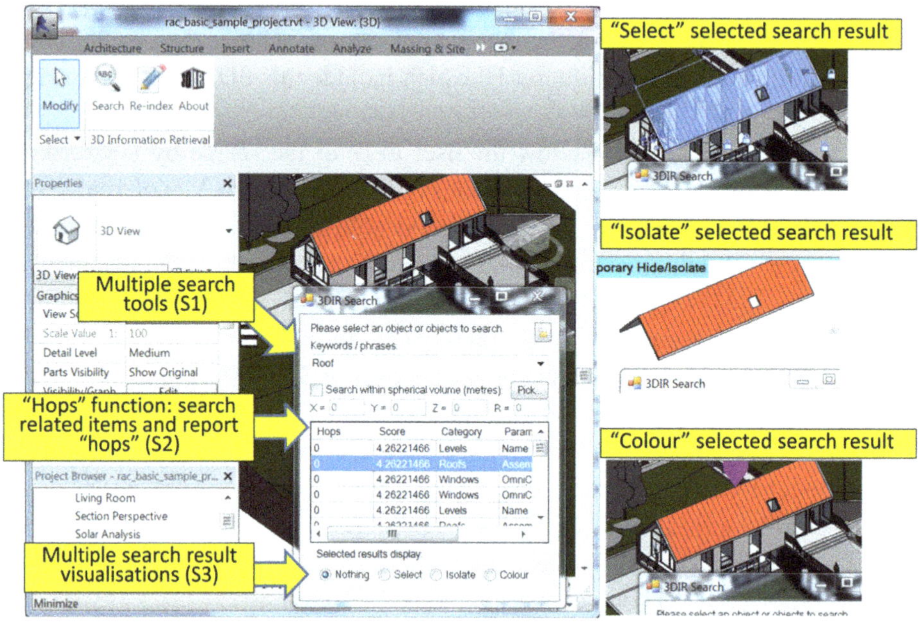

entered and search results listed related to the roof of the model. A relevance score is calculated for each V_{3D} information item using text analysis. The search can be processed either on the whole building model if nothing is selected, or limited to the selected objects. As the object selection in the model changes, the search results are updated dynamically. The user is also able to limit the search to a spherical volume. Selecting a search hit from the list (e.g. the "Roof Assembly" item) will optionally "select" (or highlight) the Revit element containing that search term (i.e. the red roof graphic from the model), "isolate" it (i.e. temporarily hide all other items in the model) or identify the object by displaying a coloured balloon next to it in the model. The size and colour of the balloon can be used to denote various attributes, such as the relevance or type of information. Figure 3 shows an example search for "roof". The three options for displaying the selected search result(s) appear on the right hand side of the figure.

Exploitation of Model Topology in 3DIR

From requirement R2 and corresponding specification S2, the exploitation of topological relationships (or E_t edges) between objects in a 3D model emerged as a promising avenue for improving information retrieval in 3D environments. The 3DIR prototype was extended to exploit such relationships. In order to establish relationships between the 3D building elements in the model, a separate list of other objects *hosting*, *touching* or *intersecting* each element in question is saved as a list of "Neighbours". In other words, E_t edges are followed to identify each V_{3D} object's neighbour V_{3D} object(s). While indexing, these Neighbours lists are indexed with their respective objects along with the object's own parameters. For each element, only one list of nearest neighbours is stored, i.e. one "hop" away. Subsequent neighbours, more than one hop away, are retrieved during the search, using a recursive function.

When searching a selected set of objects, the results are shown in the same way as when searching the whole building model. Selected objects containing the search keyword in any of their parameters are shown first in the results table with "Hops" value 0. Next, the list of Neighbours of each of the retrieved objects is also searched. Neighbouring objects containing the required keyword are listed in the results table with "Hops" value 1. The search is repeated recursively on the newly retrieved objects, each time incrementing the "Hops" value. This continues until the maximum number of Hops specified by the user. In this way, objects which are not in direct consideration but are related to the objects of the user's interest can still be retrieved but with less relevance.

Results of 3DIR Evaluation

Both the *original* (3DIR) and *modified* (3DIR+Topology) versions of 3DIR were evaluated by demonstrating the software to users and recording their feedback using a questionnaire. The samples of test subjects were drawn from different populations. 3DIR was evaluated by professional architects from the 3DIR project industry partners "Architects", each with over ten years of industrial experience. 3DIR+Topology was evaluated by a sample of postgraduate construction students from Loughborough University (11 students with an average of 1.95 years of industrial experience). The questionnaire was based on the System Usability Scale (SUS Brooke, 1996). Two additional questions were posed for 3DIR+Topology to gauge its recall and performance in averting information overload. The results are shown in Figure 4.

It can be seen from Figure 4 that although 3DIR+Topology was generally well received, without exception the new topology feature caused test subjects *to agree less strongly with the positive statements and disagree less strongly with the negative statements*. This is most pronounced in the last statement which was posed to users of both tools: "I would need to learn a lot before using 3DIR". As discussed below, it appears that a major concern for the new functionality is the added complexity it entails.

Figure 4. A comparison of the evaluation results from the original version of 3DIR and the modified version of 3DIR (3DIR+Topology). The questions are loosely based on the System Usability Scale (Brooke 1996) with three additional Likert questions for the modified version of 3DIR.

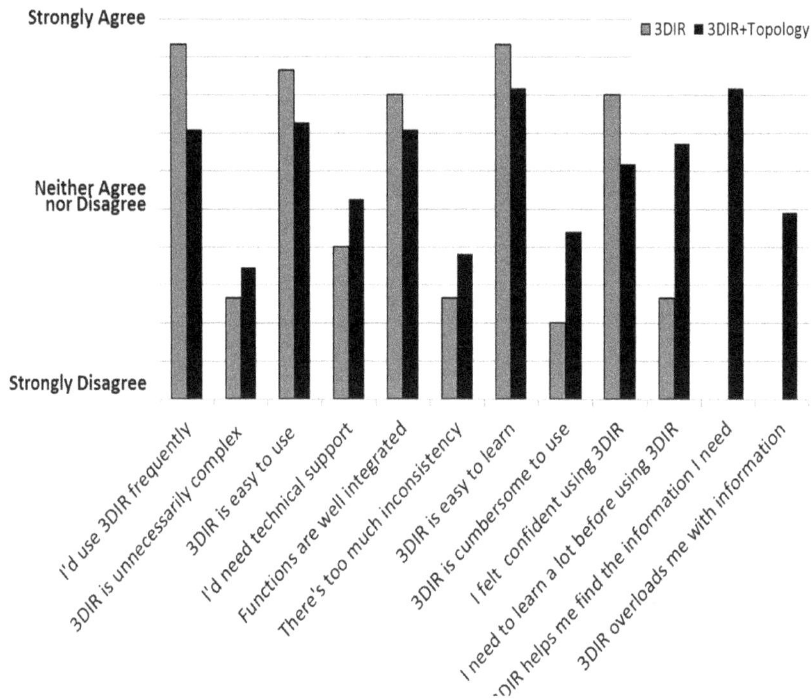

DISCUSSION

The results from the 3DIR evaluation demonstrate the promise of this approach. The results from 3DIR+Topology are encouraging but suggest that the added complexity might be difficult for users to grasp. The stronger agreement with statements of the complexity, cumbersomeness and un-learnability may also be a symptom of the speed with which the demonstration was conducted. The software demonstration lasted about twenty minutes, followed by about ten minutes of questions and answers, after which the users were asked to complete the questionnaire. It is also possible that framing the new functionality within the theoretical concept of topology added unnecessary complexity. In the demonstration session, the users used terms such as “widening the search criteria” or “finding related items” when discussing the new functionality.

A possibility that was explored was that the lukewarm assessment of 3DIR+Topology was due to the relative inexperience of the cohort of test subjects. Perhaps more experienced professionals would see the value of the functionality more clearly. This was investigated but the results showed that the support for the new functionality was roughly uniform across all levels of experience.

Although the 3DIR results reaffirm the potential of exploiting 3D data and 3D visualisation, it must be acknowledged that the slightly lower scores of 3DIR+Topology indicate that some rethinking is required about the usefulness of this extra topology functionality. During the workshop following the demonstration, test participants verbally agreed that the functionality was useful, but the SUS (System Usability Scale) questionnaire scores do not strongly support this. This functionality might be a result of "function creep", whereby the gradual widening use of a technology causes it to be unwieldy in its complexity or causes it eventually to shift away from the use for which it was originally intended. Even if this functionality is indeed useful, an improved interface design is needed to make the functionality more intuitive and facilitate the formation of more helpful mental models of the notion of topology. From discussion with test participants, the list of topological relationships exploited (touching, intersecting and hosting) needs to be expanded, and the interface design needs to convey those more clearly. The graph theoretic formulation might also provide a useful framework for users to grasp this functionality: searching text and 3D information items, all interconnected by edges.

FUTURE RESEARCH DIRECTIONS

The graph theoretic formulation proposed is intended to provide a framework for future research into information management in 3D modelling environments. Given a query text string q_i and a set of retrieved items from set V_i, it has already been shown how E_n edges can be used to retrieve linked objects from V_{3D}, and edges from E_t can be used to expand the search and rank search results by making "hops". Future research can explore more holistic relevance measures, where text matching scores are combined between q_i and multiple V_i items which are related through interlinked 3D objects in the set V_{3D}. For example, if a user is searching for "lift shaft in an atrium", such a holistic search would rate the relevance of a "lift shaft" object more highly if it was *hosted by* an "atrium" object (i.e. if there was a *hosting* E_t edge connecting the lift shaft V_{3D} object to an atrium V_{3D} object).

3DIR, so far, only allows queries based on a query text string, q_i. An important next step would be to allow 3D query objects, to enable users to search for items which are relevant to an individual 3D query object, q_{3D}. This measuring relevance between q_{3D} and those items in V_{3D} might entail complex geometric computations.

Further research is needed to evaluate the retrieval performance of 3DIR in terms of precision and recall. This would allow benchmarking against standard systems. This formal evaluation is also particularly important when considering retrieval embellishments such as one described in this chapter based on *topology*, or the holistic relevance measure proposed in this section. Would such embellishments improve retrieval performance? Of course standard measures of precision and recall require almost binary classifications of *relevant* or *not relevant* and *retrieved* or *not retrieved*. These classifications might be difficult in 3D modelling environments.

CONCLUSION

The 3DIR prototype creates an index of all text data attached to a 3D model. The user is able to search for information by selecting specific 3D objects, specifying a spherical region of the model and/or entering search keywords. Search results are displayed by highlighting 3D objects in the 3D model, isolating them or indicating them using a coloured balloon shape. The 3DIR+Topology system exploits model topology. At the indexing stage, a separate list of other objects hosting, touching or intersecting each element in question is saved as a list of "neighbours". When searching a selected set of objects, selected objects containing the search keyword in any of their parameters are shown first in the results. The list of Neighbours of each of the retrieved objects is then recursively searched until the maximum number of Hops specified by the user. In a comparative evaluation of 3DIR and 3DIR+Topology, users of the latter agree less strongly with positive statements and disagree less strongly with negative statements. This indicates that, although still useful, more careful interface design is needed to mitigate the added complexity of this functionality.

This work is distinct in its focus on retrieval of non-geometric information from 3D models, and how 3D computations can support information retrieval. It contrasts with, for example, the work of Daum & Borrmann (2014). They focus on spatial semantics, and particularly topological operators such as *Touch*, *Within* and *Contains*. They go on to propose a Query Language for Building Information Models (QL4BIM) and an efficient set of algorithms for implementing that query language. In 3DIR, 3D objects are retrieved based on the V_i items (i.e. textual information) to which they are linked. As noted above as a future research direction, it is also conceivable to query models and retrieve 3D objects based on 3D query items ("find a shape like this one") and geometrical operators. Aside from BIM, 3DIR is also aligned with some work in the domain of GIS: Ellul & Haklay 2007, Ellul & Haklay 2009, Schneider et al. 2012. The 3D computations employed in those efforts can also be applied to 3DIR in future work.

The underlying hypothesis of the 3DIR project remains compelling, that a tighter coupling between the 3D model and textual information is helpful for information retrieval. The graph theoretical formulation proposed provides a framework for this tighter coupling between 3D objects and textual information.

ACKNOWLEDGMENT

Our colleague Dr Ann O'Brien was an original member of the 3DIR team and contributed to the original 3DIR work but unfortunately passed away before this chapter was written. The 3DIR project was supported by a Brian Mercer Feasibility Award from the Royal Society, with further support from the Enterprise Office at Loughborough University and funding from University's Higher Education Innovation Fund. Autodesk provided support through membership of the Autodesk Developer Network. The 3DIR Revit app is available for free from the Autodesk App Store link at http://www.3dir.org/.

REFERENCES

Aldous, J. M., & Wilson, R. J. (2003). *Graphs and applications: an introductory approach*. Springer Science & Business Media.

Austin, S., Baldwin, A., Li, B., & Waskett, P. (2000). Analytical design planning technique (ADePT): A dependency structure matrix tool to schedule the building design process. *Construction Management and Economics*, *18*(2), 173–182. doi:10.1080/014461900370807

Borrmann, A., & Rank, E. (2009a). Topological analysis of 3D building models using a spatial query language. *Advanced Engineering Informatics*, *23*(4), 370–385. doi:10.1016/j.aei.2009.06.001

Borrmann, A., & Rank, E. (2009b). Specification and implementation of directional operators in a 3D spatial query language for building information models. *Advanced Engineering Informatics*, *23*(1), 32–44. doi:10.1016/j.aei.2008.06.005

Brilakis, I. K., & Soibelman, L. (2008). Shape-based retrieval of construction site photographs. *Journal of Computing in Civil Engineering*, *22*(1), 14–20. doi:10.1061/(ASCE)0887-3801(2008)22:1(14)

Brooke, J. (1996). SUS-A quick and dirty usability scale. *Usability Evaluation in Industry*, *189*(194), 4-7.

Caldas, C. H., Soibelman, L., & Han, J. (2002). Automated classification of construction project documents. *Journal of Computing in Civil Engineering*, *16*(4), 234–243. doi:10.1061/(ASCE)0887-3801(2002)16:4(234)

Card, S. K., Mackinlay, J. D., & Shneiderman, B. (Eds.). (1999). *Readings in information visualization: using vision to think*. Morgan Kaufmann.

Charalambous, G., Thorpe, T., Demian, P., Yeomans, S. G., Doughty, N., & Peters, C. (2013). Collaborative BIM in the Cloud and the Communication tools to support it. *Proceedings of the 30th CIB W78 International Conference on Applications of IT in the AEC industry*, 58 - 67.

Daum, S., & Borrmann, A. (2014). Processing of topological BIM queries using boundary representation based methods. *Advanced Engineering Informatics*, *28*(4), 272–286. doi:10.1016/j.aei.2014.06.001

Demian, P., & Fruchter, R. (2005). Measuring relevance in support of design reuse from archives of building product models. *Journal of Computing in Civil Engineering*, *19*(2), 119–136. doi:10.1061/(ASCE)0887-3801(2005)19:2(119)

Demian, P., & Walters, D. (2014). The advantages of information management through building information modelling. *Construction Management and Economics*, *32*(12), 1153–1165. doi:10.1080/01446193.2013.777754

Dominich, S. (2008). *The modern algebra of information retrieval*. Heidelberg, Germany: Springer.

Eastman, C., Teicholz, P., Sacks, R., & Liston, K. (2011). *BIM Handbook: A Guide to Building Information Modeling for Owners, Managers, Designers, Engineers, and Contractors* (2nd ed.). John Wiley & Sons.

Eastman, C. M. (1999). *Building product models: computer environments, supporting design and construction*. CRC Press.

Ellul, C., & Haklay, M. (2007). Requirements for Topology in 3D GIS. *Transactions in GIS*, *10*(2), 157–175. doi:10.1111/j.1467-9671.2006.00251.x

Ellul, C., & Haklay, M. M. (2009). *Using a B-rep structure to query 9-intersection topological relationships in 3D GIS–reviewing the approach and improving performance. In 3D geo-information sciences* (pp. 127–151). Springer Berlin Heidelberg.

Emmitt, S., & Ruikar, K. (2013). *Collaborative design management*. Routledge.

Gemignani, M. C. (1972). *Elementary Topology*. Courier Corporation.

Gerrish, T., Ruikar, K., Cook, M., Johnson, M., Phillip, M., & Lowry, C. (2017). BIM application to building energy performance visualisation and management: Challenges and potential. *Energy and Building*, *144*, 218–228. doi:10.1016/j.enbuild.2017.03.032

Gray, C., & Hughes, W. (2001). *Building design management*. Routledge.

Koo, B., & Fischer, M. (2000). Feasibility study of 4D CAD in commercial construction. *Journal of Construction Engineering and Management*, *126*(4), 251–260. doi:10.1061/(ASCE)0733-9364(2000)126:4(251)

Kunz, J. C., Christiansen, T. R., Cohen, G. P., Jin, Y., & Levitt, R. E. (1998). The virtual design team. *Communications of the ACM*, *41*(11), 84–91. doi:10.1145/287831.287844

Lansdale, M., & Edmonds, E. (1992). Using memory for events in the design of personal filing systems. *International Journal of Man-Machine Studies*, *36*(1), 97–126. doi:10.1016/0020-7373(92)90054-O

Leslie, H. G. (1996). Strategy for information in the AEC industry. *INCIT 96 Proceedings: Bridging the Gap*.

Lin, C. J. (2013). Smart Spatial Ontology: Bridging Semantic Ontology to Spatial Topology. *Computer-Aided Design and Applications*, *10*(3), 489–497. doi:10.3722/cadaps.2013.489-497

Lin, K. Y., & Soibelman, L. (2009). Incorporating domain knowledge and information retrieval techniques to develop an architectural/engineering/construction online product search engine. *Journal of Computing in Civil Engineering*, *23*(4), 201–210. doi:10.1061/(ASCE)0887-3801(2009)23:4(201)

OED Online. (2017). Oxford University Press. Retrieved from http://www.oed.com/view/Entry/203426?redirectedFrom=topology

Paul, N. (2009). Basic topological notions and their relation to BIM. Handbook of Research on Building Information Modeling and Construction Informatics: Concepts and Technologies, 451-472.

Pirolli, P., & Card, S. (1999). Information foraging. *Psychological Review*, *106*(4), 643–675. doi:10.1037/0033-295X.106.4.643

Rezgui, Y. (2006). Ontology-centered knowledge management using information retrieval techniques. *Journal of Computing in Civil Engineering*, *20*(4), 261–270. doi:10.1061/(ASCE)0887-3801(2006)20:4(261)

Robertson, G., Czerwinski, M., Larson, K., Robbins, D. C., Thiel, D., & Van Dantzich, M. (1998). Data mountain: using spatial memory for document management. In *Proceedings of the 11th annual ACM symposium on User interface software and technology* (pp. 153-162). ACM. 10.1145/288392.288596

Rosson, M. B., & Carroll, J. M. (2001). *Usability engineering: scenario-based development of human-computer interaction*. Elsevier.

Ruikar, K. (2014). BIM and Regulation 38. Fire Sector Summit 214.

Schneider, M., Chen, T., Viswanathan, G., & Yuan, W. (2012). Cardinal directions between complex regions. *ACM Transactions on Database Systems*, *37*(2), 8. doi:10.1145/2188349.2188350

Shaaban, S., Lockley, S., & Elkadi, H. (2001). Information visualisation for the architectural practice. In *Information Visualisation, 2001. Proceedings. Fifth International Conference* (pp. 43-50). IEEE.

Ugwu, O. O. (2005). A service-oriented framework for sustainability appraisal and knowledge management. *International Journal of IT-Construction ITCon*, *10*, 245–263.

Veeramani, D., & Russell, J. (2000). Preparing the AEC industry for the knowledge economy. *Berkeley-Stanford Construction Engineering and Management Workshop. Defining a Research Agenda for AEC Process/Product Development in 2000 and Beyond.*

Wu, I. C., & Hsieh, S. H. (2012). A framework for facilitating multi-dimensional information integration, management and visualization in engineering projects. *Automation in Construction*, *23*, 71–86. doi:10.1016/j.autcon.2011.12.010

Zhang, S., Sulankivi, K., Kiviniemi, M., Romo, I., Eastman, C. M., & Teizer, J. (2015). BIM-based fall hazard identification and prevention in construction safety planning. *Safety Science*, *72*, 31–45. doi:10.1016/j.ssci.2014.08.001

KEY TERMS AND DEFINITIONS

API: Application programming interface.

BIM: Building information model/modelling.

CAD: Computer-aided design.

E_n: The set of natural edges in the graph theoretic formulation (i.e., a link between a 3D object and one of its properties).

E_t: The set of topological edges in the graph theoretic formulation (i.e., links between related 3D objects).

IFC: Industry foundation classes.

IR: Information retrieval.

IV: Information visualization.

Q_{3D}: A query in the form of a 3D object.

Q_i: A query in the form of a text string.

R: Software "requirement" item from the scenario-based design process.

S: A software "specification" item from the scenario-based design process.

SUS: System usability scale.

V_{3D}: The set of vertices in the graph theoretic formulation representing 3D objects.

V_i: The set of vertices in the graph theoretic formulation representing information objects (i.e., properties of 3D objects as well as linked documents).

Chapter 6
Integration of BIM Work Culture for Improving Global Project Collaboration Productivity

Maszura Abdul Ghafar
Universiti Putra Malaysia, Malaysia

Zalina Shari
Universiti Putra Malaysia, Malaysia

Rahinah Ibrahim
Universiti Putra Malaysia, Malaysia

Farzad Pour Rahimian
University of Strathclyde, UK

ABSTRACT

Building information modelling is further globalizing architecture, engineering, and construction (AEC) professional partnerships. However, little is known on the effect of cultural and human factors on BIM-enabled visualization applications. This desktop study examined the extant literature on factors relating to application of BIM-enabled visualization technologies as a process that can improve, leverage, and conduct visual communication for coordination during implementation of global projects. It identifies BIM-enabled visualization having the capability in facilitating knowledge flows in complex discontinuous working environment of a property development's life cycle, and supports designers' understanding in its early working phases. This chapter presents the development of a theoretical proposition for embedding local work culture etiquette in BIM-enabled visualization application for augmenting dynamic knowledge transfer among discontinuous members in a building project. The result is expected to benefit rapidly developing countries (e.g., Malaysia) in enabling successful partnerships with counterparts from developed countries.

DOI: 10.4018/978-1-5225-5625-1.ch006

INTRODUCTION BACKGROUND

Architecture, Engineering and Construction (AEC) team members have ways of leveraging communication to coordinate their activities, although most communication traditions work across many building processes. The authors see AEC visual communication being leveraged through utilisation of technological and professional collaborative tools for managing construction activities and implementation progresses. Seminal literatures frequently advocating BIM with its potential for revolutionising the whole AEC industry by enhancing team collaboration (Gu & London, 2010); improving project integration (Woo, *et al.*, 2004); leveraging better construction information flow (Ibrahim, Krawczyk & Schipporiet, 2004); helping documentation flow (Popov, *et al.*, 2006); and providing construction simulation for teamwork planning, clash prevention and coordination interface (Fischer & Kunz, 2004).

Globalisation has made AEC industry, particularly of interest for developing country like Malaysia, utilise BIM enabled visualisation in implementing project delivery through partnerships with their respective counterparts in other countries. The need has been elevated when the Malaysian Government is promoting globalisation of services through exportation of building and professional services in the last 2012 Budget. Business Watch (2005) estimated that the AEC sector in Malaysia is one of the largest industrial employers, representing 9.8% of Malaysian Gross Domestic Product as it employs over 7.1% of workforce. The ability of BIM to depict and analyse construction schedules and logics (Heesom & Mahdjoubi, 2004), interoperability of data between applications (Fischer & Kunz, 2004), and allowing transition of *n*D CAD modelling and analysis (Lee *et al.*, 2005) are merits to AEC delivery. These benefits have proven to improve profitability, lower construction cost, and enhance time management and healthier client relationships (Azhar, 2011). Hence, the authors note that the success of AEC projects is highly dependent upon the type, level and quality of their communication exchanges between various disciplines involved in the different design and implementation phases (Pour Rahimian and Ibrahim, 2011).

This paper supports Bouchlaghem *et al.* (2005), in consigning cost as a vestige issue. It heeds their recommendation to focus towards the human issues as increasing number of professionals are embracing BIM enabled visualisation for their building projects. The authors had seen lack of emphasis given to the human factor especially the users' socio-cultural issues when using these visualisation tools. This paper agrees with Delavari *et al.* (2011) that further studies would be required on the following issues: 1) superior control over tools, 2) how fast feedback is returned to users, and 3) how much added-value are these tools to team members during the collaborative phase. In anticipation of future works in these areas, the purpose of this paper is setting up the theoretical foundation based upon identification of current literature gaps on how BIM enabled visualisation could enhance visual communication among AEC members in a complex building project process.

RESEARCH METHODOLOGY

This research conducted desktop study for examining the impacts of different visual communication approaches using technology and professional collaborative tools to manage project deliveries in the construction industry. The investigation of CAD technologies covers understanding of general conceptual perspectives regarding added-value components associated with BIM enabled visualisation. This is followed by examination from extant literature among others - the relative effect of the CAD technologies (e.g., object-oriented modelling (OOM), interoperability, and Industry Foundation Classes (IFC)) to enhance the effectiveness and efficiency in BIM enabled visualisation. Emanating from examination of the relative effect of these CAD technologies, this paper argues that professional response towards BIM enabled visualisation tools seldom receive scholarly attention.

The conducted literature led to development of theoretical understanding of basic epistemology of visualisation with respect to professional collaborative visualisation tools, such as the Virtual Design Construction (VDC) from Stanford University. The investigation of Product, Organisation and Process (POP) Models (Fischer & Kunz, 2004) as a BIM enabled visualisation process is examined in detail to identify the effect of human culture in VDC with particular focus on enhanced coordination, communication and decision making. This paper then presents how human culture has made significant impacts on the professionals' response to BIM enabled visualisation and VDC. In conclusion, the paper discusses the potential theoretical direction for professionals in the built environments to align towards advanced technological change, understanding the role of discontinuous membership in building projects, and integrating the required basic knowledge as early as the tertiary education level.

APPLICATION OF BIM ENABLED VISUALISATION TECHNOLOGIES

In this section, this paper describes the development of BIM enabled visualisation technologies in relation to technological and professionals tools. An overview on IFC and OOM as technological tools is presented followed by description of professional tools, which include collaborative tools, interoperability, and web space.

Technological Tools

Industry Foundation Classes (IFC) Standardisation

During the CAD era, 1985-1990, AutoCAD *Drawing Exchange Format* (DXF) file and the *Initial Graphics Exchange Specification* (IGES) format file are among the popular geometric format for interoperable exchange of information between different CAD programs. However, when AutoCAD software has become more powerful and started supporting many complex *object* types, DXF has become less useful. In 1990 to1995, *Standards for the Exchange of Product data* (STEP)-Exchange is created and its extension file format enabled better transfer to most CAD software. STEP file format could be exported to 3D model and used for analysis, modeling or drawings. Later in 1995 to present, IFC was invented based on the object-based data model file, being used worldwide to facilitate interoperability in AEC industry especially for collaborative format in BIM based project. It is registered by ISO and is an official International Standard ISO 16739: 2013. IFC is a platform of "open" or "common data schema that makes it possible to hold and exchange data between different proprietary software applications" (Buildingsmart, 2008). It allows team members to work from different visualisation tools with data continuity (Kam *et al.*, 2003). For instance, IFC can convert a 4D model to disseminate and improve its 4D simulation in a building project (Heesom & Mahdjoubi, 2004). This paper agrees with Froese (2003) that the implementation of IFC supports the full integration of construction projects data and processes. IFC improves "efficiency and value of information" whether they are "upstream or downstream tasks" (Kam *et al.*, 2003; pg.144). Therefore, this paper agrees with the scholars that IFC standardisation must continue to leverage and support as many AEC applications.

Object-Oriented Modelling (OOM)

According to Isikdag (2015) OOM carried parametric data about a product or an entity. Parametric data of a product is usually a measurable data such as heights, weight, depth, square footage, etc. Each of the parametric data has level of details (LoD) which reduces the complexity of a 3D object representation as it moves away from the viewer or data of other metrics such as object importance, viewpoint-relative speed or position. LoD increases efficiency of rendering by decreasing the workload on sequence of steps used, creating a faster 2D representation of a 3D scene. The level of details from OOM could facilitate visualisation into a real-time simulation (Khuan, Abdul-rahman, & Zlatanova, 2008). Combined with IFC, OOM would enable a project's team members to impart realistic budget and cost control; cut redundancy in re-entering geometric, thermal, and materials data during design and design development progresses; and facilitate design iterations, variation orders, and post-occupancy dis-satisfactions (Kam *et al.*, 2003). The same scholars (such as Kam *et al.*) noted that, the level of details in projects were insufficient then due to wide usage of the manual convention involving 2D drawings, artist impressions, and non-intelligent physical 3D models. This paper agrees with Howard and Bjork (2008) that the traditional practices do hinder team members to integrate and comprehend the project thoroughly since they inhibit the continuous widespread use of prior data for visual communication purposes. Thus, the incorporation of OOM into the current BIM technology has transformed the usage of visualisation tools in the construction industry to a new operational paradigm.

Professional Tools

Collaborative Tools

This paper agrees with McKinney and Fischer (1998) that architects, engineers, and contractors needed a comprehensive tool that could allow them to simulate and visualise construction sequences as part of their interactive experience. The method of collaboration enhances self-examination of one's behavior and communication. Collaborative method increases the success of a team in engaging problem solving. A collaborative tool is a medium that helps a team to collaborate while effective collaborative tools aid predefined tasks together and solves problem. 4D visualisation and *n*D CAD (computer-aided design) are the keys for visualising construction scheduling and activities, improving collaboration between team members, and detecting problems prior to construction (Koo & Fischer, 2000; *Kam et al.*, 2003). This paper concurs with Bouchlaghem *et al.* (2005) that prudent decisions can be made accurately and reasonably without depending merely on an individual's professional experience and assumptions. This paper sees potential improvement within the construction industry for better coordinated work, efficient communication, and effective use of information resources to support decision making. In this instance, the authors posit that the use of collaborative tool would remain crucial to attain effective decision-making between the building professionals.

Interoperability

Hamil (2012) acknowledges the fragmentation of AEC and describes that conventional AEC teams do work in silos. The "siloness" attitude is creating coordination problem, communication challenges and risk of argumentative decisions between team members at a higher level. This kind of work culture is certainly not collaborative and is resulting in bad coordinated documentation. Hamil concludes with this situation per se, interoperability is critical. The issue of incompatibility among IT/ICT technologies is improving in AEC, where previously it had inhibited seamless data visualisation among team members. Interoperability is needed to reduce repetitive information transformations. With respect to interoperability, Froese (2003) suggested a Model-based system, Jezernik and Hren (2003) proposed a better compatible format, Schreyer *et al.* (2005) suggested a "middleware controlled messaging infrastructure", openBIM framework (O'Keeffe, 2016) and Common Data Environment (CDE) (Comiskey *et al.*, 2017; Comiskey *et al.*, 2016) to integrate the much fragmented AEC industry applications. The reluctance of AEC team members to use new systems, because they are accustomed to their own visualisation techniques, have created further difficulties. For instance, when many documents are sent out to other team members, they become unreadable and incompatible on their respective machines. In essence, Kam *et al.* (2003) asserted that interoperability continues contributing to delay and cause inefficiency of information delivery and acceptance. Nonetheless, this paper strongly believes that despite many inadequacies of interoperability mentioned above, interoperability is key to further making the required improvements and support the fragmented AEC industry.

Web Space

The Web-space is another beneficial approach for non-collocated AEC team members to communicate. Rohrer and Swing (1997) recommended to integrate visualisation components with web media to give continuous related information sharing for end users. Kamat and Martinez (2001) proposed use of Virtual Modelling Language (VRML) to universally interchange format within the World Wide Web and integrate 3D graphics and multimedia in visualisation models. In the same vein, Jezernik and Hren (2003) advocated use of Extensible Markup Language (XML) to extend interface between VRML and CAD models. Kam et al. (2003, p.g 164) asserted that both models have helped "defined the ownership of each item of information, support dynamic and collaborative approach to data sharing, and facilitate better access and privilege controls." Therefore, the web space is vital to aid future visualisation process and support dynamic collaboration between non-collocated team members to communicate, collaborate and get real-time feedbacks.

Figure 1 demonstrates how 4D CAD, *n*D technologies, interoperability, and web space have merged to create the platform to effectively collaborate between AEC professionals from our survey summarised above so far. Technologically, 4D CAD advances *n*D models to depict construction activities, MEP coordination and communication between AEC professional. Therefore, this paper posits that the application of 4D CAD, *n*D and interoperability in BIM enabled visualisation

Figure 1. Development of BIM-CAD visualisation utilisation in AEC industry

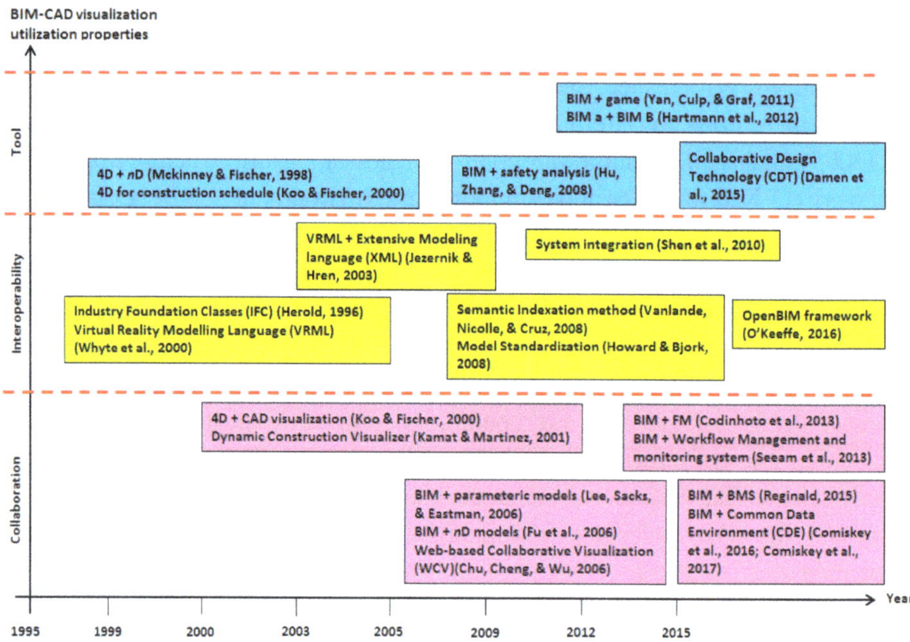

are significant to facilitate visual communication between AEC professionals in building projects.

As such, this study expects that IFC, OOM, 4D visualisation, and *n*D development will become more eminent to leverage and bring visualisation tools to a new paradigm in construction. The study also anticipates that interoperability and web space will further support the dynamic collaboration and effective decision-making between the building professionals in the complex working environment. Therefore, this paper recommends a review of continuing professional development programs and their respective educational programs to include awareness and skill development due to technological advancements in AEC applications.

VISUAL COMMUNICATION

Visualisation is better known in the conceptual design field and can be classified into two main disciplines: *Information* and *Scientific*. Both visualisation disciplines interpret data into meaningful result if it fulfils criteria, such as: 1) data used is a non-visual data and is used for communication purposes; 2) the data has the ability to produce additional modalities and information for communication purposes; and 3) the transformation of the data is readable and recognisable. Information visualisation is further subdivided into two categories, namely artistic and pragmatic visualisation factors (Kosara, 2007). Figure 2 illustrates the subtle divisions of information and scientific visualisation disciplines under the visualisation field.

Artistic visualisation is a visualisation technique which presents data as its information. It usually has no visual affection, yet consists of sublime and contemplative quality, such as found in paintings. On the other hand, pragmatic visualisation explores, analyses or presents information in a way that allows users to gain new insights into typical 1D or 2D format, such as chart, table, or figure (Kosara, 2007). Contrarily, scientific visualisation involves the exploration and analysis of complex datasets from real world experience for gaining understanding and insight into the data (Brodlie, 1995). Generally, scientific visualisation is subdivided into four categories, which are 1D, 2D, 3D, and animation. Animation (Friendly & Denis, 2001) is the rapid display of a sequence of static images and/or objects to create an illusion of movement. The most common method of presenting animation is as a motion picture or video program. In scientific visualisation, e.g. computer animation, it becomes an art, technique, and science of creating moving images via the use of computers. For instance, the computer simulation is a computer program, or network of computers, that attempts to simulate an abstract model of a particular system.

VIRTUAL DESIGN AND CONSTRUCTION (VDC)

This section discusses details of the VDC concept and its three main components to see how it fits into the visualisation field. Continuing the concept of analysing and exploring data, this paper refers to works by the Centre for Integrated Facility

Figure 2. Tree structure of the visualisation theory; adapted from Kosara (2007), and Friendly and Denis (2001)

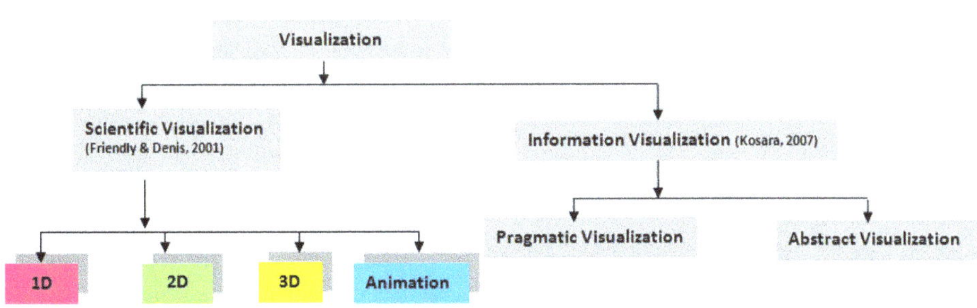

Engineering (CIFE) who has developed the Virtual Design and Construction (VDC) for the design-construction-operation team. VDC supports multidisciplinary performance models for product, organisation and process—later called POP models by Fischer and Kunz (2004). Sacks *et al.* (2010) extended VDC into a BIM representation due to similarities in their fundamental tenets, components, and procedures. Intrinsically, VDC provided an integrated project framework in managing tracks and changes in the POP models over time for better project management (Kunz & Fischer, 2012). Tools, such as Revitt, NavisWorks, SimVision, etc. have been developed under the VDC framework for analysing the effectiveness of multi-stakeholder meetings in order to meet the business objectives of the project and the client. The inference of the tools is summarised below in Table 1.

Table 1. Tools developed under VDC framework (adapted from Khanzode et al. 2006)

Categories	Item	Function
Visualisation tools	3D object modelling technology such as AutoCAD ADT, Revit, ArchiCAD II, Tekla	Visualisation tools will generate collective understanding by simulating building when it is completed and aid coordination among project members especially the mechanical, electrical and plumbing, and construction operations (Fischer & Kunz, 2004).
Process modelling and visualisation tools	4D visualisation tools such as CommonPoint Project 4D and NavisWorks Timeliner	Project members can visualise 3D model of the building and enhance further understanding on how building will be constructed over time (Koo & Fischer, 2000).
Organisational and process modelling tools	Such as SimVision	The tool simulates organisational effort to complete project and detect potential risks that may suspend project progress. (Fischer & Kunz, 2004).
Online collaboration tools	iRoom, Project Based Learning Lab	These tools allow collocated and non-collocated team members to collaborate using POP models (Schreyer *et al.*, 2005).
Techniques to analyse the effectiveness of multi-stakeholder meetings, sustainable issues, latency, etc. in order to meet the business objectives of the project and the client.	VCG (Vickrey- Clarke-Groves) mechanism Integrated Concurrent Engineering (ICE). The six criterion framework	The mechanism values meeting activities in construction (Garcia, Kunz, & Fischer, 2003). The tool offer coordination latency as a unifying, intuitive, descriptive performance metric and reducing it to near-zero as a project design goal (Chachere, Kunz, & Levitt, 2009) Predictive project performance tools to use on design-construction projects and support business objectives (Ho, Fischer, & Kam, 2009b)

Table 2 illustrates the three types of virtual POP building models that are widely used in the construction industry. These models are complimenting each other in facilitating their stakeholders' perspectives for making better decision-making throughout a project's life-cycle.

POP models have proven their capabilities in enhancing coordination, visualisation, and planning. However, they lack the formalisation for interacting between themselves at various levels of details, across disciplines, and project phases (Kam & Fischer, 2004). It is here that the authors would like to propose to include the effect of human culture in VDC models through enhanced coordination, communication, and decision making functions. This paper supports Kam and Fischer's (2004) recommendation for further investigating how these POP models could assist communication in BIM enabled visualisation for the following segments. The following sections could provide potential areas for human culture integration.

Product Models

Product models have elements such as: Floors, Walls and Beams (Kunz & Fischer, 2012) facilitating 4D CAD visualisation to coordinate, analyse and aid decision making of spatial and planning between multidisciplinary specialist professionals effectively (Kam & Fischer, 2004). Stakeholders can cut cost and adhere to construction progress time without any disruptions (Heesom & Mahdjoubi, 2004; Fischer & Kunz, 2004). However, insufficient level of details in planning and design hampers any analytical operation of 4D CAD visualisation during construction process (Kamat & Martinez, 2001). For instance, when a plan shows activities and their logic relationship, their corresponding activities do not show specific start and end dates due to missing product assembly (Kamat & Martinez, 2001). Ibrahim and

Table 2. Three types of POP model in the AEC industry (source: Kunz and Fischer 2005)

Modelling type	Function
Visual 3D and 4D models (Product models)	• Involve stakeholder in project's facility, schedule and organisation • Improve coordination in project lifecycle
Knowledge-based models that support automation (Organisational models)	Lean professional practices increasing consistency, mitigate error and reduce time to perform tasks.
Building Information Models (Process Models)	• Supports data exchange between tools • Speed up analysis time • Reduce data input and transfer errors

Paulson (2008) highlighted that the early stage of product development is dominated by tacit-dominated knowledge areas, the early design phase benefits when sufficient visualised details in them helps better prediction analysis and decision-making.

Organisational Models

The second VDC component is organisation. The organisational model would define the organisation of different experts in planning a project delivery (Kam & Fischer, 2004). Inputs are gained from two main categories: 1) reporting hierarchy structure and 2) project's work tasks (Fischer & Kunz, 2004). The Virtual Design Team (VDT) developed by Jin *et al.* (1995) visualises qualitative predictions of organisational performance concurrent with their schedule and cost diagnostic, and qualified risks that associated with the planned configuration of the project. Project managers could have clear visions how to construct, to manage organisation and tools similarly as engineers to designing bridges (Kam & Fischer, 2004). The visualised process based on the Critical Path Method (CPM) assumption through SimVision®—the commercial model of VDT—is capable to depict a project member's information and communication capacity in a project at macro and micro level (Kam & Fischer, 2004). Therefore, the application of VDT has the potential to increase coordination activities and encourages an organisation to incorporate information processing and communication in its operations. To date, there are several organisational tools developed under the VDC context to achieve organisational objectives. These are surmised in Table 3.

Process Models

The third component of VDC is process. The process models ranges from macro to micro schedules, delivery dates, and milestone activities of a project in 4D CAD visualisation. Unfortunately, conventional social methods of construction inputs to design still transpire in current practice (Fischer, 2006). For example, professionals still sit together in meetings to discuss the preliminary design phase and gain manual inputs. This can be difficult in large scale project integration, when time, budget, and stakeholders' attention are not available in the early project phase. Professionals tended to interpret materials in their mind and share predictions verbally using other professionals' sketches resulting slow, incompetent and error-prone design (Fischer, 2006). Therefore, the application of 4D CAD visualisation in VDC would allow multiple stakeholders to visualise design intention and helps in the decision making process. Table 4 shows the utilisation and application of work process developed in the VDC frameworks.

Table 3. Organisational tools in VDC to achieve organisational objectives

Tools	Organisational objectives
TEAM Analysis Framework (Team interaction, Emotional interaction, Artifact interaction and Model interaction	The tool visually compares digital and paper artifacts, and factors of material that can impact process and meeting outcome (Liston *et al.*, 2007).
Latency Theory- ICE (Integrated Concurrent Engineering)	The tool uses simulated "Just in Time" approach to reduce latency near-zero by unifying coordination and performance matrix. The theory indicates that all collaborative arranged operations are quantified, efficient and reliable (Chachere *et al.*, 2009).
IOI method (Identify occupant interaction)	A visual method to identify occupants' interactions automatically in relation to space, organisation and temporal aspect of renovation planning (Ho, Fischer, & Haymaker, 2009).
6 Criterion framework	The framework is to validate a new route evaluating new VDC methods, evaluate different validation methods and select the top method to achieve the purpose of validation (Ho, Fischer, & Kam, 2009).
IVL (Integrated VDC and Lean) method	A simulated method that project teams can adopt to produce consistent results for MEP coordination (Khanzode, 2010).

Table 4. Application of work process in VDC framework

Method	Objectives
Requirement Model Specification	To enable active connections between the requirements for a building project and the Building Product Model based design solution (Kiviniemi *et al.*, 2004).
Horse shoe method	To formalise construction concepts to make AEC self-aware in the context of virtual computer models of facilities and their construction schedules and organisations (Fischer, 2006).
MACDADI (Multi-Attribute Collaborative Design, Assessment and Decision Integration)	A method of structured collaboration with social and technical elements intended to build consensus on AEC decisions by improving transparency, precision, and comprehensiveness of rationale (Haymaker, Chachere, & Senescu, 2011).

In summary, bringing POP models during the early design phase requires building professionals to have sufficient levels of detail, thus, enabling 4D CAD visualisation to visualise better predicted analysis and make better decision making while the process progresses. The application of VDT has proven to increase coordination activities and encourages organisation to incorporate information processing and communication in their operations. Therefore, the authors agree that interoperability, 4D CAD visualisation, and VDT could facilitate dynamic collaboration, communication, information processing, and effective decision making. More studies are recommended to determine the different levels of details required to support each operational phase since Ibrahim and Paulson (2008) found the unique regressive nature of tacit operational knowledge areas as a project progresses towards its implementation.

UTILISATION OF VISUALISATION FOR PROFESSIONAL COMMUNICATION

This study generally finds that VDC has the potential breath for visualising a building project. The exploration of the VDC framework shows that visualisation is very important for design teams to comprehend and bring forward the design intent to construction implementation. In fact, Ibrahim and Paulson (2008) argued that during the preliminary design phase, critical information such as costs and special requirements/ annotations should get across other team members' knowledge even when it is felt "irrelevant" in some cases. The failure to ensure the flow of that particular knowledge could gradually leads to the "knowledge loss" phenomenon, which tend to manifest during on-site construction. They further posited that in most western practices, cost and design are alienated and not considered in pre-construction phase and its emergence later could impact the project's schedule and cost. 4D CAD visualisation, therefore, has the potential to help team members to visualise whilst having logical, temporal and spatial understanding of a building project. Coordination, decision making, and analysis of problems can also be potentially augmented when a single platform is used in AEC while in building project. In this instance, the authors argue that visualisation enhances the operational knowledge and helps stabilise the knowledge flows in such discontinuous environment of the property development life cycle.

There is urgent need to understand visual tools in order for designers to support the design in the early phase of a project. When models such as 4D CAD, *n*D, and cost are introduced in the early design phase, VDC can demonstrate the POP advantages in simulating construction activities, organisational behaviour, and product modelling tools. The paper foresees the distant cry from the traditional practice when building professionals need to rely on their past experiences to understand and comprehend a complex integration issue, such as the MEP. Below are several recommendations towards utilising visualisation for professional communication purposes.

Alignment of Professional Towards Technology Changes

The development of 4D CAD visualisation certainly shows benefits to mitigate risks, reduce construction time and cost. However, it is becoming increasingly difficult to ignore how these visualisation tools have effects on human and work culture of professionals during AEC collaborations. VDC framework has shown many potential benefits in building project. However, professionals reluctance to implement VDC technology is predicted because of some cultural factors(Abdul Ghafar, 2016; Abdul Ghafar, Ibrahim, & Shari, 2014). Paulson and Fondahl (1980) had earlier found that the above problem stem from individuals' attitudes towards risks involved; difficulty in executing; and other participants' acceptance in implementing new technologies. Delavari *et al.* (2011) have highlighted the importance of engaging the human side of the building professionals when they involve IT-supported procedures during the design stage. They found three factors engaging team members with their visualisation tools effectively, which are having control over the tools, obtaining speedy feedback to users besides their existing application functionality. Therefore, this paper is recommending to integrate professional control, speedy feedbacks and cultural factors within the highly functional tools in the VDC framework.

Understanding the Role of Discontinuous Member

The notion of discontinuous membership during the construction phases was initially brought about by Ibrahim and Paulson (2008) in their study of the knowledge loss phenomenon in the construction industry. The scholars found that new team members have difficulties in retrieving and gaining available information from other team members and from the developer's database due to the inherent dynamic operating environment in its lifecycle process. One aspect is the different culture of each team member while different workflows would further contribute to the individual professional's limited knowledge retrieval. Thus, resulting in poor decision-makings, which could lead to misunderstanding, conflict and poor organizational performance (Hofstede, 1997). Ramsey and Levitt (2005) discovered that team members cultivate "a set of beliefs" and are fused with individual skills and experiences to form one's knowledge in the operating environment. For example, "sticky" (Szulanski, 2000) tacit knowledge phenomenon (Nissen, 2006) is hampering discontinuous membership learning of necessary project information leading to misunderstanding, variants and rework. Members do not know what information is required to perform their task and what is needed for the next task. The situation is worsen when the project's knowledge is tacit (Flanagan, Eckert & Clarkson, 2007) and further socialization (Nonaka, 1994) is required to internalize that knowledge. For example, a typical project team depends on the continuous attribution of the prominent person in that team. The absence of this continuity could lead to missing information in between the property development phases when new team members are unacquainted to the project's history. Shumate, Ibrahim, and Levitt (2010) affirm the existence of this scenario and sought to promote knowledge-flow through the efficiency of tacit knowledge transfer during a workflow process. For example, when a design decision knowledge skips the fabrication task, the mistake will only manifest when the builder realises a building component has arrived at site with a wrong specification. Due to potential expensive reworks, this paper recommends further studies on different methodologies of knowledge transfer tailored towards serving the discontinuous membership operating environment in the property development industry.

Reforming Education Curriculum

In line of improving Integrated Project Delivery (AIA California Council, 2007) practice, this paper foresees the need to improve building professional's education. Integrated project delivery (IPD) is "a collaborative alliance of people, systems, business structures and practices into a process that harnesses the talents and insights of all participants to optimise project results, increase value to the owner, reduce waste, and maximise efficiency through all phases of design, fabrication, and construction" (AIA California Council, 2007). The foreseen changes to the method of advanced manufacturing into conventional building methodology in the construction industry would now forces architectural and engineering schools in universities to review some fundamentals of their professional curriculums. The need to learn how to use new design tools and to do design differently than present curriculum will require experimentation in the educational process of future graduates. This is supported by Comiskey, et al. (2017) study that AEC students prove to gain functional advantages while using BIM during their learning processes. While sketching has been a relevant tool to convey ideas and understanding during the design exercises, the authors agree with Ibrahim and Rahimian (2010) in expecting major barriers when dealing with complex design projects due to its technical visualisation limitation. In a later study, Rahimian and Ibrahim (2011) found the main obstacle for design communication was in the very nature of tacit knowledge in the architectural design activities. They elucidated that with the use of virtual visualisation techniques and improved communication tools, these approaches have potential to make implicit knowledge become explicit while giving better collaboration with stakeholders simultaneously. Here, the authors recommend the integration of advanced visualisation and collaboration tools in the professional education programs for accelerating a project's comprehension.

DISCUSSION AND CONCLUSION

This paper analysed three aspects of BIM enabled visualisation culture including technology and professional application: epistemology, VDC and POP models development, and utilisation for professional communication. Figure 3 summarises

Figure 3. Theoretical development of work culture embedment in BIM enabled visualisation for educating building professionals

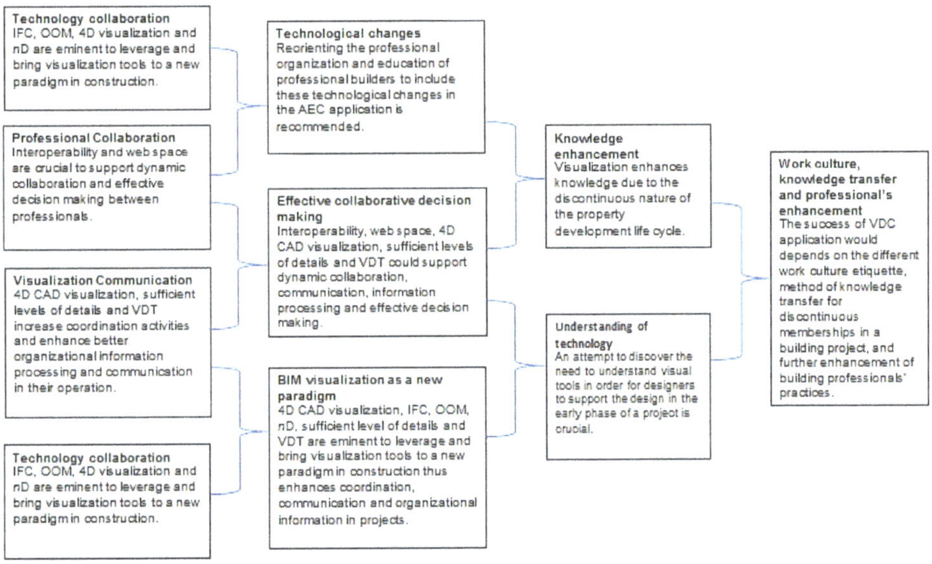

the paper's key discussion points. In lieu of implementing VDC in rapidly developing countries, such as Malaysia, the need to increase professional control and providing speedy feedbacks are necessary over the existing functionalities of technologies. The potential for extending VDC to include specific human traits during operational process is an area of interest to the authors because it will support globalisation across different countries. This paper proposes that the success of VDC application depends on the different work culture etiquette, method of knowledge transfer for discontinuous memberships, and further enhancement of building professionals' practices. Hence, the authors support in improvising current professional education programs for similar purpose.

Specifically, a CAD's visualisation culture is proposed to coordinate local industry stakeholders, while supporting the development of a new generation of construction stakeholders especially in developing countries. Additionally, discontinuous members, e.g. specialist contractor can comprehend a project's goal without having difficulty to find the project information from other team members. The inventions of various tools from various scholars proved to enrich communication barriers and supported professional collaborative environment among stakeholders. This will certainly benefit AEC industry across the globe especially developing countries like Malaysia. In conclusion, the success of VDC application would depend on the different work culture etiquette, method of knowledge transfer for discontinuous memberships in a building project, and further curriculum enhancement of professionals' education programs. The benefits in improving these factors would allow stakeholders in both developed and developing countries to partner successfully in implementing joint global projects.

ACKNOWLEDGMENT

This paper is part of the first author's PhD dissertation at Universiti Putra Malaysia sponsored by the SLAI scholarship. The authors would like to thank Dr. Mohd Yazah Md Raschid for his contributions in this paper.

REFERENCES

Abdul Ghafar, M. (2016). *Professional Collaborative Cultural Model*. Universiti Putra Malaysia. Retrieved from https://upm.academia.edu/MaszuraAbdulGhafar

Abdul Ghafar, M., Ibrahim, R., & Shari, Z. (2014). Embedding cultural knowledge in Building Information Modeling (BIM) for fabrication efficicncy to reduce industrialized construction waste. In *15th International Conference on Computing in Civil and Building Engineering (ICCCBE)*. Orlando, FL: ASCE. 10.1061/9780784413616.025

AIA California Council. (2007). *Integrated Project Delivery–A Working Definition*. Retrieved from http://www.aia.org/aiaucmp/groups/aia/documents/pdf/aias076695.pdf

Azhar, S. (2011). Building Information Modeling (BIM): Trends, Benefits, Risks, and Challenges for the AEC Industry. *Leadership and Management in Engineering*, *11*(3), 241–252. doi:10.1061/(ASCE)LM.1943-5630.0000127

Bouchlaghem, D., Shang, H., Whyte, J., & Ganah, A. (2005). Visualisation in architecture, engineering and construction (AEC). *Automation in Construction*, *14*(3), 287–295. doi:10.1016/j.autcon.2004.08.012

Brodlie, K. (1995). Scientific visualization - Past, present and future. *Nuclear Instrument and Methods in Physics Reserach A*, 104–111.

Buildingsmart. (2008). *Industry Foundation Classess (IFC) data model*. Retrieved from http://www.buildingsmart.org/standard/ifc

Business Watch. (2005). ICT and Electronic Business in the construction Industry, IT adoption and e-business activity in 2005. *The European E-Business Market Watch. European Commission, Enterprise and Industry Directorate General.*

Chachere, J. M., Kunz, J., & Levitt, R. (2009). The Role of Reduced Latency in Integrated Concurrent Engineering. *CIFE Working Paper#WP116.*

Codinhoto, R., Kiviniemi, A., Kemmer, S., & Gravina da Rocha, C. (2013). BIM-FM Implementation: An Exploratory Invesigation. *International Journal of 3-D Information Modeling, 2*(2), 1–15.

Comiskey, D.; McKane, M.; Eadie, R.; & Goldberg, D. E. (2016). Providing Collaborative Education with an International Dimension: An Ulster University and Pennsylvania State University Case Study. *International Journal of 3-D Information Modeling, 5*(2), 1–15.

Comiskey, D., McKane, M., Jaffrey, A., Wilson, P., & Mordue, S. (2017). An analysis of data sharing platforms in multidisciplinary education. *Architectural Engineering and Design Management*, 1–18.

Damen, T., Sebastian, R., MacDonald, M., Soetanto, D., Hartmann, T., Di Giulio, R., Bonsma, P., & Luig, K. (2015). The Application of BIM as Collaborative Design Technology for Collective Self-Organised Housing. *International Journal of 3-D Information Modeling, 4*(1), 1–18.

Delavari, N., Said, N. S., Ibrahim, R., & Abdullah, M. (2011). HCI to Engage Design Team Members in IT-integrated Design Collaboration Process. *WSEAS Transaction Information Science and Applications*, *8*(9), 341–355.

Fischer, M. (2006). Formalizing Construction Knowledge for Concurrent Performance-Based Design. In *Intelligent Computing in Engineering and Architecture* (pp. 186–205). Berlin: Springer. doi:10.1007/11888598_20

Fischer, M., & Kunz, J. (2004). The Scope and Role of Information Technology in Construction. In *Proceeding-Japan Society of Civil Engineers* (pp. 1–32). DOTOKU GAKKAI. doi:10.2208/jscej.2004.763_1

Flanagan, T., Eckert, C., & Clarkson, P. J. (2007). Externalizing tacit overview knowledge: A model based approach to supporting design teams. *Artificial Intelligence for Engineering Design, Analysis and Manufacturing*, *21*(03), 227–242. doi:10.1017/S089006040700025X

Friendly, M., & Denis, D. J. (2001). *Milestones in the history of thematic cartography, statistical graphics, and data visualization*. Retrieved from http://www.math.yorku.ca/SCS/Gallery/milestone/milestone.pdf

Froese, T. (2003, April). Future Directions For IFC-Based Interoperability. *ITcon: Special Issue IFC- Product Models for the AEC Arena, 8*, 231–246.

Garcia, A. C. B., Kunz, J., & Fischer, M. (2003). *Meeting Details : Methods to Instrument Meetings and Use Agenda Voting to Make Them More Effective*. CIFE Technical Report #147.

Gu, N., & London, K. (2010). Understanding and facilitating BIM adoption in the AEC industry. *Automation in Construction, 19*(8), 988–999. doi:10.1016/j.autcon.2010.09.002

Hamil, S. (2012). *Building Information Modeling and Interoperability*. Retrieved from https://www.thenbs.com/topics/BIM/articles/bimAndInteroperability.asp

Haymaker, J., Chachere, J. M., & Senescu, R. R. (2011). Measuring and improving rationale clarity in a university building design process. *Journal of Architectural Engineering, 17*(3), 97–11. doi:10.1061/(ASCE)AE.1943-5568.0000041

Heesom, D., & Mahdjoubi, L. (2004). Trends of 4D CAD applications for construction planning. *Construction Management and Economics, 22*(2), 171–182. doi:10.1080/0144619042000201376

Ho, P., Fischer, M., & Haymaker, J. (2009). *Automated Identification of Occupant Interactions in Renovations of Occupied Buildings*. CIFE Working Paper#WP122.

Ho, P., Fischer, M., & Kam, C. (2009). *Prospective Validation of Virtual Design and Construction Methods: Framework, Application, and Implementation Guidelines*. CIFE Working Paper#WP123.

Hofstede, G. (1997). *Cultures and Organizations: Software of the mind*. McGraw Hill International.

Howard, R., & Bjork, B. (2008). Building information modelling – Experts' views on standardisation and industry deployment. *Advanced Engineering Informatics, 22*(2), 271–280. doi:10.1016/j.aei.2007.03.001

Ibrahim, M., Krawczyk, R., & Schipporiet, G. (2004). No Title. Two Approaches to BIM: a Comparative Study. Proceedings of eCAADe 2004.

Ibrahim, R., & Paulson, B. C. (2008). Discontinuity in organizations: Identifying business environment affecting of knowledge flow in PLM. *International Journal of Product Lifecycle Management*, *3*(1), 21–36. doi:10.1504/IJPLM.2008.019972

Ibrahim, R., & Pour, F. (2010). Comparison of CAD and manual sketching tools for teaching architectural design. *Automation in Construction*, *19*(8), 978–987. doi:10.1016/j.autcon.2010.09.003

Isikdag, U. (2015). Building Information Models: An Introduction. In Enhanced Building Information Models. In Enhanced Building Information Models (pp. 1–12). Springer International Publishing.

Jezernik, A., & Hren, G. (2003). A solution to integrate computer-aided design (CAD) and virtual reality (VR) databases in design and manufacturing processes. *International Journal of Advanced Manufacturing Technology*, *22*(11–12), 768–774. doi:10.100700170-003-1604-3

Jin, Y., Levitt, R., Kunz, J., & Christiansen, T. R. (1995). The Virtual Design Team : A Computer Simulation Framework for Studying Organizational Aspects of Concurrent design. *Simulation*, *64*(3), 42–57. doi:10.1177/003754979506400304

Kam, C., & Fischer, M. (2004). Capitalizing on early project decision-making opportunities to improve facility design, construction, and life-cycle performance — POP, PM4D, and decision dashboard approaches. *Automation in Construction*, *13*(1), 53–65. doi:10.1016/j.autcon.2003.08.004

Kam, C., Fischer, M., Manager, G., Oy, O. G., Karjalainen, A., Manager, C., ... Solutions, C. (2003). The product model and fourth dimension project. *ITcon*, *8*, 137–166.

Kamat, V. R., & Martinez, J. C. (2001). Visualizing Simulated Construction Operations in 3D. *Journal of Computing in Civil Engineering*, *15*(4), 329–337. doi:10.1061/(ASCE)0887-3801(2001)15:4(329)

Khanzode, A. (2010). *An Integrated, Virtual Design and Construction and Lean (IVL) Method for Coordination of MEP.* CIFE Technical Report #TR187.

Khanzode, A., Fischer, M., Reed, D., & Ballard, G. (2006). *A Guide to Applying the Principles of Virtual Design & Construction (VDC) to the Lean Project Delivery Process.* CIFE Working Paper#093.

Khuan, C. T., Abdul-rahman, A., & Zlatanova, S. (2008). 3D Solids and Their Management. In *DBMS. In Advances in 3D geoinformation systems* (pp. 279–311). Springer Berlin Heidelberg. doi:10.1007/978-3-540-72135-2_16

Kiviniemi, A., Fischer, M., Bazjanac, V., & Paulson, B. (2004). *PREMISS - Requirements Management Interface to Building Product Models: Problem Definition and Research Issues.* CIFE Working Paper#092.

Koo, B., & Fischer, M. (2000). Feasibility Study of 4D CAD in Commercial Construction. *Journal of Construction and Management*, *126*(4), 251–260. doi:10.1061/(ASCE)0733-9364(2000)126:4(251)

Kosara, R. (2007). Visualization Criticism - The Missing Link Between Information Visualization and Art. In *2007 11th International Conference Information Visualization (IV '07)* (pp. 631–636). IEEE.

Kunz, J., & Fischer, M. (2012). *Virtual Design and Construction: Themes, Case Studies and Implementation Suggestions.* CIFE Working Paper#097, 14.

Lee, A., Marshall Pointing, A., Aouad, G., Tah, J., Cooper, R., & Fu, C. F. (2005). *nD modeling- a driver or enabler for construction improvement?* RICS Research paper Series, University of Salford.

Liston, K., Fischer, M., Kunz, J., & Dong, N. (2007). *Observations of Two MEP iRoom Coordination Meetings: An Investigation of Artifact Use in AEC Project Meetings.* CIFE Working Paper#WP106.

Mckinney, K., & Fischer, M. (1998). Generating, evaluating and visualizing construction schedules with CAD tools. *Automation in Construction*, *7*(6), 433–447. doi:10.1016/S0926-5805(98)00053-3

Nissen, M. E. (2006). *Harnessing Knowledge Dynamics Principled Organizational Knowing and Learning* (Vol. 2). IRM Press. doi:10.4018/978-1-59140-773-7

O'Keeffe, S. E. (2016). OpenBIM Framework for a Collaborative Historic Preservation System. *International Journal of 3-D Information Modeling, 5*(4), 1–11.

Paulson, B. C., & Fondahl, J. (1980). *Towards Improved Transportation Construction Through Research.* Stanford, CA: The Construction Institute, Stanford University.

Popov, V., Mikalauskas, S., Migilinskas, D., & Vainiunas, P. (2006). Complex Usage of 4D Information Modelling Concept for Building Design, Estimation, Scheduling and Determination of Effective Variant. *Technological and Economic Development of Economy*, *12*, 91–98.

Rahimian, F. P., & Ibrahim, R. (2011). Impacts of VR 3D sketching on novice designers' spatial cognition in collaborative conceptual architectural design. *Design Studies*, *32*(3), 255–291. doi:10.1016/j.destud.2010.10.003

Ramsey, M. S., & Levitt, R. E. (2005). *A computational framework for experimentation with edge organizations*. Academic Press.

Reginald, A. I. (2015). Integrating BIM with BMS in Energy Performance Assessment: Case Study of a University Building in UK. *International Journal of 3-D Information Modeling, 4*(1), 19–44.

Rohrer, R., & Swing, E. (1997). Web-Based Information Visualization. *IEEE, 17*(4), 52–59.

Sacks, R., Koskela, L., Dave, B. A., & Owen, R. (2010). Interaction of Lean and Building Information Modeling in Construction. *Journal of Construction Engineering and Management, 136*(9), 968–980. doi:10.1061/(ASCE)CO.1943-7862.0000203

Schreyer, M., Scholar, V., Hartmann, T., Engineer, S., & Fischer, M. (2005). Supporting Project Meetings With Concurrent Interoperability in a Construction Information Workspace. *ITcon, 10*, 153–167.

Seeam, A., Zheng, T., Lu, Y., Usmani, A., & Laurenson, D. (2013). BIM Integrated Workflow Management and Monitoring System for Modular Buildings. *Journal of 3-D Information Modeling, 2*(1), 17–28.

Shumate, M., Ibrahim, R., & Levitt, R. (2010). Dynamic Information Retrieval and Allocation Flows in Project Teams With Discontinuous Membership. *European Journal of International Management, 4*(6), 556–575. doi:10.1504/EJIM.2010.035589

Szulanski, G. (2000). The process of knowledge transfer: A diachronic analysis of stickiness. *Organizational Behavior and Human Decision Processes, 82*(1), 9–27. doi:10.1006/obhd.2000.2884

Woo, J.-H., Clayton, M. J., Johnson, R. E., Flores, B. E., & Ellis, C. (2004). Dynamic Knowledge Map: Reusing experts' tacit knowledge in the AEC industry. *Automation in Construction, 13*(2), 203–207. doi:10.1016/j.autcon.2003.09.003

Chapter 7
Demystifying Collaboration in BIM-Based Projects Under Design-Build Procurement:
Clash Detection as a Use Value

Benjamin R. Jowett
Northumbria University, UK

Malak Al Hattab
American University of Beirut, Lebanon

Mohamad Kassem
Northumbria University, UK

ABSTRACT

Building information modelling (BIM) tools and workflows, new procurements methods, and emerging management practices are being adopted on projects to overcome collaboration barriers and improve project performance within the architecture, engineering, construction, and operation (AECO) sector. Academic literature and industry reports recommend the use of collaborative procurement methods such as design and build (DB) procurement and integrated project delivery (IPD) when adopting BIM workflows. However, to date there are little operationalization and empirical evidence of the value realization potential when using BIM in conjunction to these procurement methods. This chapter draws upon five case studies of BIM-based DB projects to analyze and quantify the potential of value realization using clash detection as a use value. The results reveal potential hurdles inhibiting BIM from reaching its full potential. Accordingly, recommended changes to the current processes are suggested to facilitate BIM in enhancing value on DB projects.

DOI: 10.4018/978-1-5225-5625-1.ch007

INTRODUCTION

The construction sector of the US, UK, and other countries has been suffering from a constant decline in performance while manufacturing and other industries have been continuously experiencing a boom in productivity (Sveikauskas et al., 2014; Teicholz et al., 2001). The construction sector has consistently been scrutinized for its inefficiency and failure to meet stakeholder expectations and values. The Construction Industry Training Board (CITB) (2016) reveals that both time and cost for design and construction processes are inconsistently predicted and are adversely affecting project stakeholder satisfaction.

The construction sector is built on the interactions and collaboration of multi-disciplinary teams whose processes and information are intertwined. Communication, integration and alignment of values are key success factors given the growing interdependence and complexity of design and construction tasks (Knotten et al., 2015). Reports spanning across several decades (e.g. Latham, 1994; Egan, 1998; Constructing Excellence, 2009; Cabinet Office, 2011) identified the need for improving project communication and changing adversarial contractual and procurement structures as critical strategies for the construction sector.

One of key process, technology and policy innovations that emerged within the construction sector is Building Information Modeling (BIM). BIM tools and workflows enable project stakeholders to digitally model facility, simulate its performance, and manage information flows across the whole project lifecycle. BIM is increasingly adopted or mandated by central government around the world. For example, the UK government's construction strategy stated that all centrally funded public sector projects needed to achieve Level 2 BIM by April 2016. According to this strategy, the adoption of Level 2 BIM processes is expected to promote the full alignment of supply chains with the people responsible for operating and maintaining the assets (Her Majesty's Government [HMG], 2013). The idealized benefits of BIM in many strategy documents and industry reports (GCCG, 2011) prompted many scholars to investigate the benefits of BIM (Bryde et al., 2013, Love et al., 2014). The collaboration benefits from BIM have been partially operationalised or measured using either quantitative or qualitative key performance indicators (Liu et al., 2016, Oraee et al., 2017, Papadonikolaki et al., 2017, Papadonikolaki and Wamelink, 2017). However, there is still a dearth of studies that investigate collaboration benefits of BIM in conjunction to the procurement framework of projects.

Motivated by Eastman et al. (2008)'s proposition that the use of BIM is clearly advisable in conjunction to Design-Build (DB) procurement, this study aims to address whether the value realization potential of BIM is achieved within DB projects. DB procurement involves a client procuring design and construction services through a single organisation, thereby shifting risk to the supplier whilst also having to only manage a single contract. Under DB delivery, the DB firm subcontracts specific design and construction elements to their suppliers while retaining the single contractual link to the client (Hickethier et al., 2013).

The theoretical approach used in this study to link these two strands (i.e. BIM, and DD procurement) adopts some basic Value Management concepts. Value Management is a structured approach to determine what value means to the organization and the project, then delivering to those requirements (Association of Project Management [APM], 2012). The Office of Government Commerce [OGC] (2007) identified that the benefits of having successful value management include more effective team working and the reduction of unnecessary project costs. Both BIM and Value Management are considerably growing within the industry; however, the realisation of their actual benefits remains limited due to the prominence of traditional procurement methods (Eadie et al., 2014; Lindblad, 2013).

While BIM, Value Management, and collaborative procurement strategies concur towards the same outcomes, there is a dearth of studies that combines concepts from across the three subjects. The aim of this chapter is to investigate how DB projects can enhance Value, specifically Use Value, through the utilization of BIM. The specific objectives are:

- Highlight any alignments or misalignments between BIM, Value, and DB/IPD processes by reviewing related literature on these subject areas;
- Determine the effectiveness of BIM processes for enhancing Use Value on DB projects by focusing on the transition from the Preconstruction design phase into the Construction design phase; and
- Identify any necessary improvements or alterations to the design and construction processes to facilitate BIM adoption and its ability to enhance Value on DB projects.

KEY CONCEPTS AND DEFINITIONS

This section introduces the key concepts and definitions from across BIM, DB and Value Management that are relevant to the proposed study. The understanding of these concepts is important to justify the linking proposed across the three subjects (i.e. BIM, DB, and value management), the proposed research methodology, and the empirical analysis of the case studies.

Building Information Modeling

BIM is the current expression of digital innovation in the construction sector (Succar and Kassem, 2015). BIM is a value creating collaboration through the entire lifecycle of an asset, underpinned by the creation, collation and exchange of shared three dimensional (3D) models and intelligent, structured data attached to them (UK BIM Task Group, 2013). It helps stakeholders make educated decisions and execute the project with reduced costs, schedules, rework, and better quality (Azhar, 2011; Eastman et al., 2008; Redmond et al., 2012). The key value proposition of BIM lies in enabling collaboration between participants throughout the project's lifecycle while achieving stakeholder requirements with improved predictability (Demian & Walters, 2013; Eisenmann & Park, 2012).

In BIM workflows there are different levels that determine varying levels of BIM implementation among the project participants and project stages. In industry, four levels are proposed by the UK BIM Task Group: Level 0, Level 1, Level 2, and Level 3 and are summarized in Figure 1 (BIM Industry Working Group, 2011). Level 2 is the mandated level by the UK government. Level 2 BIM is a collaborative way of working, in which 3D models with the required data are created in separate discipline models according to a set of guides, standards and specifications (Kassem et al., 2016). A research-based concept that implicitly embeds the level of implementation among project teams and stages is the BIM capability stages of Succar (2009). The three capability stages are: modelling (BIM Stage 1), collaboration (BIM Stage 2) and integration (BIM Stage 3) and their effect on the project lifecycle phases and corresponding project teams is shown in Figure 2.

Figure 1. BIM implementation levels (adapted from British Standards Institution [BSI], 2013)

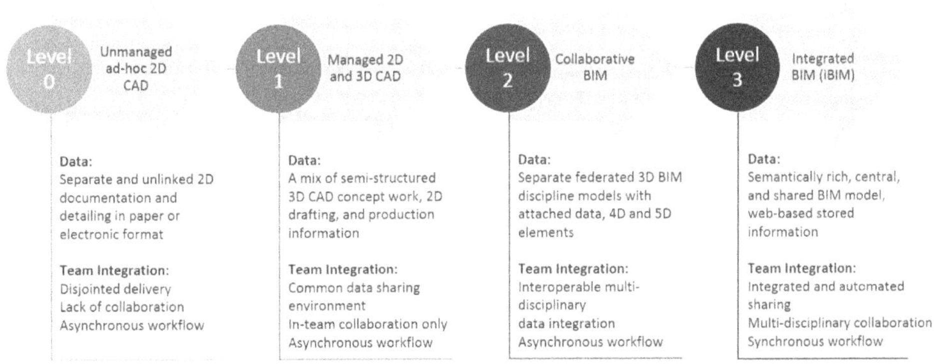

Figure 2. BIM Capability Stages and their effect on project stages and teams (© Copyright 2009, Succar. Used with permission)

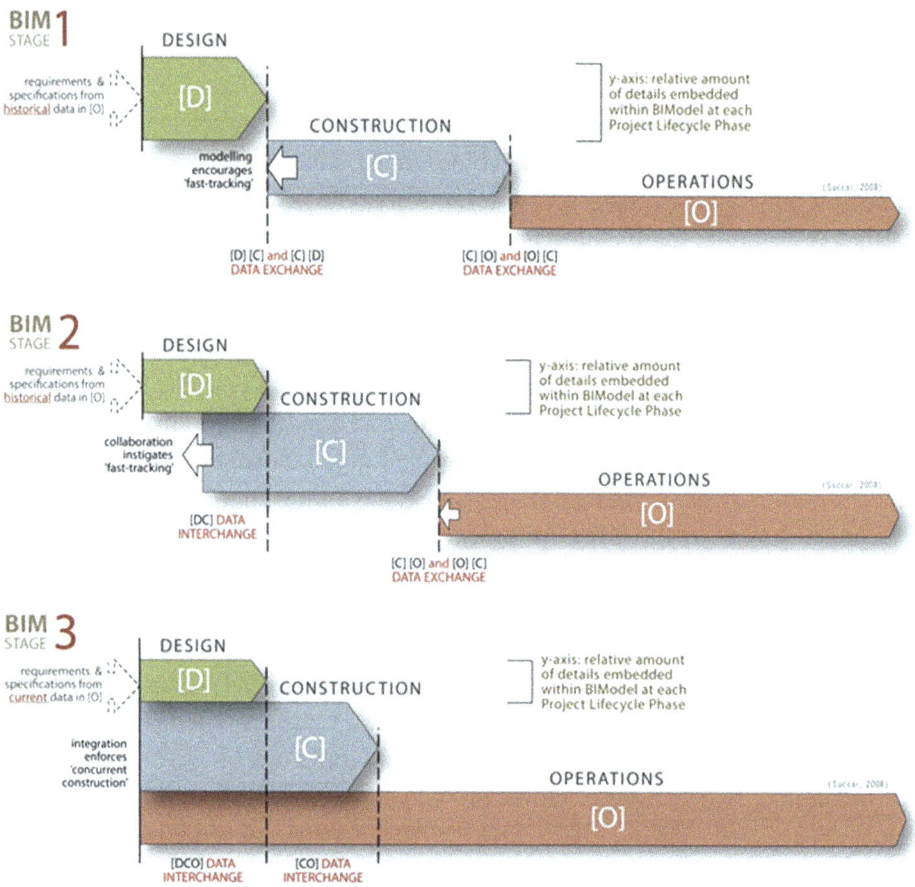

Value Management

Value has varying definitions dependent upon an individual's core beliefs, morals, and ideals (Thyssen et al. 2008; Thomson et al., 2003) and human interest (Korsgaard, 1986; Thomson et al, 2003). Contemporary attempts defining value tend to be increasingly mathematical (Thyssen et al. 2008) and argue that many values are objective and can be quantified (Moore, 1998; Thyssen et al., 2008). The most commonly adopted definition of value in construction is expressed as the ratio between Function and Cost (Park, 1999). Table 1 presents various definitions of value as presented in current literature.

Table 1. Definitions of value

Definitions of Value	Source
Value = Time, Cost, Quality	Best & De Valence (1999)
Value = the relationship between the contribution of the function and the satisfaction of the need and the cost of the function	BSi (1997)
Value = Functionality, Build Quality & Impact (additionally: Finance, Time, Environment & Resources)	Construction Industry Council [CIC] (2006)
Value = Benefits / Price	Fallon (1971)
Value = Capital Cost, Operating Cost, Time, Exchange (earning potential or sale worth of completed project), Environmental Impact, Utility & Esteem	Kelly (2007)
Value = Function / Cost	Park (1999)
Value = Function, Form, Economy & Time	Pena & Parshall (2001)
Value = Use Value, Esteem Value & Exchange Value	Thiry (1997)

Identifying and bringing all key stakeholders together at the correct time in the project is necessary for an effective Value Management process (Male et al., 2007). MacLeamy's curve (2008) states the need for more integration and effort investment at the early design phase to prevent costly and time-consuming changes. Information sharing and communication is key in all proposed Value Management processes. In the construction sector, the traditionally low and ineffective use of information technology contributes to poor communication across construction projects and affects value realisation. Empirical evidence from case studies were found about the positive impact of BIM on communication and consequently on value realisation (Shahrin & Johansen, 2013). Levels of supply chain integration (McCormack & Lockamy III, 2004) within a BIM workflow were also found to be linked with a suppliers' BIM maturity (Papadonikolaki et al., 2015).

Procurement Methods: DB and IPD

Traditional construction projects are procured by a client who engages consultants to deliver design work which is then subject to competitive tender by main contractors. The main contractors collate prices from subcontractors and submit a lump sum bid. This traditional procurement structure inhibits cooperation between project teams and diminishes their ability to maximize value on a project (Matthews & Howell, 2005). These limitations of the traditional procurement method called upon more integrated and collaborative forms for procuring assets within the construction sector.

IPD emerged to address the shortcomings of the adversarial relationships associated with traditional procurement. Under IPD, key stakeholders from each discipline are present from project inception to ensure that the project design meets the needs of all stakeholders (American Institute of Architects [AIA], 2007; AIA California Council, 2008). Although several professional organisations have supported the further development of IPD and research has demonstrated its benefits and challenges (Matthews & Howell, 2005; Cohen, 2010), the uptake of this approach by the construction industry remains limited (Becerik-Gerber & Kent, 2010; Looi, 2013).

Design and Build contracts are more popular primarily due to having a longer track-record than IPD (Darrington, 2011). Under DB contracts, the main contractor is responsible for both the design and construction work on a project for an agreed lump-sum price. Being responsible for design, the main contractor can appoint design consultants to carry out this work if they lack this expertise in-house. Furthermore, some designers will become appointed at different stages where, for example, construction designers only become involved until later stages, whereas consultants (e.g. Architect; Mechanical and Electrical (or services) Engineer; Quantity Surveyor or Costs Manager; Structural engineer, and Project manager) may be involved much sooner (Design-Build Institute of America [DBIA], 2017). DB procurement allows for various levels of integration through its different procurement forms, namely, qualifications-only selection, best value selection, and price driven selection.

Although both IPD and DB aim to achieve enhanced integration compared to traditional structures, the underlying concepts and contractual arrangements differ. These differences can be summarized into the following categories: risk allocation, team selection methods, degree of owner involvement, and accountability and risk management. Despite these differences, DB is contractually well-suited for increasing the collaboration among project team members at the project's start, specifically the designers and constructors, which makes the implementation of IPD principles possible. However, a greater level of involvement is required on the client's behalf to achieve IPD's benefits (AIA, 2007).

Under IPD, the stakeholders work collaboratively and make collective decisions while adjusting target cost to achieve the desired value. The wider involvement of the project team on the decision making improves the team's understanding of the project requirements (Ballard et al., 2012) while keeping aligned the interests of stakeholders.

On the other hand, within the DB the client sets the project requirements and then the main contractor determines how to design, build, and manage the construction process to stay within the target price and value (Cohen, 2010; Ballard et al., 2012; Karasulu et al., 2013). This means that DB does not provide the same level of transparency IPD brings and is unable to achieve the alignment of parties' interests achievable within IPD (Ballard et al., 2012; Karasulu, 2013). However, it does offer a project delivery solution for less educated clients (Karasulu et al., 2013).

RELATED LITERATURE

The studies that have addressed the intersections between BIM, Value, DB, and IPD subject areas are listed in Table 2. A few studies have explored the alignment between BIM and DB but they did not address their impact on value realization.

With the absence of any industry standards for the measurement of BIM benefits, it is important to redefine project delivery and management processes to harness the potential of BIM (Ilozor & Kelly, 2011; Bockstael & Issa, 2016; Leite et al., 2010). Reported benefits can be reaped through the alignment between BIM and IPD methods around one integrated model as suggested by Ilozor and Kelly (2011). Mcgraw-Hill Construction (2012) presented the benefit of specifically using Clash Detection and Avoidance as one of the immediate benefits of utilizing BIM on projects. They further noted the value of identifying significant clashes prior to commencing construction work to avoid substantial costs in reworking. The findings of Papadonikolaki et al. (2015) indicated that there are benefits to be derived from the utilization of BIM with an integrated supply chain, which could be applied to the DB scenario.

The utilization of Clash Detection and Avoidance on commercial construction projects was further investigated by Bockstael and Issa (2016). They suggested that the avoidance of design conflicts can result in a reduction of Requests for Information (RFI) and design-related change orders and subsequently in costs and wastes saving. Increasing the Level of Definitions (LOD) within a design model can improve the design accuracy and the ability of BIM tools to detect clashes, and subsequently enhance decision making while reducing conflicts (Leite et al., 2010). Yet, it is necessary to filter irrelevant clashes when investigating and assessing the results of the clash detection process (Leite et al., 2010).

The realignment of business processes with technology use should entail an understanding of the individualities of the construction process (Koskela & Kazi, 2003). For example, the integration of BIM and lean processes requires any process change to be rooted in the conceptual understanding of the theory of production in construction (Koskela et al., 2009). Similarly, much of the literature surrounding Value Management seeks to align with an IPD structure and little considerations is given to DB contracts despite the assertions of Looi (2013) and Darrington (2011) that clients are still more disposed to pass risk down the supply chain in methods such as DB rather than pursuing an IPD structure. Owen et al. (2010) highlighted that IPD, BIM, and Value are commonly developed in isolation from one another and there are limited studies investigating their alignments.

Some market-wide processes proposing some alignment between project delivery processes and BIM workflows have been proposed. For example, the Royal Institute

Table 2. Literature review interaction matrix

Source	BIM	Value	IPD	DB	Literature Summary
AIA (2007)	x		x		IPD Guide book
Azhar et al. (2011)	x		x		Case study approach to investigate time and cost savings
Bercerik-Gerber & Kensek (2010)	x		x		Survey approach to investigate BIM research trends and directions
Bouazza & Greenwood (2017)	x	x	x		Literature review paper to discuss opportunities linking BIM and Knowledge Management
Bryde et al. (2013)	x		x		Case study approach to investigate benefits of BIM on 35 projects using secondary data
BSi (2013)	x				Level 2 BIM Standard
Darrington (2011)			x	x	Paper discussing possibility of using DB contract in an IPD approach
Eastman et al. (2011)	x		x		Guide book on BIM
Froese (2010)	x		x		Paper discussing integration of ICT & BIM into project management
Ilozor & Kelly (2011)	x		x		Literature review on BIM, IPD & partnering
Shahrin & Johansen (2013)	x	x			Interviews within 2 case study projects investigating meeting client requirements using BIM
Karasulu et al. (2013)	x	x	x		Literature Review and interview to investigate advancement of TVD in IPD using BIM
Kassem et al. (2014)	x		x		Literature Review and design competitions to test adoption of BIM protocols for collaborative work processes
Kelly (2007)		x			Action research in workshops to investigate client values
Matthews & Howell (2005)			x	x	Case study to demonstrate benefits of using IPD
Owen et al. (2010)	x		x		Literature Review and paper discussing challenges for adopting integrated design
Papadonikolaki et al. (2015)	x		x		Case Study research to investigate use of BIM and integrated supply chains
SEC (2007)		x	x		Memorandum by SEC advocating use of IPD
Smith & Tardif (2009)	x				Guide book on implementing BIM
Talebi (2014)	x	x	x		Literature review on benefits and challenges of BIM
Thyssen et al. (2008)		x			Case study and workshop to investigate early engagement for value creation
Aibinu & Papadonikolaki (2016)	x			x	Case study research on the relationship between DB procurement and coordination from BIM

of British Architects (RIBA) updated their standard plan work (i.e. RIBA Plan of Work 2013) to include the alignment with the Information Exchanges that occur within a BIM process. However, there is still a dearth of studies investigating such an alignment and in particular, the transition from one of its stages to the next. Project stakeholders become involved in a project at different phases depending on the procurement method adopted which may also affect value realization. Hence, it is necessary to understand: the interaction between stakeholders across the project lifecycle; how this interaction varies within BIM workflows under different procurement methods, and its effect on value realization. This paper addresses this gap by focusing on the transition from the preconstruction design stages (Stages 0 - 4) to the Construction stage (Stage 5).

BIM USE TO ENHANCE VALUE ON DB PROJECTS: EVIDENCE FROM PRACTICE

Research Methods

The lack of studies on the use of BIM for enhancing value on DB projects necessitates the collection of relevant data and evidence from practice. Cross-sectional case study research design is used to explore the effect on value when adopting BIM workflows under DB procurement. Quantitative measures were utilized to evaluate specific project data and support themes captured through qualitative data collection. The adopted research methods for fulfilling the posed aim are discussed in this section.

Concepts and Measures

The three concepts underpinning this study are: Value, BIM, and project team integration. It is important to establish a consistent objective view of what Value means within construction projects while also accounting for value's subjective aspects. The definition provided by Kelly (2007) is used in this study and detailed in Table 3. The Utility as a Value Criteria, identified by Thiry (1997) and Kelly (2007) as being a key component of delivering value, was selected for this study as it is a concept that can be transposed and applied across multiple projects compared to other project specific Value Criteria such as Capital Cost, Operational Cost, etc.

Defining BIM and clarifying the level of implementation is necessary when evaluating its impact on enhancing value as previous studies (e.g. Papadonikolaki et al; 2015) showed that these can be correlated. This study adopts:

Table 3. Value criteria on construction projects (adapted from Kelly, 2007)

Value Criteria	Measure
Capital Cost (CAPEX)	All investment costs incurred prior to project completion
Operational Cost (OPEX)	All costs incurred after project completion until the client's time horizon
Time	Time from initial project workshop until completion
Exchange	The earning potential or sale worth of the completed project
Environmental Impact	Impact on Land, Amount of Carbon utilized during project life-cycle
Utility	Use Value
Esteem	Regard/Respect benefits to the client from the world at large due to the project

- The definition established by BSi (2013) for the BIM levels (Figure 1) to represent the levels of BIM implementation within the practical context of the selected projects; and
- Succar (2009)'s three BIM capability stages (i.e. Modelling, Collaboration, and Integration) – each inferring a certain level of fast tracking or overlap between project stages/stakeholder – to (1) analyze and visualize the current effect of BIM use and the procurement method (i.e. DB) on value realization (i.e. Use Value), and (2) subsequently visualize the proposed improvement.

The selected case study projects are UK-based and adopted the RIBA Plan of Work 2013 (RIBA, 2013). Table 4 shows the mapping between RIBA Plan of Work stages and the corresponding stages of Succar's framework. The study will focus on investigating the transition from RIBA Stage 4 to RIBA Stage 5 within DB BIM-based projects. The corresponding area of investigation within the Succar's framework is highlighted in Figure 3.

Data Collection

Primary project data were collected from five construction projects across the UK. This data helps identify both the benefits and challenges encountered when adopting

Table 4. Alignment of project stages (Succar, 2009; RIBA, 2013)

Succar (2009) Project Phases	Equivalent RIBA 2013 Stages
Design	0, 1, 2, 3, and 4
Construction	5
Operations	6 and 7

Figure 3. Project transition phase to be investigated (adapted from Succar, 2009)

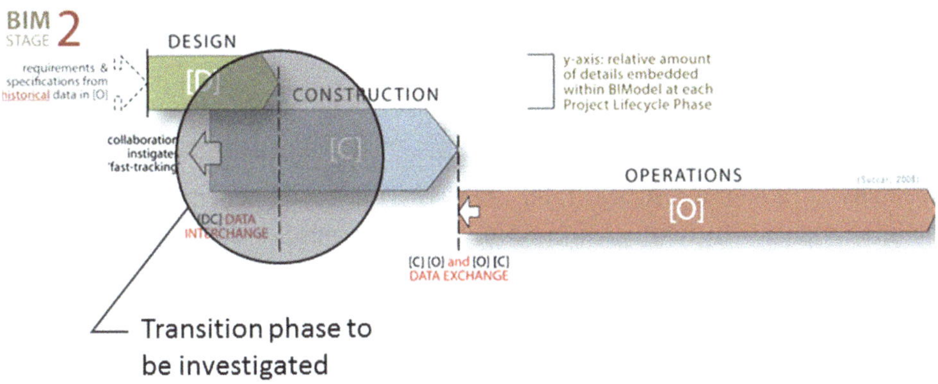

BIM processes in DB projects. Utility will be referred to as the Use Value meaning that the item under investigation must have a specific use or benefit to deliver Utility (Kelly, 2007). The key areas of Use Value that can be realized from a construction project employing BIM are defined by Bryde et al. (2013) as: cost and time reduction or control, communication improvement, coordination enhancement, quality increase or control, risk reduction, and scope clarification.

Building on the findings of the literature review, Clash Detection and Avoidance is identified as one of the main BIM uses that generate cost and time reductions and enhanced coordination (Mcgraw-Hill Construction, 2012; Leite et al., 2010; Bockstael & Issa, 2016). This BIM use, therefore, formed a fundamental data source for measuring the Use Value of BIM across the five selected projects.

One of the objectives of the research is to investigate the transition from the Preconstruction Design (Stage 4) into Construction Design (Stage 5) and the corresponding integration of the supply chain. Hence, collecting data from both phases is necessary to investigate such transition. The cross-sectional nature of this study helped this approach by enabling the collection of such data from across five projects.

The selection of the case studies included a representative sample of Design and Build projects to ensure reliability and generalization of the research. The selection was based on the contract value of the projects ranging between £5m and £25m in value and delivered under DB procurement. An overview of each case study is provided in the following section.

Case Study Overviews

An overview on each case study is presented in Figure 7. Background information about each project includes the contract value, the procurement method, start and end dates, and the project team structure.

Case Study A comprises 7600m^2 of teaching and workshop space including specialist rail equipment such as 150m of external track and catenary. The completed building is intended to train engineers to meet the future needs of the wider HS2 project as well as those in the rail sector. The project was procured via the Scape Framework which supports the DB procurement structure. The direct client for the project was a representative of the end user which added a further line of communication to an otherwise conventional DB team structure as shown in Figure 4.

Case Study B involves the new build construction of a Swimming Pool with associated changing rooms, gymnasium and fitness studio area. The completed building is intended to replace a similar facility that existed on the same site due to the previous building deteriorating beyond repair. It was procured via the Scape Framework, aligning with the DB procurement structure. The client for the project was also the end user, thus simplifying project communications in terms of project requirements.

Case Study C involves the new build construction of a Leisure Centre consisting of a learner and main pool, gymnasium, changing areas, fitness studios, and sports courts. The building will replace an equivalent building on an adjacent site which had become costly to maintain and run whilst deteriorating in condition. It was also procured via the Scape Framework and has the same project team structure as Case Study B.

Case Study D consists of a 5000m^2, three-storey building which is being constructed to BREEAM Excellence standards. The building will comprise of office, laboratory, and workshop space for up to 700 people. It is designed to be a central hub for the science park that it is located on and will become an important regional center for a range of businesses from start-ups to large corporate companies. Case Study D was procured via the North Wales Construction Framework, which also coincides with the DB structure and has the same project team structure as Case Study B and C.

Case Study E involves the construction of a 450-place secondary school including class rooms, sports hall, conferencing facility, as well as public arts space and gallery. The project has been constructed adjacent to an existing sister Primary School facility and designed to offer open and flexible learning spaces. It was procured via the EFA Framework, complementing the DB procurement structure. The client was a representative of the end user while also involving the use of a Project Management consultant which added further lines of communication to the team structure shown in Figure 4.

Figure 4. Case study projects background information

ANALYSIS OF CASE STUDY DATA

Data pertaining to the two Use Value items, (1) clash detection and avoidance and (2) project team integration, has been collected on each project and aggregated across the five case studies to facilitate the analysis. Such data will be presented in this section for each project and will be analyzed across all projects in the next section.

Clash Detection and Avoidance

The number of clashes from the federated BIM models is recorded at the end of Stage 4 (Design) and at the onset of Stage 5 (Construction). Clashes are investigated between each pair of disciplines: architectural (denoted by A) and MEP (denoted by M); architectural and structural (denoted by S) packages, and structural and MEP packages. It is also necessary to filter the clashes for actual and irrelevant/non-clashes as suggested by Leite et al. (2010). These values are summarized in Table 5 for the five case studies for each pair of trades which also shows differences in the number of clashes during the transition from Stage 4 and Stage 5. All projects had an increase in the number of clashes between at least one pair of trades during the transition from Stage 4 to Stage 5. In particular, clashes between Structural and MEP increased in four of the case studies during the transition from Stage 4 to Stage 5.

Table 5. Actual clashes at Stages 4 and 5 for the five case studies

Project	S4 (A&M)	S5 (A&M)	Difference	S4 (A&S)	S5 (A&S)	Difference	S4 (S&M)	S5 (S&M)	Difference
Project A	19	8	-11	7	11	4	46	36	35
Project B	1	11	10	5	14	9	4	4	4
Project C	32	18	-14	24	11	-13	1	19	17
Project D	14	164	150	11	14	3	5	67	62
Project E	16	7	-9	9	4	-5	21	97	97

The number of model components for each case study at Stage 4 and Stage 5 are summarized in Table 6. The table also shows whether there has an increase or decrease in the number of model components for each discipline during the transition from preconstruction to construction. The number of model components increased for all trades across all projects when passing from Stage 4 to Stage 5 with the exception of the Architectural disciplines in Projects B and C.

The BIM authoring tools of each design package are listed in Table 7. For the structural trade, the data shows that there has been a change in the adopted design package across all projects when transitioning from Stage 4 to Stage 5.

Project Team Integration

The stage at which teams get involved on the project and the level of team integration depends on the procurement structure and contractual arrangements. However, the level of BIM implementation may either induce different engagement dynamics

Table 6. Disciplines showing increase in number of model components between Stage 4 and Stage 5

Case Study	Increase in number of model components from Stage 4 to Stage 5				
	Architectural	Structural	MEP	Total Component count Stage 4	Total Component count Stage 5
A	Yes	Yes	Yes	17085	24699
B	No	Yes	Yes	5339	16938
C	No	Yes	Yes	23725	49405
D	Yes	Yes	Yes	17398	34914
E	Yes	Yes	Yes	14717	18704

Table 7. Case studies entailing a change in the type of authoring tools used between Stages 4 and 5

Case Study	Native Authoring Tool					
	Stage 4 Architecture Format	Stage 5 Architecture Format	Stage 4 Structure Format	Stage 5 Structure Format	Stage 4 MEP Format	Stage 5 MEP Format
A	Archicad	Archicad	Revit	Tekla	Revit	Revit
B	Revit	Revit	Revit	Strucad	Revit	Revit
C	Revit	Revit	Revit	Strucad	Revit	Revit
D	Revit	Revit	Revit	StruMIS	Revit	Revit
E	Revit	Revit	Revit	Strucad	Revit	Revit

or be distorted by default engagement structure of such procurement methods. Exploring the involvement of project teams can help determine the differences in the input of organizations between the Preconstruction and Construction stages of a project. It can also provide insights to enhance integration and the Use Value of the implemented BIM capability stage.

The level of organizational integration and involvement at Stages 4 and 5 is summarized in Table 8. The data shows that across all projects there has been at least one specialist trade who get involved prior to being awarded a contact.

Table 8. Organizations involvement at Stages 4 and 5

Organization	Case Study A			Case Study B			Case Study C			Case Study D			Case Study E		
	S4	Main Contract Award	S5	S4	Main Contract Award	S5	S4	Main Contract Award	S5	S4	Main Contract Award	S5	S4	Main Contract Award	S5
Main Contractor	2		2	2		2	2		2	2		2	2		2
Architectural Engineer	2		2	2		2	2		2	2		2	2		2
Structural Engineer	2		2	2		2	2		2	2		2	2		2
Steelwork Consultant	2			2			2			2			2		
MEP Consultant	2			2			2			2			2		
Steelwork Supply/ Design	2*		2	2*		2			2	2*		2	2*		2
MEP Supply/Design	2*		2			2			2	2*		2			2
Cladding and Roofing	2*		2			2			2	2*		2			2
Curtain Walling			2			2			2	2*		2			2
Filtration	NA		NA	2		2	2*		2*	NA		NA	NA		NA

X* denotes that an organization was involved prior to being awarded a subcontract
S4=Stage 4, S5=Stage 5

Table 9 presents the level of details (LOD) [the geometric part of the Level of Definitions] of each BIM model element at Stages 4 and 5. The data shows that (1) different LODs co-exist at the same stage to reflect the actual decision making and the client requirements, and (2) across all projects there has been an increase in the LOD for LOD4 to LOD5 for several work packages.

Table 10 shows the responsibility for the design of different elements at Stage 4. At this stage the responsibility for all components, with the exception of filtration systems, reside with the consultant.

Table 11 shows the responsibilities for the design of different elements and their corresponding BIMs (models). From this data, there appears to be a shift between the BIMs (models) authoring responsibility and the corresponding design responsibility between stages 4 and 5. In other words, the player responsible for the design is not

Table 9. Lod for each bim model element at stages 4 and 5

BIM Model Element	Case Study A			Case Study B			Case Study C			Case Study D			Case Study E		
	S4		S5	S4		S5	S4		S5	S4		S5	S4		S5
Substructure	4	Main Contract Award	5	4	Main Contract Award	5	4	Main Contract Award	5	4	Main Contract Award	5	4	Main Contract Award	5
Steelwork	4		5	4		5	4		5	4		5	4		5
Floor slabs	4		5	4		5	4		5	4		5	4		5
Cladding	4		4	4		4	4		4	4		4	4		4
Roof	4		3	4		4	4		4	4		4	4		4
Internal walls	4		4	4		4	4		4	4		4	4		4
Ceilings	4		4	4		4	4		4	4		4	4		4
Curtain walls	4		4	4		5	4		4	4		4	4		5
Doors	4		4	4		4	4		4	4		4	4		4
MEP	4		5	4		5	4		5	4		5	4		5
Filtration	NA		NA	4		5	4		5	NA		NA	NA		NA

Table 10. Model and design responsibility at stage 4

BIM Model Element	Model and Design Responsibility at Stage 4									
	Case Study A		Case Study B		Case Study C		Case Study D		Case Study E	
	Mod	Des	Mod	Des	Mod	Des	Mod	Des	Mod	Des
Substructure	Con	Con	Con	Con	Con	Con	Con	Con	Con	Con
Steelwork	Con	Con	Con	Con	Con	Con	Con	Con	Con	Con
Floor slabs	Con	Con	Con	Con	Con	Con	Con	Con	Con	Con
Cladding	Con	Con	Con	Con	Con	Con	Con	Con	Con	Con
Roof	Con	Con	Con	Con	Con	Con	Con	Con	Con	Con
Internal walls	Con	Con	Con	Con	Con	Con	Con	Con	Con	Con
Ceilings	Con	Con	Con	Con	Con	Con	Con	Con	Con	Con
Curtain walls	Con	Con	Con	Con	Con	Con	Con	Con	Con	Con
Doors	Con	Con	Con	Con	Con	Con	Con	Con	Con	Con
MEP	Con	Con	Con	Con	Con	Con	Con	Con	Con	Con
Filtration	NA	NA	Con	Con	Con	Con	NA	NA	NA	NA

*Con = Consultant

Table 11. Model and design responsibility at stage 5

BIM Model Element	Model and Design Responsibility at Stage 5									
	Case Study A		Case Study B		Case Study C		Case Study D		Case Study E	
	Mod	Des	Mod	Des	Mod	Des	Mod	Des	Mod	Des
Substructure	Con	Con	Con	Con	Con	Con	Con	Con	Con	Con
Steelwork	SC	SC	SC	SC	SC	SC	SC	SC	SC	SC
Floor slabs	Con	Con	Con	Con	Con	Con	Con	Con	Con	Con
Cladding	Con	SC	Con	SC	Con	SC	Con	SC	Con	SC
Roof	Con	SC	Con	SC	Con	SC	Con	SC	Con	SC
Internal walls	Con	Con	Con	Con	Con	Con	Con	Con	Con	Con
Ceilings	Con	Con	Con	Con	Con	Con	Con	Con	Con	Con
Curtain walls	Con	SC	SC	SC	Con	SC	Con	SC	SC	SC
Doors	Con	Con	Con	Con	Con	Con	Con	Con	Con	Con
MEP	SC	SC	SC	SC	SC	SC	SC	SC	SC	SC
Filtration	NA	NA	SC	SC	SC	SC	NA	NA	NA	NA

**Con = Consultant, SC = Subcontractor*

responsible for authoring the corresponding BIM (model) which is ultimately used in clash detection for coordination purpose. All the data about team integration from across the five case studies is aggregated in Table 12 to provide a clearer analysis of the changes occurring between Preconstruction and Construction.

DISCUSSION OF FINDINGS: CROSS-PROJECT ANALYSIS

To enable a cross-project analysis, the data is aggregated for each pair of disciplines from across all the five studies. Tables 13, 14 and 15 show the aggregated data for

Table 12. Summary analysis of case studies for team integration use value area

	Model and design author change from Stage 4 to 5					Early integration of Stage 5 supplier in Stage 4					Increase in model LOD from Stage 4 to 5				
Case Study / BIM Element	A	B	C	D	E	A	B	C	D	E	A	B	C	D	E
Substructure	No	No	No	No	No	No	No	No	No	No	Yes	Yes	Yes	Yes	Yes
Steelwork	Yes*	Yes*	Yes	Yes*	Yes*	Yes	Yes	No	Yes	Yes	Yes	Yes	Yes	Yes	Yes
Floor slabs	No	No	No	No	No	No	No	No	No	No	Yes	Yes	Yes	Yes	Yes
Cladding	No	No	No	No	No	Yes	No	No	No	No	No	No	No	No	No
Roof	No	No	No	No	No	Yes	No	No	No	No	No	No	No	No	No
Internal walls	No	No	No	No	No	No	No	No	No	No	No	No	No	No	No
Ceilings	No	No	No	No	No	No	No	No	No	No	No	No	No	No	No
Curtain walls	No	No	No	No	No	No	No	No	No	No	No	Yes	No	No	Yes
Doors	No	No	No	No	No	No	No	No	No	No	No	No	No	No	No
MEP	Yes*	Yes	Yes	Yes*	Yes	Yes	No	No	Yes	No	Yes	Yes	Yes	Yes	Yes
Filtration	NA	Yes	Yes*	NA	NA	NA	Yes	Yes	NA	NA	NA	Yes	Yes	NA	NA

X denotes that a subcontractor organization was integrated into Stage 4 prior to being awarded a subcontract*

Table 13. Architecture and MEP grouping data

Case Study	Architectural and MEP										
	S4-S5 software change		S4-S5 increase in components count		S4-S5 responsibility change		Supply chain involvement before Stage 5		S4-S5 increase in LOD		S4-S5 increase in clashes
	Arch	MEP	Arch	MEP	Arch	MEP	Arch	MEP	Arch	MEP	Arch vs MEP
A	No	No	Yes	Yes	No*	Yes	NA	Yes	No	Yes	No
B	No	No	No	Yes	No*	Yes	NA	No	No	Yes	Yes
C	No	No	No	Yes	No*	Yes	NA	No	No	Yes	No
D	No	No	Yes	Yes	No*	Yes	NA	Yes	No	Yes	Yes
E	No	No	Yes	Yes	No*	Yes	NA	No	No	Yes	No
%	0%	0%	60%	100%	0%*	100%	NA	40%	0%	100%	40%

Some architectural elements changed in design responsibility; however, the whole model did not and was regarded as No.

Table 14. Architecture and structural grouping data

Case Study	Architectural and Structural										
	S4-S5 software change		S4-S5 increase in components count		S4-S5 responsibility change		Supply chain involvement before Stage 5		S4-S5 increase in LOD		S4-S5 increase in clashes
	Arch	Struct	Arch	Struct	Arch	Struct	Arch	Struct	Arch	Struct	Arch vs Struct
A	No	Yes	Yes	Yes	No*	Yes	NA	Yes	No	Yes	Yes
B	No	Yes	No	Yes	No*	Yes	NA	Yes	No	Yes	Yes
C	No	Yes	No	Yes	No*	Yes	NA	No	No	Yes	No
D	No	Yes	Yes	Yes	No*	Yes	NA	Yes	No	Yes	Yes
E	No	Yes	Yes	Yes	No*	Yes	NA	Yes	No	Yes	No
%	0%	100%	60%	100%	0%*	100%	NA	80%	0%	100%	60%

Table 15. MEP and structural grouping data

Case Study	MEP and Structural										
	S4-S5 software change		S4-S5 increase in components count		S4-S5 responsibility change		Supply chain involvement before Stage 5		S4-S5 increase in LOD		S4-S5 increase in clashes
	MEP	Struct	MEP	Struct	MEP	Struct	MEP	Struct	MEP	Struct	MEP vs Struct
A	No	Yes	Yes	Yes	Yes	Yes	Yes	Yes	Yes	Yes	No
B	No	Yes	Yes	Yes	Yes	Yes	No	Yes	Yes	Yes	Yes
C	No	Yes	Yes	Yes	Yes	Yes	No	No	Yes	Yes	Yes
D	No	Yes	Yes	Yes	Yes	Yes	Yes	Yes	Yes	Yes	Yes
E	No	Yes	Yes	Yes	Yes	Yes	No	Yes	Yes	Yes	Yes
%	0%	100%	100%	100%	100%	100%	40%	80%	100%	100%	80%

'Architecture and MEP', 'Architectural and Structural', and 'MEP and Structural' respectively. This cross-project analysis aims to improve the understanding of potential associations between these factors or independent variables and value realization when transitioning from Stage 4 to 5.

The clash detection and avoidance was identified through the literature as an important Use Value due to its potential contribution to time and cost savings. Research work discussed earlier identified that a reduction in the number of clashes can result in both a reduction in RFI's and change orders, subsequently reducing delays and costs (Bockstael & Issa, 2016; Mcgraw-Hill Construction, 2012).

The data analysis pertaining to this use value area revealed that the most significant changes during the transition from Preconstruction to the Construction phase occurred within the Structural and MEP grouping. Four of the case studies showed an increase in the number of detected clashes in this grouping compared to the other two groupings involving the Architectural model. In conjunction with these findings, both the Structural and MEP disciplines demonstrated an increase in the number of model components when comparing the preconstruction and construction model. This observation suggests that more information was being added to a model after a contract has been awarded rather than prior to it. The noted findings provide a balanced view about the idealized impact associated with combining DB and BIM in terms of anticipating the effort and investing more knowledge into the upstream phases of a project's life cycle (AIA, 2007; Macleamy, 2008). Clashes still either persist or grow in number at construction phase (i.e. Stage 5).

However, there are some independent variables in these case studies that may be associated with this increase in clash number. One of such variables is the increase in the level of information and the number of components when transitioning from Stage 4 to Stage 5. In addition to the increase the number of components, most disciplines consistently showed an increase in LOD from Preconstruction to Construction across all five case study projects, aligning with the suggestions of Leite et al. (2010). The latter study indicates that the LOD influences the accuracy of design elements and subsequently the volume of identified clashes.

Four of the five case studies revealed an association between the increase in LOD, the increase in the number of model components, and the increase in the number of detected clashes. When adding the supply chain involvement variable, it is evident that there are varying levels of organizational involvement at different stages: only in a few instances the MEP and structural consultants were involved at Stage 4 (Tables 14 and 15); the majority of subcontractors / work packages do not get involved in Stage 4 but only in Stage 5 after a contract award (Table 12). In fact, the formal contractual stance states that any use of digital information prior to a formal contract is completely at the risk of the supply chain organization until such a subcontract and supporting BIM protocol is formally established (AIA, 2013). However, despite the established contractual risks that face the supply chain under a traditional DB structure, some packages – MEP and Steelwork mainly – showed early supply chain involvement in Stage 4.

In the unique instance where the Structural and MEP disciplines did not show an increase in clashes (Table 15, Case Study A), both disciplines were involved on the project at Stage 4 prior to being awarded a contract. This project did not show an increase in the number of clashes despite most of the other independent variables were present (i.e. increase in components count; change of responsibility for design and modelling; increase in LOD). This is in line with the ideas presented by Macleamy (2009) that earlier involvement of key project stakeholder and integration of knowledge can decrease the changes needed at later stages by detecting any clashes earlier in the project. This also provides empirical evidence about a theoretical notion proposed in Succar (2009) who inferred the potential occurrence of increased fast tracking between stages/stakeholders in transitioning across the BIM capability stages (i.e. modelling, collaboration, and integration)

Considering the design and model responsibility during the transition from Preconstruction to Construction, both the Structural and MEP disciplines experienced a change in responsibility where different organizations handled the design of these disciplines at different stages. This shift in responsibility is primarily due to the team integration structure dictated by the DB procurement method. The initial analysis of the project team during Stage 5 showed a disconnection between the responsibility of organizations for the construction design and BIM representation of certain design elements, in addition to a change in the authoring tool used. For example, all five case studies showed a change in the assigned organization responsible for designing and 3D modeling for the Cladding and Roofing elements within the Architecture discipline. It is suggested that levels of BIM collaboration on projects depends upon the level of supply chain maturity (Papadonikolaki et al., 2015). This is consistent with the observations from the presented case studies where some of the supply chain organizations responsible for Stage 5 design were collaborating using 2D design only. This shift in modeling and designing responsibility and the discrepancy in maturity levels yield inconsistencies in the produced components and a distortion of transferred knowledge, thus can be also associated the observed increase in detected conflicts.

Despite the capabilities of automating the detection of clashes with BIM, the persistence of conflicts between different design disciplines and their increase during the later stages of a project are a sign of underlying issues within the project delivery process itself. The findings reveal potential associations between project team structure and the amount of detected conflicts. Although the DB procurement structure integrates the design and construction processes under the responsibility of a single entity, the fragmentation of the involved supply chain and its varying maturity levels inhibit the realization of BIM's potential in enhancing value for the project stakeholders. Traditional management strategies, isolated team work, and risk adverse mentalities pose great challenges for harnessing the continuously claimed

BIM merits. Delaying knowledge integration till later project stages and varying the assignment of responsibility to different organizations (i.e. model responsibility vs. design responsibility) can not only result in more conflicts and inconsistencies, but also delay their detection beyond the point of effective resolution.

The findings also reveal that where early full integration of the supply chain occurred prior to contract award, the organizations were required to operate at risk until a formal subcontract has been awarded by the main DB firm. This affects the willingness of the supply chain to participate in a more integrated process given that they would not be covered under any insurances should any errors occur (Looi, 2013). In conclusions, the five case studies provided empirical evidence that value realization cannot be fully attained through combining BIM and DB procurement methods unless the entire supply chain is integrated earlier and unified towards common project goals under contractual arrangements that cater for these needs.

SOLUTIONS AND RECOMMENDATIONS

To represent the current actual level of attainable collaboration in BIM-based projects under DB procurement, the model depicted in Figure 5 is proposed.

Clash Detection and Avoidance was identified to be one of the key areas that could generate cost and time reduction whilst improving coordination (Mcgraw-Hill Construction, 2012; Leite et al, 2010; Bockstael &

Issa, 2016). When this potential value enhancing area is placed in tandem with the principles of Macleamy (2008) and AIA (2007), detecting clashes later in the project stages entails detrimental impact on time and cost during the construction stage. In alignment with the objectives of this chapter, it is necessary to identify any

Figure 5. Current state of BIM Stage 2 in DB projects

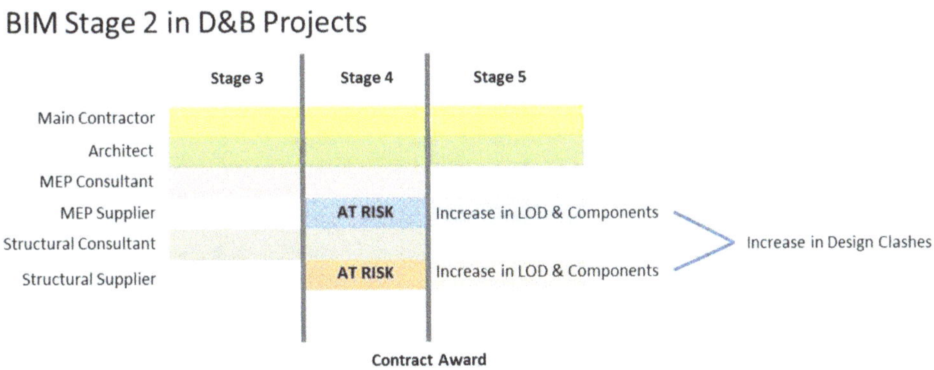

alterations that are required to the DB process to facilitate BIM and its influence on delivering enhanced value. In this regard, a proposed BIM Stage 2 model is provided in Figure 6. The model proposes that the appointment of key construction design packages of Structural and MEP be procured in preconstruction prior to construction contract award. Supply chain selection can be based upon success criteria rather than a lump sum cost which would then place more emphasis on generating valuable outcomes (SEC, 2007; Integrated Project Initiatives [IPI], 2014). This process could also be applied to the DB structure for construction design critical packages, such as Structural and MEP. In engaging with the supply chain sooner during preconstruction while also having a contractual agreement in place to support this involvement and protect organizations against unnecessary risk, the anticipated increase in the number of components, LOD, and consequently the potential increase in clashes could be identified and addressed much sooner in the DB process (AIA, 2007; Macleamy, 2008).

FUTURE RESEARCH DIRECTIONS

The limited literature works surrounding the use of BIM on DB projects necessitates further research to be conducted on this topic to expand the existing body of knowledge and enhance the performance of BIM-based DB projects. The following are suggested areas that are worthy of exploration:

- Increasing the number of case studies and explore potential statistical correlations between the identified independent variables affecting value realization;

Figure 6. Proposed changes for BIM Stage 2 in DB projects

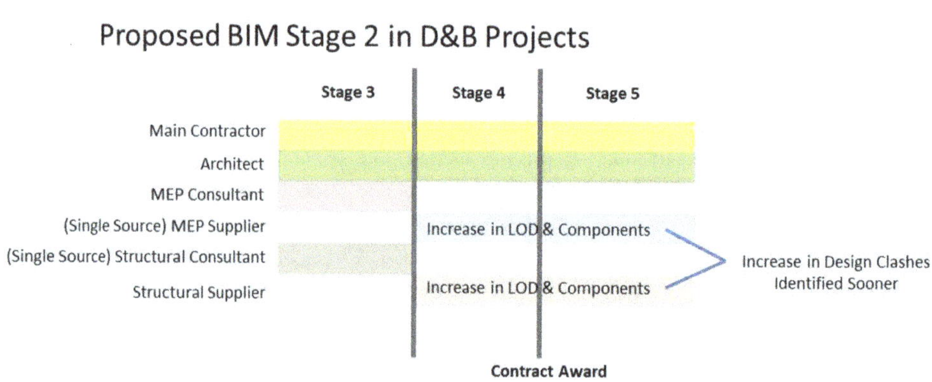

- Investigating the maturity capability of the supply chain organization and its effect on collaboration and value realization;
- The development of metrics for the Utility Value of BIM for main contractors that consider the subjective nature of value; and
- The investigation of contractual structures and BIM protocols for DB projects that eliminate the requirement for supply chain members to operate at risk prior to a subcontract agreement.

CONCLUSION

Prior studies suggest that the realization of value in BIM-based projects depends on the integration and maturity of the involved supply chain as well as the contractual arrangements that accommodate this integration. While existing research constantly regards the virtues of BIM in achieving enhanced value for stakeholders through its collaborative approach, little evidence and relevant data is available in the literature, in particular on BIM-based project using DB procurement. This chapter therefore explored the opportunities and challenges facing value realisation in BIM-based projects procured through Design and Build method. Two utility criteria (i.e. Clash Detection and Avoidance, and Project Team Integration), deemed useful in enhancing use value on projects, were investigated. Analyzing the transition from the preconstruction to the construction stage of five BIM-based DB projects revealed a disconnection between the two stages. Much of the detailed construction design and the appointment of construction designers still occur at the construction stage which is in contradiction to the early engagement and integration principles that are idealized in the literature. The element of risk attached to the current state of the supply chain's early involvement in DB projects is worthy of note. The case studies showed that as the LOD and number of components increased in supply chain packages, the number of clashes followed suit in most of the case study projects. Additionally, the results showed a change in the organizations responsible for the design and the corresponding 3D BIMs (models) and an incomplete integration of organizations in the preconstruction stage before the contract award.

Accordingly, an early involvement of the supply chain within preconstruction whilst eliminating the need for consultancy cost in the same stage has the potential to provide a structure that can deliver an enhanced value. However, two other factors need to be considered when adopting this approach: (1) the willingness of employers to invest more in the upfront design to accrue potential benefits during construction is still not widely proven; and (2) the earlier selection of supply chain members in preconstruction has the potential to compromise commercial competitiveness, therefore selection of such subcontractors need ensure that both value and commercial competitiveness are achieved.

Based on the findings from the five case studies and notions within the literature review, a model was proposed to address the identified challenges affecting the value realization in BIM-based DB projects. The proposed model suggests anticipating the appointment of key construction designers from current stage 5 (construction) to the earlier design phases (stage 4) in DB projects. This will eliminate the current at-risk-operation for the concerned stakeholders; increase their contribution to collaborative workings, and enhance value realization. However, challenges related to the BIM maturity of these stakeholders and the commercial competitiveness of procurement remain to be addressed.

ACKNOWLEDGMENT

We acknowledge the support of Wilmott Dixon Construction Ltd in providing data for this research and supporting the lead author in conducting this study.

REFERENCES

AIA California Council. (2008). *Integrated project delivery frequently asked questions*. Retrieved March 12, 2017, from http://aiacc.org/wpcontent/uploads/2010/07/AIACC_1108FAQ.pdf

Aibinu, A., & Papadonikolaki, E. (2016). BIM implementation and project coordination in Design-Build procurement. In *Proceedings of the 32nd Annual ARCOM Conference* (pp. 15-24). Manchester, UK: Association of Researchers in Construction Management.

American Institute of Architects. (2007). *Integrated Project Delivery: A Guide*. New York: American Institute of Architects.

Association of Project Management. (2012). *APM body of knowledge* (6th ed.). Buckinghamshire, UK: Association of Project Management.

Azhar, S. (2011). Building Information Modeling (BIM): Trends, benefits, risks, and challenges for the AEC industry. *Leadership and Management in Engineering*, *11*(3), 241–252. doi:10.1061/(ASCE)LM.1943-5630.0000127

Ballard, G., Jung, W., Kim, Y., & Han, S. (2012). Understanding of target value design for integrated project delivery with the context of game theory. In *Proceedings of Construction Research Congress 2012*, (pp. 556-563). West Lafayette, IN: American Society of Civil Engineers

Bercerik-Gerber, B., & Kent, D. C. (2010). Understanding construction industry experience and attitudes toward integrated project delivery. *Journal of Construction Engineering and Management*, *136*(8), 815–825. doi:10.1061/(ASCE)CO.1943-7862.0000188

Best, R., & De Valence, G. (Eds.). (1999). *Building in value: predesign issues*. Sydney, Australia: Routledge.

BIM Industry Working Group. (2011). *A report for the government construction client group Building Information Modelling (BIM) working party strategy paper*. Retrieved from http://www.bimtaskgroup.org/wp-content/uploads/2012/03/BIS-BIM-strategy-Report.pdf

Bockstael, D., & Issa, M. (2016). A methodology for contractor clash detection using Building Information Modelling on commercial construction projects. *Journal of Information Technology in Construction*, *21*, 233–249.

Bouazza, T., & Greenwood, D. (2017). Revolutionizing the AEC Industry through the use of BIM and KM. In *Proceedings of The Ninth International Conference on Construction in the 21st Century (CITC-9)*. Dubai, UAE: Academic Press.

British Standards Institution. (2013). *PAS1192-2:2013 Specification for information management for the capital/delivery phase of construction projects using building information modelling*. London, UK: British Standards Institution.

Bryde, D., Broquetas, M., & Volm, J. M. (2013). The project benefits of Building Information Modelling (BIM). *International Journal of Project Management*, *31*(7), 971–980. doi:10.1016/j.ijproman.2012.12.001

Cabinet Office. (2011). *Government Construction Strategy*. Retrieved November 5, 2017 from: https://www.gov.uk/government/uploads/system/uploads/attachment_data/file/61152/Government-Construction-Strategy_0.pdf

Chang, C. Y., Pan, W., & Howard, R. (2015). *Will the application of Building Information Modelling increase acceptance of integrated delivery systems in China?* London, UK: University College London.

Cohen, J. (2010). *Integrated Project Delivery: Case studies*. Sacramento, CA: AIA California Council.

Constructing Excellence. (2009). *Never waste a good crisis: a review of progress since rethinking construction and a thought for our future*. London, UK: Quick Brown and Fox Ltd.

Construction Industry Council. (2006). *Design quality indicators*. Retrieved May 14, 2017 from www.dqi.org.uk/DQI/Common/history.html

Construction Industry Training Board. (2016). *UK industry performance report based on the UK construction industry key performance indicators*. Watford, UK: Construction Industry Training Board.

Darrington, J. (2011). Using a design-build contract for lean integrated project delivery. *Lean Construction Journal*, *7*, 85–91.

Demian, P., & Walters, D. (2013). The advantages of information management through Building Information Modelling. *Construction Management and Economics*, *32*(12), 1153–1165. doi:10.1080/01446193.2013.777754

Design-Build Institute of America. (2017). *What is Design-Build?* Retrieved April 21, 2017 from https://www.dbia.org/about/Pages/What-is-Design-Build.aspx

Eadie, R., Odeyinka, H., Browne, M., McKeown, C., & Yohanis, M. (2014). Building Information Modeling adoption: An analysis of the barriers to implementation. *Journal of Engineering and Architecture*, *2*(1), 77–101.

Eastman, C., Teicholz, P., Sacks, R., & Liston, K. (2008). *BIM handbook: a guide to Building Information Modeling for owners, managers, designers, engineers, and contractors*. John Wiley & Sons. doi:10.1002/9780470261309

Egan, J. (1998). *Rethinking construction: the report of the construction task force*. London, UK: HMSO.

Eisenmann, S., & Park, B. (2012). Building Information Modeling's impact on team performance. In *Proceedings of the 48th ASC Annual International Conference*. Normal, IL: Associated Schools of Construction.

Fallon, C. (1971). *Value analysis* (2nd ed.). Washington, DC: Wiley-Interscience.

GCCG. (2011). Government Construction Client Group: BIM Working Party Strategy Paper. GCCG.

Her Majesty's Government. (2013). *Construction 2025: industrial strategy for construction – Government and industry in partnership*. London, UK: Her Majesty's Government.

Hickethier, G., Tommelein, I. D., & Lostuvali, B. (2013). Social network analysis of information flow in an IPD-project design organisation. In Proceedings of the 21st Annual Conference of the International Group for Lean Construction (IGLC). Fortaleza, Brazil: International Group for Lean Construction.

Ilozor, B., & Kelly, D. (2011). Building Information Modelling and Integrated Project Delivery in the commercial construction industry: A conceptual study. *Journal of Engineering, Project, and Production Management*, *2*(1), 23–36.

Karasulu, Y., Pishdad-Bozorgi, P., & Moghaddam, E. H. (2013). Advancing target price and target value design process in IPD using BIM and risk-sharing approaches. In *Proceedings of the 49th ASC Annual International Conference*. San Luis Obispo, CA: Associated Schools of Construction.

Kassem, M., Iqbal, N., Kelly, G., Lockley, S., & Dawood, N. (2014). Building Information Modelling: Protocols for collaborative design processes. *Journal of Information Technology in Construction*, *19*, 126–149.

Kassem, M., Jenaban, M., Craggs, D., & Dawood, D. (2016). A tool for the assessment of project compliance with Level 2 BIM. *16th International Conference on Construction Applications of Virtual Reality*, Hong Kong.

Kelly, J. (2007). Making client values explicit in value management workshops. *Construction Management and Economics*, *25*(4), 435–442. doi:10.1080/01446190601071839

Knotten, V., Svalestuen, F., Hansen, G. K., & Laedre, O. (2015). Design management in the building process - a review of current literature. *Procedia Economics and Finance*, *21*, 120–127. doi:10.1016/S2212-5671(15)00158-6

Knutt, E. (2016). *Low confidence on level 2 BIM rests on weak uptake of '8 pillars': BIM + survey*. Retrieved April 21, 2017 from http://www.bimplus.co.uk/news/bim-survey-lowconfid3enc4e-limi1ted-experience/

Korsgaard, C. M. (1986). Aristotle and Kant on the source of value. *Ethics*, *96*(3), 486–505. doi:10.1086/292771

Koskela, L., & Kazi, A. S. (2003). Information technology in construction: how to realize the benefits? In S. Clarke, E. Coakes, M. G. Hunter, & A. Wenn (Eds.), Socio-technical and human cognition elements of information systems (pp. 60-75). Hershey, PA: Idea Group Inc (IGI) Publishing. doi:10.4018/978-1-59140-104-9.ch004

Latham, M. (1994). *Constructing the team*. London, UK: HMSO.

Leite, F., Akcamete, A., Akinci, B., Atasoy, G., & Kiziltas, S. (2011). Analysis of modeling effort and impact of different levels of detail in building information models. *Automation in Construction*, *20*(5), 601–609. doi:10.1016/j.autcon.2010.11.027

Lindblad, H. (2013). *Study of the implementation process of BIM in construction projects: analysis of the barriers limiting BIM adoption in the AEC industry* (Master's Thesis). KTH Royal Institute of Technology, Stockholm, Sweden.

Looi, M. (2013). *Integrated project insurance – an inevitable step to BIM implementation*. Retrieved April 21, 2017 from: http://www.lexology.com/library/detail.aspx?g=249c6f9c-c364-4364-a3b6-eb4881e903f3

Love, P. E., Matthews, J., Simpson, I., Hill, A., & Olatunji, O. A. (2014). A benefits realization management building information modeling framework for asset owners. *Automation in Construction*, *37*, 1–10. doi:10.1016/j.autcon.2013.09.007

MacLeamy, P. (2008). *BIM, BAM, BOOM! How to build greener, high-performance buildings*. Urban Land Green Magazine.

Male, S., Kelly, J., Gronqvist, M., & Graham, D. (2007). Managing value as a management style for projects. *International Journal of Project Management*, *25*(2), 107–114. doi:10.1016/j.ijproman.2006.09.001

Matthews, O., & Howell, G. (2005). Integrated Project Delivery: An example of relational contracting. *Lean Construction Journal*, *2*(1), 46–61.

McCormack, K., & Lockamy, A. III. (2004). The development of a supply chain management process maturity model using the concepts of business process orientation. *Supply Chain Management*, *9*(4), 272–278. doi:10.1108/13598540410550019

McGraw-Hill Construction. (2012). *SmartMarket report - the business value of BIM in North America*. Bedford, MA: McGraw-Hill Construction.

Moore, G. E. (1998). The conception of intrinsic value. In J. Rachels (Ed.), *Philosophical Studies*. Oxford, UK: Oxford University Press.

National Building Specification. (2014). *BIM management handbook*. Newcastle, UK: National Building Specification.

Office of Government Commerce. (2007). *Management of risks: Guidance for practitioners*. Office of Government Commerce.

Owen, R., Amor, R., Palmer, M., Dickinson, J., Tatum, C., Kazi, A. S., ... East, B. (2010). Challenges for Integrated Design Delivery Solutions. *Architectural Engineering and Design Management*, *6*(4), 232–240. doi:10.3763/aedm.2010.IDDS1

Papadonikolaki, E., Vrijhoef, R., & Wamelink, H. (2015). BIM adoption in integrated supply chains: a multiple case study. In *Proceedings of the 31st Annual ARCOM Conference* (pp. 631-640). Lincoln, UK: Association of Researchers in Construction Management.

Park, R. J. (1999). *Value engineering: a plan for invention.* CRC Press.

Pena, W., & Parshall, S. (2001). *Problem seeking: an architectural programming primer* (4th ed.). New York: John Wiley and Sons Inc.

Project Management Institute. (2013). *The project management body of knowledge* (5th ed.). Project Management Institute Inc.

Raisbeck, P., Millie, R., & Maher, A. (2010). Assessing Integrated Project Delivery: a comparative analysis of IPD and alliance contracting procurement routes. In *Proceedings of the 26th Annual ARCOM Conference* (pp. 1019-1028). Leeds, UK: Association of Researchers in Construction Management.

Redmond, A., Hore, A., Alshawi, M., & West, R. (2012). Exploring how information exchanges can be enhanced through cloud BIM. *Automation in Construction, 24*, 175–183.

Royal Institute of British Architects. (2013). *Plan of work.* Retrieved April 14, 2017 from http://www.ribaplanofwork.com/Download.aspx

Saxon, R. (2013). *Growth through BIM.* London, UK: Construction Industry Council.

Shahrin, F., & Johansen, E. (2013). Challenges in engaging the client during the capture, translation, transformation, and delivery (CTTD) of client requirements (CR) within the Building Information Modelling (BIM) environment. In *Proceedings of the 7th Nordic Conference on Construction Economics and Organisation*, (pp. 469-479). Trondheim, Norway: Academic Press.

Smith, D., & Tardiff, M. (2009). *Building Information Modeling: a strategic implementation guide for architects, engineers, constructors, and real estate asset manager.* John Wiley and Sons. doi:10.1002/9780470432846

Specialist Engineering Contractors. (2007). *Supplementary memorandum submitted by Specialist Engineering Contractors' Group.* Retrieved April 22, 2017 from http://www.publications.parliament.uk/pa/cm200708/cmselect/cmberr/127/127we57.htm

Succar, B. (2009). Building information modelling framework: A research and delivery foundation for industry stakeholders. *Automation in Construction, 18*(3), 357–375. doi:10.1016/j.autcon.2008.10.003

Succar, B., & Kassem, M. (2015). Macro-BIM adoption: Conceptual structures. *Automation in Construction*, *57*, 64–79. doi:10.1016/j.autcon.2015.04.018

Sveikauskas, L., Rowe, S., Mildenberger, J., Price, J., & Young, A. (2014). *Productivity growth in construction*. U.S. Bureau of Labor Statistics.

Talebi, S. (2014). Exploring advantages and challenges of adaptation and implementation of BIM in project life cycle. In Proceedings of the 2nd BIM International Conference on Challenges to Overcome (pp. 1–20). Lisbon, Portugal: BIM Forum.

Teicholz, P., Goodrum, P. M., & Haas, C. T. (2001). U.S. Construction labor productivity trends. *Journal of Construction Engineering and Management*, *127*(5), 427–429. doi:10.1061/(ASCE)0733-9364(2001)127:5(427)

Thiry, M. (1997). *Value management practice*. Sylva, NC: PMI Publications.

Thomson, D., Austin, S., Devine-Wright, H., & Mills, G. (2003). Managing value and quality in design. *Journal of Building Research & Information*, *31*(5), 334–345. doi:10.1080/0961321032000087981

Thyssen, M. H., Emmitt, S., Bonke, S., & Christoffersen, A. K. (2008). Managing client values in construction design. *Proceedings of Joint CIB Conference W102 – Information and Knowledge Management in Building / W096 - Architectural Management "Improving Performance"*.

ADDITIONAL READING

Akponeware, A., & Adamu, Z. (2017). Clash detection or clash avoidance? An investigation into coordination problems in 3D BIM. *Buildings*, *7*(3), 75. doi:10.3390/buildings7030075

Al Hattab, M., & Hamzeh, F. (2015). Using social network theory and simulation to compare traditional versus BIM–lean practice for design error management. *Automation in Construction*, *52*, 59–69. doi:10.1016/j.autcon.2015.02.014

Eadie, R., Browne, M., Odeyinka, H., McKeown, C., & McNiff, S. (2013). BIM implementation throughout the UK construction project lifecycle: An analysis. *Automation in Construction*, *36*, 145–151. doi:10.1016/j.autcon.2013.09.001

Hickethier, G., Tommelein, I. D., & Gehbauer, F. (2012). "Reducing rework in design by comparing structural complexity using a Multi Domain Matrix." Proc. 20th Ann. Conf. Int'l. Group for Lean Constr., San Diego, CA.

Kassem, M., & Succar, B. (2017). Macro BIM adoption: Comparative market analysis. *Automation in Construction*, *81*, 286–299. doi:10.1016/j.autcon.2017.04.005

Parn, E., Edwards, J., & Sing, M. (2018). Origins and probabilities of MEP and structural design clashes within a federated BIM model. *Automation in Construction*, *85*, 209–219. doi:10.1016/j.autcon.2017.09.010

Solihin, W., Eastman, C., & Lee, Y. C. (2016). A framework for fully integrated building information models in a federated environment. *Advanced Engineering Informatics*, *30*(2), 168–189. doi:10.1016/j.aei.2016.02.007

Tsai, M., Md, A. M., Kang, S., & Hsieh, S. (2014). Workflow re-engineering of design-build projects using a BIM tool. *Zhongguo Gongcheng Xuekan*, *37*(1), 88–102. doi:10.1080/02533839.2012.751302

KEY TERMS AND DEFINITIONS

Building Information Modeling: A set of technologies, processes, and policies enabling multiple stakeholders to collaboratively design, construct, and operate a Facility in virtual space. As a term, BIM has grown tremendously over the years and is now the "current expression of digital innovation" across the construction industry.

Clash Detection: A process of digitally identifying, inspecting and reporting conflicts between disciplines' models through the use of BIM technologies and workflows.

Collaboration: A process of jointly working with others towards a common vision or goal.

Design-Build: A project delivery method whereby design and construction are performed by one entity.

Integrated Project Delivery: A collaborative approach for delivering a project where all team members work as one entity to deliver value.

Level of Definitions: A BIM metric encapsulating the levels of detail (LOD; graphical content of models) and the levels of information (non-graphical content of models).

LOD 300: The model element is graphically represented within the model as a specific system, object, or assembly in terms of quantity, size, shape, location, and orientation. Non-graphic information may also be attached to the model element.

LOD 400: The model element is graphically represented within the model as a specific system, object, or assembly in terms of size, shape, location, quantity, and orientation with detailing, fabrication, assembly, and installation information. Non-graphic information may also be attached to the model element.

LOD 500: The model element is a field verified representation in terms of size, shape, location, quantity, and orientation. Non-graphic information may also be attached to the model elements.

Value Management: A management approach for satisfying stakeholders' requirements and achieving their perceived values.

Chapter 8
GML-Based nD Data Management With a Big Geo Data Semantic World Modeling Approach

Juergen Rossmann
RWTH Aachen University, Germany

Martin Hoppen
RWTH Aachen University, Germany

Arno Buecken
RWTH Aachen University, Germany

ABSTRACT

3D simulation applications benefit from realistic and exact forest models. They range from training simulators like flight or harvester simulators to economic and ecological simulations for tree growth or succession. The nD forest simulation and information system integrates the necessary methods for data extraction, modeling, and management of highly realistic models. Using semantic world modeling, tree data can efficiently be extracted from remote sensing data – even for very large areas. Data is modeled using a GML-based modeling language and a flexible data management approach is integrated to provide caching, persistence, a central communication hub, and a versioning mechanism. Combining various simulation techniques and data versioning, the nD forest simulation and information system can provide applications with historic 3D data in multiple time dimensions (hence nD) as well as with predicted data based on simulations.

DOI: 10.4018/978-1-5225-5625-1.ch008

INTRODUCTION

At 3D GeoInfo 2012, we presented an innovative and efficient way to generate "Virtual Forests" from remote sensing data (Bücken & Rossmann, 2013). Individual trees are delineated from normalized digital surface models and annotated with height and species. This approach is the first step towards various forestall simulation applications based on real-world data like the simulation of forest machines (Figure 1), a flight simulator, a tree growth or a succession simulation. To provide a basis for an efficient and modern data management of such vast datasets, a database-driven method for 3D simulation systems previously presented at 3D GeoInfo 2010 is used (M Hoppen, Rossmann, Schluse, & Waspe, 2010). It provides a persistence layer and a common data schema for simulation systems. Now, it is enhanced by techniques for database-driven, distributed data management and simulation, for data versioning and for the use of big, heterogeneous geo data.

In this revised work, we focus on the integration, enhancement, and on future trends regarding these two core technologies of a large-scale nD forest simulation and information system. In particular, algorithms for the attribution of the individual tree, details on the GML-based (Open Geospatial Consortium (OGC), n.d.), object-oriented schema family ForestGML for forestry data, and the concept of database-driven communication are presented. Overall, a shared world model is

Figure 1. Driver training by means of an advanced forest machine simulator

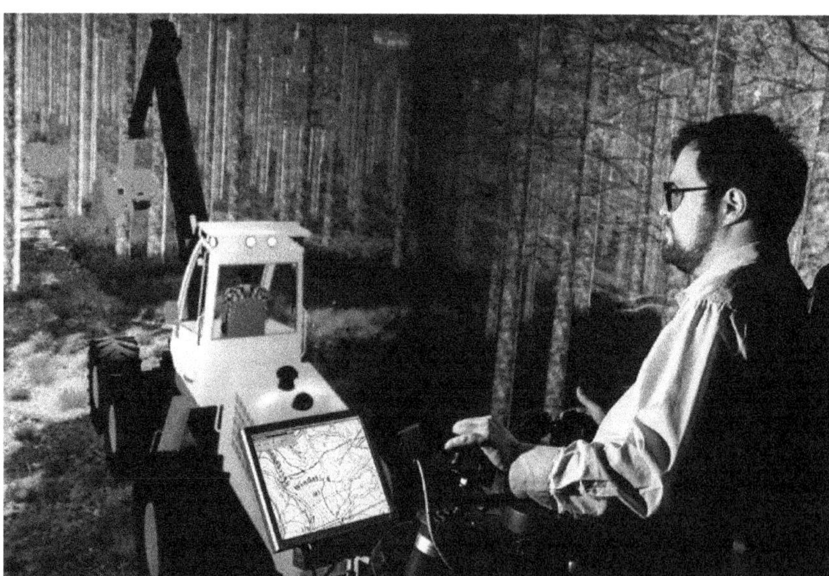

efficiently managed in a geo database and filled using modern techniques of semantic world modeling. The latter transform remote sensing data into a semantic object representation that can be used for the various simulation scenarios as mentioned above. Furthermore, data versioning can be used to analyze past scenarios like a windthrow, where the corresponding storm loss must be calculated. In this context, multiple time dimensions (hence nD) are introduced to the system. Furthermore, even simulated or predicted future states can be managed in a database for conservation, analysis, and comparison. These two concepts – simulation and versioning – add multiple time dimensions yielding an nD forest simulation and information system. Furthermore, given the performance of today's database systems, it even becomes feasible to use the presented system for a multi-client simulation. Here, different clients are simultaneously working with the shared world model, while their actions' effects are distributed over the very same active geo database system.

The work at hand is organized as follows. In the next section, we give an overview of related work. In Section "Single Tree Delineation and Attribution", the tree extraction approach is introduced and current results are presented. Subsequently in Section "Database Interface", the database interface is introduced, including database versioning and data streaming. Here, we give an insight into the systems nD capabilities as well as current developments, and show how the database interface and the tree extraction interact. Subsequently, we give details on the new ForestGML schema family used for data modeling in the Virtual Forest in Section "Data Modeling". A selection of applications benefitting from realistic tree data is presented in Section "Applications". Finally, we conclude this work in the last section.

RELATED WORK

There are several approaches for the delineation and attribution of individual trees from remote sensing data documented in literature. (Garcia, Suárez, & Patenaude, 2007) compare four common algorithms for single tree delineation from nDSM (normalized digital surface model) datasets and point out their advantages and disadvantages. (Reitberger, 2010) provides algorithms that work on full waveform data. The volumetric algorithm used in this paper focuses on nDSM data. It can detect up to 90 percent of the trees in a spruce forest that is ready to harvest. (Hyyppä & Inkinen, 1999) estimate the diameter at breast height (DBH) of a tree depending on its height and crown diameter, but do not specify the parameters of their heuristic as it might vary for different areas.

Concerning the database synchronization between a 3D system and a central database presented in this paper, there are comparable approaches whose motivation is, in parts, similar. Yet, none of them focuses on all the aspects relevant for this work. In various rather basic approaches regarding database usage, databases store additional information (meta information, documents, films, positions, hierarchical structure …) on scene objects (Borgatti et al., 2004; Damer et al., 2004; Forte & Kurillo, 2010; Guan, Ren, & Zhong, 2012; Guarnieri, Pirotti, & Vettore, 2010; Manoharan, Taylor, & Gardiner, 2002; Martina & Bottino, 2012; Pacheco & Wierenga, 2014; Richards-Rissetto et al., 2012; Sendler, 2009; Shiratuddin & Thabet, 2011; Verein Deutscher Ingenieure (VDI), 2002; S. Wang et al., 2010; Zhao et al., 2013).

While this approach suffices for many scenarios, for our application, the whole 3D model should be managed within databases to obtain all their benefits for all objects and their properties. This allows to query models, distribute them between databases, realize versioning on them, and enforce their schema throughout all participating sub systems to make it available as a common language. While other approaches do store the scene data itself in databases, they often lack an object-oriented data modeling approach, which is ideally suited for 3D model data. Moreover, 3D model data should be modeled in application specific schemata (e.g., ForestGML for forest models or CityGML for city models). Yet, many other approaches use a scene-graph-like, generic geometric model, mostly with attributes (Damer et al., 2004; Haist & Coors, 2005; Kamiura, Oisol, Tajima, & Tanaka, 1997; Schweber, 1998; Vakaloudis & Theodoulidis, 1998; Van Maren, Germs, & Jansen, 1998; Walczak, 2012). Here, making the data management approach flexibly adapt to different schemata is limitedly supported using import and export to different file formats (Doboš & Steed, 2012; Haist & Coors, 2005; Scully, Doboš, Sturm, & Jung, 2015; Z. Wang, Cai, & Bu, 2014). In contrast, others support different or flexible schemata, e.g., a schema alteration approach that adds attributes to generic base objects (Van Maren et al., 1998) or approaches supporting different static (Haist & Coors, 2005) or dynamic (Kamiura et al., 1997; Schmalstieg et al., 2007) schemata. Most of these approaches focus on one specific field of application and support only a corresponding set of (fixed) schemata. This very common approach is described in many publications, especially from the field of Building Information Modeling (BIM) (Beetz, van Berlo, de Laat, & van den Helm, 2010; Domínguez-Martín, 2014; Hoerster & Menzel, 2015; Kang & Lee, 2009; Malaikrisanachalee & Vathananukij, 2011; Nour, 2016; Tarandi, 2011). Similarly, Product Data Management (PDM) systems (Sendler, 2009; Verein Deutscher Ingenieure (VDI), 2002) support arbitrary schemata. However, they are not explicitly established within a database schema. They use a so-called "black box integration" approach that manages only file metadata in a database while the data itself is stored in secured file vaults. Such vaulting approaches can also extensively be found for pure 3D applications (Eisenmann, Fuchs, De Wilde, & Basso, 2012; Fang, Yan, Wenhui, & Sen, 2012; Iliescu, Ciocan, & Mateias, 2014; Mahdjoub, Monticolo, Gomes, & Sagot, 2010; Roberts, Ducheneaut, & Smith, 2010). Again, while such approaches fulfill the needs of the respecting applications, the presented work needs to store the whole 3D models using database technology. Regarding data distribution, many other approaches also have a distributed architecture. Usually, this includes multiuser support, a client-server model, or access control and rights management. Only few of them use distributed database technology (Schmalstieg et al., 2007; Schweber, 1998; Takemura, Kitamura, Ohya, & Kishino, 1993) with databases for the client, which is advantageous for distributing the common data schema or for local querying. Temporal data management or versioning is also supported by some approaches and mostly requires special schema structures or is based on file vaulting (Beetz et al., 2010; Doboš & Steed, 2012; Eisenmann et al., 2012; Iliescu et al., 2014; Lobur, Matviykiv, Kernytskyy, & Dobosz, 2011; Malaikrisanachalee & Vathananukij, 2011; Scully et al., 2015; Sendler, 2009; Shahabi, Banaei-Kashani, Khoshgozaran, Nocera, & Xing, 2010; Stadler, Nagel, König, & Kolbe, 2009; Tarandi, 2011; Vakaloudis & Theodoulidis, 1998; Verein Deutscher Ingenieure (VDI), 2002; Z. Wang et al., 2014). These restrictions limit their usage for arbitrary data schemata.

Thus, while there are many similar applications integrating database technology, none of them is sufficiently comprehensive for managing nD geo data for a forest simulation and information system.

SINGLE TREE DELINEATION AND ATTRIBUTION

The approach presented at 3DGeoInfo 2012 (Bücken & Rossmann, 2013) efficiently generates individual trees with very limited user interaction. The idea of this algorithm is to associate a volume to each local maximum in the nDSM and to assign the local maxima to the groups "tree" and "lateral branch" based on their volumes. There are several ways how the total volume in the nDSM can be distributed among the local maxima. Figure 2 shows some possibilities. In the left distribution, the total volume of each local maximum mainly depends on the depth of the intersection between the peaks. Therefore, this distribution is not suitable, because for a tree with a split crown both parts will be classified as "lateral branch". The distribution in the middle leads towards a behavior similar to the watershed algorithm (Diedershagen, Koch, Weinacker, & Schütt, 2003). While the watershed algorithm judges only based on the ground area of a local maximum, this variant would base its decision on the ground area and the height of a segment. The disadvantage is that the volumes of both segments are quite similar to each other. If we consider the size distribution between the trees in the forestry unit, it is difficult to find a suitable threshold for all trees when using this distribution. The distribution on the right hand side assigns the common basal volume of two peaks to the dominant one. Therefore, the dominant

Figure 2. Several possible distributions of the volume

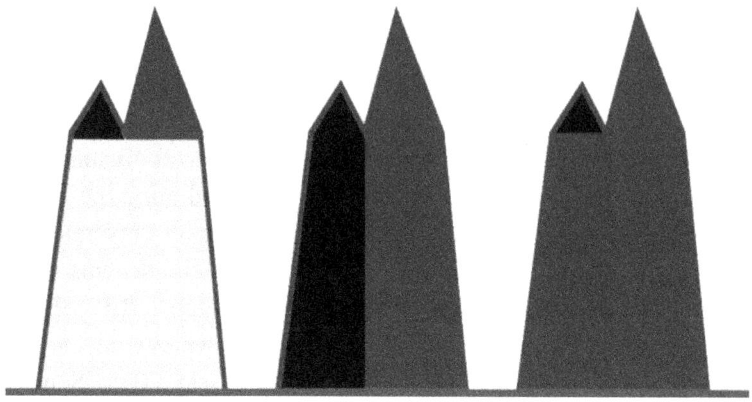

peak will be classified as a tree and the smaller peak will only be classified as an additional tree, if its own volume is significant enough.

Figure 3 shows an example of how the nDSM for a sample tree is assigned to the local maxima. A flow simulation that starts with the highest maximum can calculate this distribution. The typical runtime of this algorithm is $O(n^2)$, but with an implementation that takes advantage of the limited height of a tree and the denumerable amount of height levels in a tree (typically centimeter levels), it is possible to accelerate this algorithm to $O(n)$. This more advanced implementation belongs to the class of plane sweep algorithms.

Figure 3. The distribution for the nDSM of a single tree

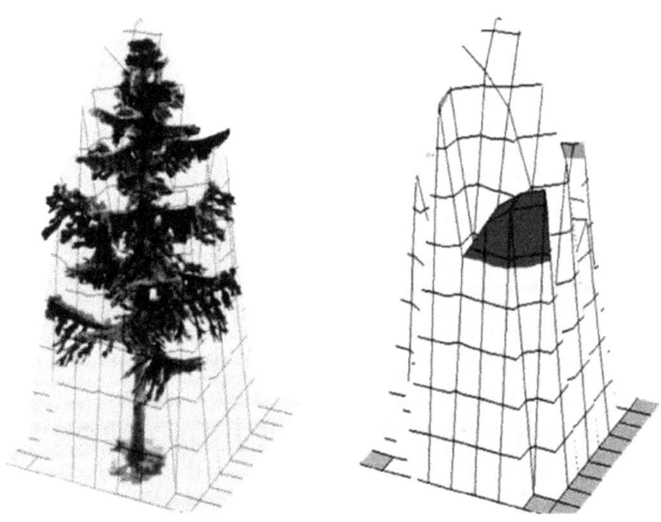

In our previous work, we also showed how the threshold required to distinguish between the classes "tree" and "lateral branch" can be estimated with a heuristic based on the receiver operator characteristic.

The trees generated with the volumetric algorithm carry only the information directly available from the given data layers, e.g., their height, their crown area and their species. Other information like the DBH is still missing. However, realistic forest models require the horizontal dimension of the trees to prevent a virtual environment from looking like an endless repetition of the same scene.

Figure 4. Comparison of two nDSM for the same forest unit

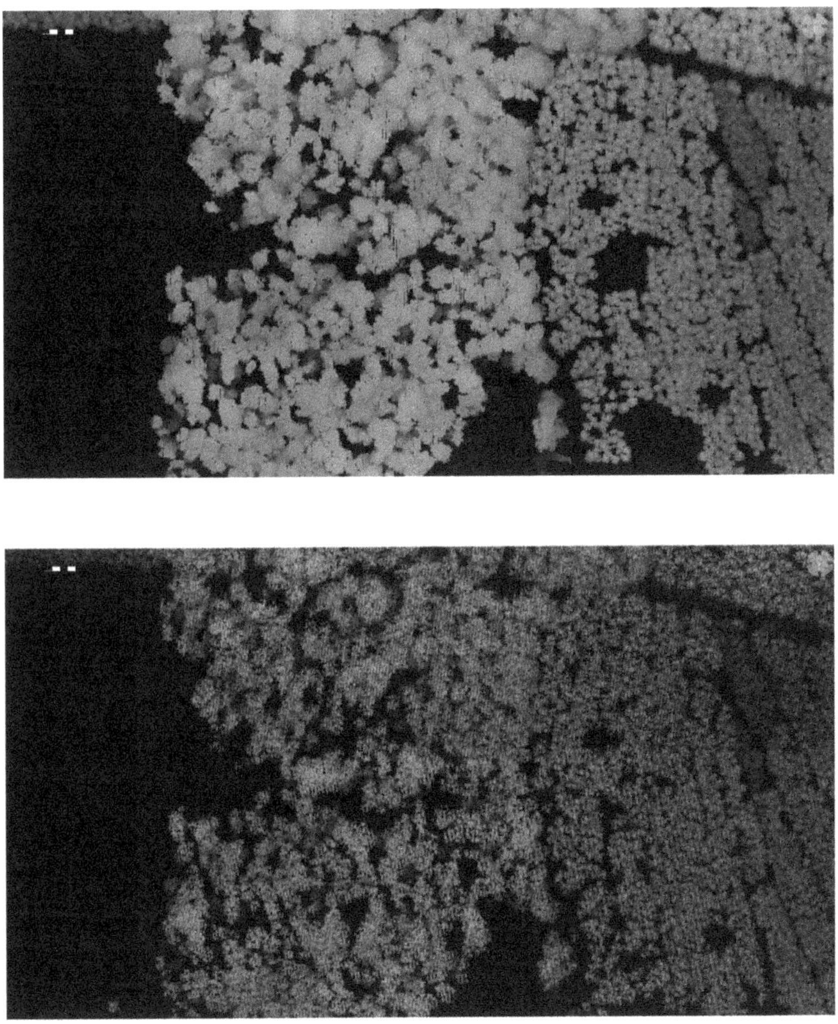

But how does this algorithm scale when examining whole states? When we designed the approach, the term "big geo data" referred to an area of a couple of hundred square kilometers of high resolution data. Nowadays, it refers to high resolution data for multiple 10,000 square kilometers and even larger areas. This data is inhomogeneous as it is produced by different sensors with different resolutions, different optical properties and most important different times of flight. Figure 4 contains an example of LIDAR data for one forestry unit which origins from two different flight campaigns and already shows how the data differs. The bottom example shows a clearly visible noise as more measurement points are located on the ground. There are several reasons leading to these effects: A smaller footprint of the laser or a different threshold for the first echo can produce these effects, as the laser better penetrates the canopy. Also, a measurement at a different time of the year could lead to the same effect, especially in broadleaf forests.

A state is not covered by a single reading. For example, North Rhine-Westphalia renews a third of its aerial images and one sixth of its LIDAR information per year. It is even possible that multiple companies collect the data within the same year and use different sensor systems.

The introduced approach already handles this inhomogeneous data as it is based on the nDSM and not on the point cloud. The only question to be solved is, how we can generate a consistent nDSM for large scale areas considering multiple flight campaigns from several years. Therefore, we established a high performance tool chain, which can process more than 100km^2 of data per hour on a standard desktop computer. This is possible by using the optimized fractal interpolation algorithm introduced in (Buecken & Rossmann, 2011) instead of a nearest neighbor, Voronoi or Delaunay interpolation. This algorithm is able to produce results free from artifacts, can consider inhomogeneously distributed gaps of varying sizes and uses height information from all around a gap to interpolate missing content. In addition to this algorithm, we implemented heuristics to detect small gaps that produce unwanted noise and mark them for interpolation. With this approach, we produce a consistent nDSM for a whole federal state, which is suitable for the proposed individual tree delineation. The calibration approach based on the receiver operator characteristic is used per flight area, as this procedure can eliminate noise effects, but still allows a variation in the width of trees. Figure 5 illustrates this effect. A wide beam diameter can produce measured points well outside of the tree crown.

The flight date is often provided as a meta information for a whole dataset or as a comma separated list, which contains the date for each individual tile of 1km^2. To handle this important data, we embed it into the geo data itself. The GeoTIFF format does not feature EXIF information as it is common in most photography image

Figure 5. Wide laser beams can produce LIDAR points outside of the tree crown

formats as JPEG or PNG. However, there are data fields reserved for metadata, one of which is the "date of production." We used this metadata field to annotate each GeoTIFF tile with the exact flight date.

This information is important when producing a consistent model. As the flight date of the used remote sensing data within one model can vary and the trees in a forestry unit will have grown in this time, it is important to address the period of time since the recording the data and to calculate the growth in between. We use taper curves or yield tables to estimate the change within this time. This way, it is possible to have a consistent model, even if the geo data is recorded at different times.

When a high resolution full waveform scan of a forestry unit is available, it is possible to detect the stems and their diameter directly from the remote sensing data (Reitberger, 2010). However, when modeling large areas, there are usually only rather lowly resolved nDSMs available. Therefore, we needed to find a heuristic for the DBH that uses attributes that can be monitored directly from remote sensing data, and that provides realistic estimates. In this work, we present a heuristic for the class of spruces in our test areas in the federal state of North Rhine-Westphalia. Laser scanner data with approximately 6 points per square meter, as well as ground truth data recorded during the last state forest inventory was used.

In this sample plot forest inventory, 10,489 spruces were measured including individual heights and DBH values at the height of 1.30m above the ground. From those trees, the ones located in state-owned forestry units were selected because the state inventory provides additional data for these units, including the yield class required in the process. Unrealistic datasets were manually filtered from this result (e.g., trees with a height of 3m and a stem diameter of 90cm). 971 trees remained for the heuristic's parameterization. These trees were located in all areas of North Rhine-Westphalia.

Crown areas of the sample trees were not recorded during the state inventory. Thus, a simple heuristic was used to estimate them. The basal area for the forestry unit can be calculated from the age and the yield class, which are both recorded in the state forest inventory. The basal area of the individual sample tree was set into relation with the basal area of the stand. We suppose that the share of the crown area in relation to the area of the forestry unit equals this percentage. Together with this information, a complete dataset is available, where each sample tree is annotated with its individual tree height, crown area and DBH. DataFit 9 ("DataFit Product Features," n.d.) was used to calculate a relation between these parameters. The strong correlation is shown in Table 1.

DataFit delivers the model given in Equation (1) for the relation between height *h*, crown area *A* and diameter at breast height *DBH* with the parameters given in Table 2.

$$\mathrm{DBH}(h,A) = a + bh + cA + dh^2 + eA^2 + fhA + gh^3 + hA^3 + ihA^2 + jh^2A \quad (1)$$

DBH values calculated with this model show a standard deviation of 3.26cm when compared to the recorded DBH values. Figure 6 shows a plot of the generated function. The sample points from the recorded data are marked with small black dots and a line to show the difference between the calculated and the recorded value.

The residual graph for the generated function (Figure 7) also shows only a small difference between the calculated and the measured DBH.

A different dataset was used to verify the generated heuristic. It was recorded in the project Virtual Forest II in forestry unit 121B1 in Schmallenberg-Schanze located in the Sauerland, North Rhine-Westphalia. The dataset contains more than 1000 spruces measured during a full record of the unit. We segmented the individual trees for this unit (see (Bücken & Rossmann, 2013) for the segmentation results),

Table 1. The distribution for the nDSM of a single tree

	Height	Crown Area	DBH
Height	1	0.7478184412	0.8851391766
Crown Area	0.7478184412	1	0.8968172646
DBH	0.8851391766	0.8968172646	1

Table 2. Parameters for Equation (1)

Parameter	Value
a	0.134758834716376
b	$-6.38872262480706*10^{-3}$
c	$5.37391200303616*10^{-3}$
d	$4.10037527413944*10^{-4}$
e	$-3.26290891771389*10^{-4}$
f	$6.43778614242134*10^{-4}$
g	$-7.6951423469473*10^{-6}$
h	$1.48843462166253*10^{-6}$
i	$1.38137983019296*10^{-6}$
j	$-6.28555527339635*10^{-6}$

Figure 6. Graph of the generated function

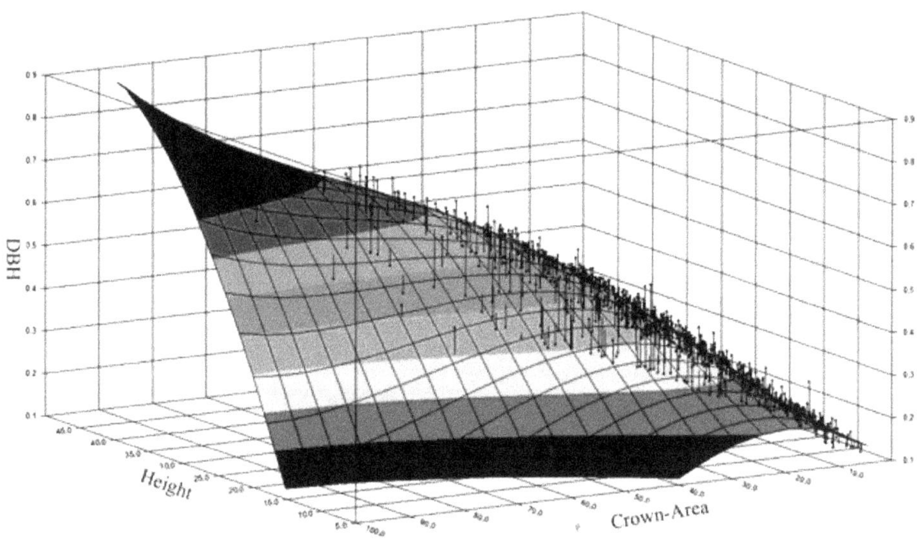

calculated the DBH with the formula given above and compared the distribution of breast height for the measured (dark gray) and calculated (light gray) dataset in a histogram (Figure 8). It turns out that the generated heuristic models the attribute DBH in a quality more than sufficient for simulation purposes.

As the data used to calibrate this curve is not based on remote sensing data but on ground truth, it is not required to adapt this to the parameters of flight campaigns. The different beam diameters can produce varying crown areas, but this only

Figure 7. Residual graph of the generated function

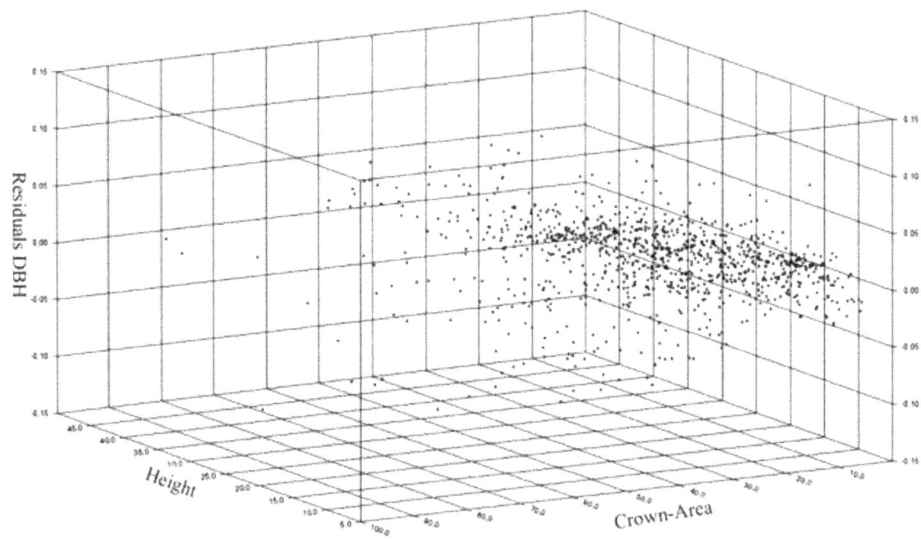

happens to trees, which are at the border of a unit or which stand individually as the boundaries between crowns can still be accessed even with larger beam diameters. For this limited number of trees, we have to accept the resulting error, especially as the beam diameter is usually not provided and the effect on the result of inventories is very limited due to the small amount of trees with these properties.

DATABASE INTERFACE

In applications from the forestry sector, vast amounts of geo data and other data has to be dealt with. Thus, a modern forest simulation and information system can only be successful when it can efficiently manage, access, and model such data. This makes the usage of database technology essential.

Database Synchronization

As previously introduced in (M Hoppen et al., 2010; Martin Hoppen, Schluse, Rossmann, & Weitzig, 2012), an approach for database synchronization can be used to connect a 3D simulation system with a central database (Figure 9). The 3D simulation system's runtime database is used as real-time capable cache that drives the 3D simulation itself. In contrast, the central database provides persistence and

Figure 8. Comparison of DBH values calculated from remote sensing data (light gray) with the terrestrial data (dark gray)

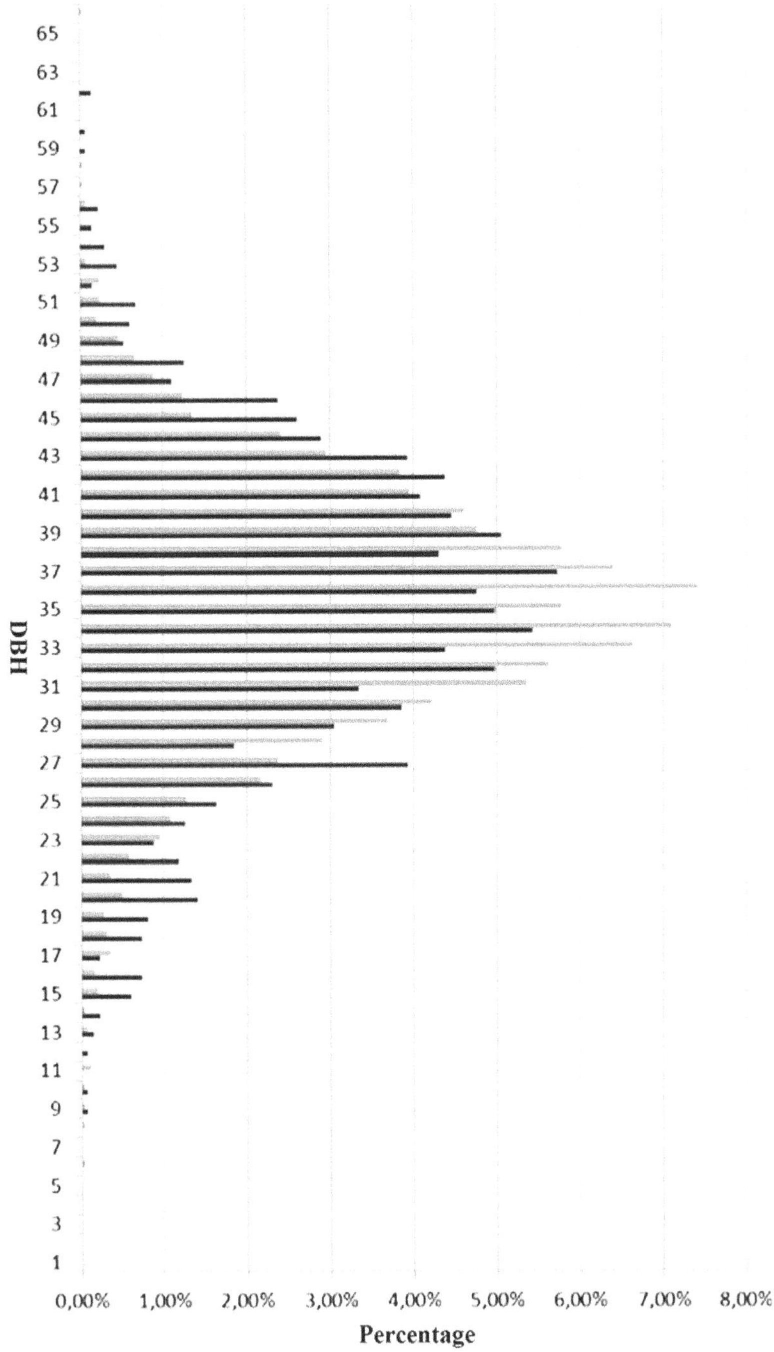

versioning, a common application schema, and an integrated communication hub for collaboration or distributed simulation.

Synchronization is performed on three levels: schema, data and functional level. Thus, not only objects are replicated from the central database to the internal simulation database and kept "in sync". At first, the simulation system adopts the data schema of the central database by synchronization to its internal simulation database. This offers an advantage over the immediate usage of objects from the central database. By using replicated objects in the simulation database, details of the central database are hidden from all internal components of the simulation system (particles, physics, sensor data processing, rendering, etc.) allowing for transparent data access, real-time visualization and simulation. As mentioned above, the internal simulation database works as a cache, speeding up repeating access patterns. These would otherwise lead to repetitive queries to the central database. The central database also provides a persistence layer to the simulation system. This is because changes made to replicate objects can be resynchronized back to the original counterpart objects within the central database.

An important part of data synchronization is change tracking and bidirectional resynchronization. Here, both databases' notification services are used to track changes of original (central database) or replicate (simulation database) objects. Upon resynchronization, the collected information is used to bring such pairs of objects back "in sync". Thus, not only changes in the simulation database are resynchronized

Figure 9. Schema and data synchronization between a central (geo) database and a simulation database

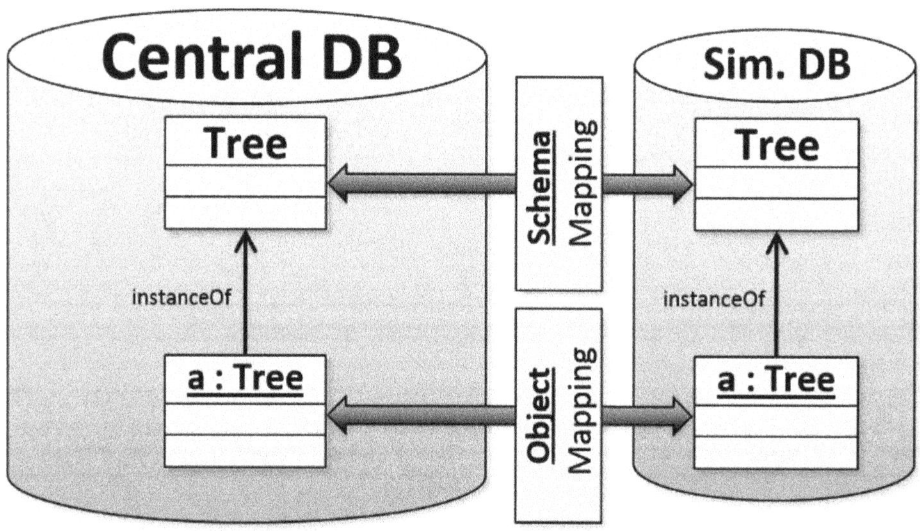

to the central database, but also vice versa. This way, the central database can now also be used as an active communication hub. It can drive a distributed simulation or be used to implement distributed algorithms. Here, data changes can trigger different clients to compute a certain part of a distributed computation. More details on database synchronization and the used databases Versatile Simulation Database (VSD) and SupportGIS Java (SGJ) are given in (Martin Hoppen et al., 2012).

Database Versioning

Another feature of the presented approach for database synchronization is versioning (M Hoppen, Schluse, Rossmann, & Averdung, 2015). Thus, the fourth and even more dimensions of time are not only supported by the several simulation techniques of the underlying simulation system. Additionally, the central SGJ database provides a mechanism to store timestamped object lifespans and versions of attribute values. Versioned data can be accessed by the database interface in terms of temporal snapshots. User interfaces or other components of the simulation system can set a reference time and the database interface takes care of updating the corresponding replicates in the simulation database. The change of reference time leads to an update of the replicated data to match the chosen time.

Using database versioning, change processes within the central database and its connected simulation clients can be recorded over time, archived, replayed, and analyzed. An important basis for this is the change distribution mechanism mentioned above. When a simulation client, connected to the central database, changes a local replicate copy of an object, it synchronizes the change back to the original object within the central database. This change is tracked by the aforementioned versioning mechanism. It is also reported to all other connected clients that will adopt the change in their corresponding local replicate copies. Thus, by managing all objects in the central database, all changes can be globally recorded. A simple example is given in Figure 10. The central database contains a tree object with an attribute "felled" initially set to "false". The object is synchronized to two simulation clients. On client #1, the tree has just been felled ("felled" becomes "true"). The change was notified by its simulation database and synchronized to the central database. There, the previous value "false" is versioned with the current timestamp. Finally, the central database notified the change. In the next step, client #2 would adopt the change from the central database, thus also marking its replicate copy of the tree as felled.

After a simulation, one of the simulation clients can be used to access the historic data as mentioned above. For example, to replay a recorded movement of a vehicle where each database change represents a version of a position attribute, the client will iterate through the corresponding time frame. At each step, it will query its database interface to update all its replicate copies to the respective point in time.

Figure 10. Example for DB-driven communication including versioning

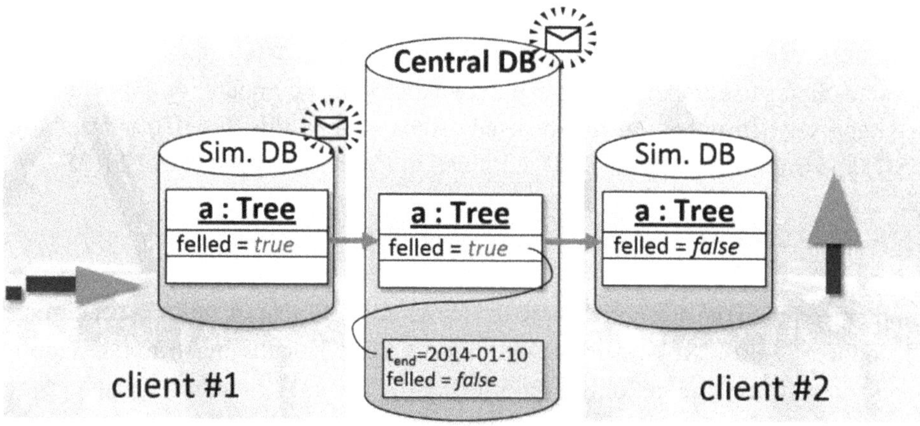

Another usage scenario is managing forest inventory data. Here, the central database holds a forest model while the simulation clients are used to access and update the data for inventory measures. Local changes are synchronized to the central database and the objects' previous states are versioned.

In a simple versioning mechanism, a standard approach using one time dimension, the system time of the database management system, is used. Here, the corresponding timestamps cannot be chosen, but always correspond to the point in time a change is committed to the central database. This is called the "transaction time" and the database a "transaction time database" (Elmasri & Navathe, 2010). Our novel approach for nD versioning (M Hoppen et al., 2015) extends this to additional time dimensions, allowing to specify arbitrary points in time. One of these time concepts is "valid time" t_V. A database supporting only this time dimension is called a "valid time database". A database combining both concepts is called a "bitemporal database". A third time dimension t_E can be used to represent the effectiveness of a value in terms of its official availability to other users. This notion is often used in the context of forest inventory data were properties of the managed objects are typically updated step-by-step during the year but the new values are not available until an effective date, e.g., December 31. Nevertheless, the captured values shall be associated with the corresponding point in (valid) time of the real world.

Another usage for this more flexible versioning approach is to insert simulated or predicted values with future timestamps into a database. This makes results persistent and allows them to be analyzed using database techniques. By using a second, orthogonal time dimension like the transaction time or a dedicated simulation time, multiple simulation runs can be inserted side by side into the database and compared. As an example, results from multiple runs of a tree growth simulation (see Applications) with different parameters could be compared.

Besides timestamps, other metadata is identified to be necessarily managed along with each version. Usually, the user is interested in the question "Who has changed what, when and why?". The "when" and the "what" are recorded by the versioning mechanism itself. The "who" and the "why" have to be integrated into the metadata in terms of a process specification.

To define the necessary behavior of the versioning system, requirements are specified using the example of a tree database. An exemplary sequence of database operations is specified:

- CREATE: On 2013-06-01, user "A" inserts a tree object with height 10m (and a derived DBH of 10cm) for t_V 2014-01-01, t_E 2014-01-01
- UPDATE: On 2014-06-01, user "B" updates the tree height to 12m (and a derived DBH to 12cm) for t_V 2015-01-01, t_E 2015-01-01
- UPDATE: User "B" made a mistake. On 2014-06-02, he updates the tree height to 12.1m (and a derived DBH to 12.1cm) for t_V 2015-01-01, t_E 2015-01-01
- DELETE: The tree is felled on 2015-06-01. Driver "C" of the harvester measures a height of 14m and a DBH of 14cm. Subsequently, the tree is deleted. Both events are stored for t_V 2015-06-01, t_E 2015-06-01
- UPDATE: Forestry scientists find a new formula for the relation between tree height and DBH that is applied to the database. The update is performed on 2015-08-01, t_E 2016-01-01. For all values of t_V, scientist "D" calculates the following DBH values: 2014-01-01: 10.5cm, 2015-01-01: 12.5cm
- UPDATE: An old data sheet is found containing exact measurements for 2014-02-01: On 2016-06-01, official in charge "E" updates the tree height to 10.1m (and a derived DBH of 10.1cm) for t_V 2014-02-01, t_E 2016-06-01.

Based on this example, queries and their intended results are specified. Standard access patterns ("day-to-day business"): Get the current value of the reference tree as created and updated above:

- SELECT: t_V 2013-01-01, t_E 2013-01-01: NULL
    - (i.e., no such object exists)
- SELECT: t_V 2014-01-01, t_E 2014-01-01: (10m, 10cm)
- SELECT: t_V 2014-03-01, t_E 2014-01-01: (10m, 10cm)
- SELECT: t_V 2015-01-01, t_E 2015-01-01: (12.1m, 12.1cm)
- SELECT: t_V 2015-06-01, t_E 2015-06-01: (14m, 14cm)
- SELECT: t_V 2016-01-01, t_E 2016-01-01: NULL

Besides these queries for the tree's state at a certain tuple of points in time, the versioning log itself can also be queried. Queried on 2017-01-01: Get metadata on all changes of "C":

- UPDATE, on 2015-06-01, t_V 2015-06-01, t_E 2015-06-01:
- "C", (14m, 14cm), job 4711, harvesting measure
- DELETE, on 2015-06-01, t_V 2015-06-01, t_E 2015-06-01:
- "C", (14m, 14cm), job 4711, harvesting measure

Altogether, this can be formalized in terms of precedence rules for the different time dimensions (M Hoppen et al., 2015). An exemplary graphical representation of these rules is given in Figure 11.

Figure 12 shows a user interface to access the versioning mechanism in a forest inventory component of the nD forest simulation and information system. The user provides the system with reference date values. Forest data is loaded from the central database using these timestamps and changes to the data are connected with these timestamps. Thus, the reference dates (for each time dimension) are

Figure 11. Exemplary precedence rule for the nD versioning approach (M Hoppen et al., 2015): Expansion of a three-time-dimensional scope stops in direction of valid time axis (V) to allow retroactive updates relative to the V-axis to "slide in between" existing values.

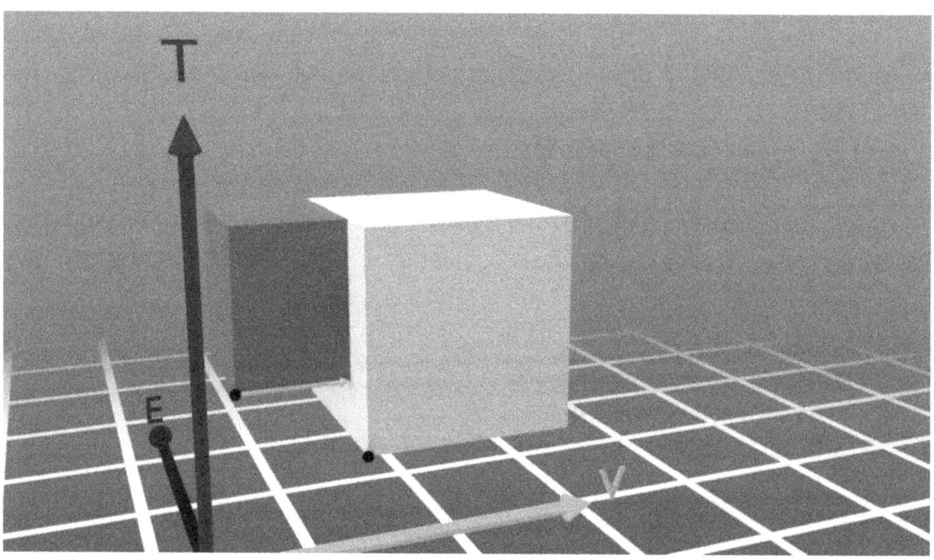

Figure 12. User interface for the versioning mechanism (reference date and change history)

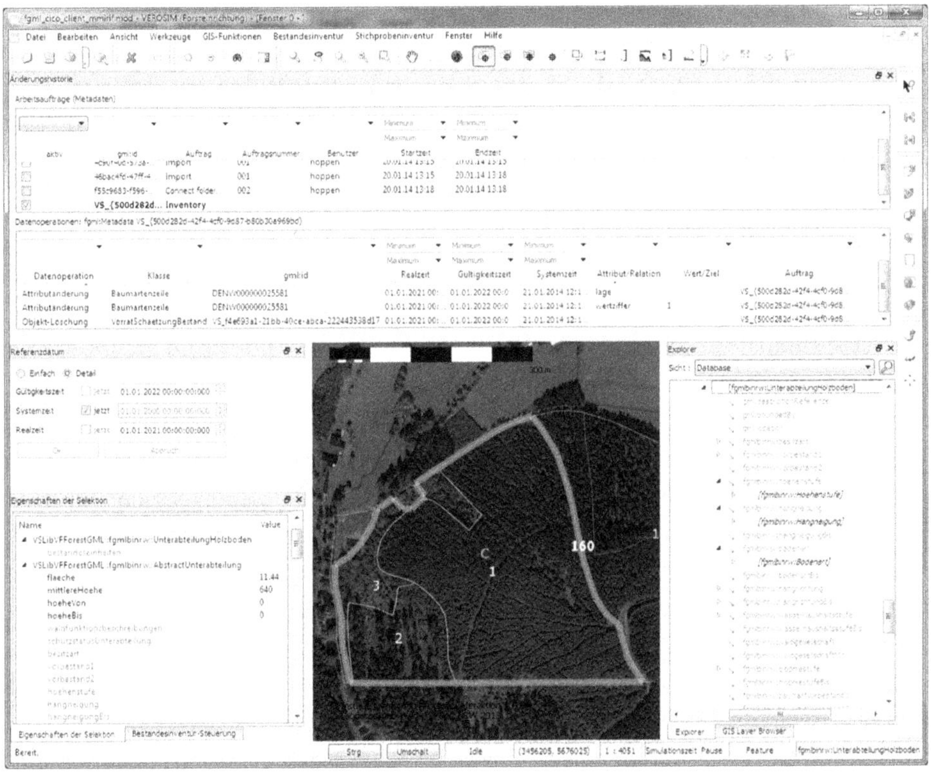

used for reading and writing data. Besides timestamps, changes are also connected with process metadata, e.g., describing a survey or a felling measure. In Section "Data Modeling," the usage of ForestGML for modeling such process metadata is presented. Querying process metadata and associated changes is also integrated into the system's user interface.

Data Streaming

Another feature of the database interface is data streaming. Parts of huge datasets are loaded into the simulation database when required and are unloaded when not required anymore. It is used to save memory in the simulation database and speed up data access by only loading required data. We distinguish tree types of streaming: spatial (or geometrical) streaming, hierarchical streaming and streaming for data processing.

Spatial streaming can be used to virtually move through huge nD datasets like city or forest models. Here, the area is subdivided into tiles. A strategy when moving may be to load all objects within the bounding rectangle of nearby tiles and unload the other objects. Hierarchical streaming is used to access hierarchically structured, object-oriented data with a GUI element like a tree view. While opening a relation between objects will trigger a loading of all associated objects, closing the relation will trigger their unloading. For example, in a CityGML scenario, opening a Building's association to its BuildingParts will load all of them into the simulation database, closing it will unload them.

The third type is streaming for data processing purposes. This approach is used when large amounts of data in the central database shall be processed in smaller subsets within the simulation database. The idea is similar to an iterator (from programming languages) or a cursor (from database technology). Each subset is loaded, processed in some way, resynchronized if necessary, and unloaded. Resynchronization is only applied when either the subset itself is changed or additional objects are created by data processing. In the latter case, such newly created objects may be unloaded as well. Streaming for data processing can use different iteration strategies. It may apply spatial streaming to process data spread over an area, hierarchical streaming to process a subtree structure of objects, or other specialized approaches. For example, the approach is used to apply the algorithms for tree extraction mentioned above to large forest areas. Forest stand objects are incrementally loaded from a stand inventory database. For each stand unit, the corresponding rasterized remote sensing data is retrieved (using another mechanism), tree data is extracted and corresponding tree objects are extracted. These tree objects are synchronized to the tree inventory database. Finally, the current stand object and the newly created tree objects are unloaded before the next stand object is loaded.

DATA MODELING

In the context of the research project Virtual Forest, several GML-3.1.1-based data schemata have been developed to model the different types of inventory or other data. They were mostly derived from existing relational or flat schemata. One of their key properties is the combination of geometry and numerical data into integrated object-oriented data models. This facilitates data integrity, as most other current data models in forest inventory store these two types of data in separate data models as well as data management systems. They inherently cover the hierarchical structure present in most inventory data by allowing to explicitly model objects and their interrelations. The use of GML-based schemata guarantees the data to be open for different applications rather than being locked in proprietary formats. Furthermore, standard OGC web services like WFS can directly be applied.

Based on these initial data models, a follow-up schema family called ForestGML (based on GML-3.2.1) has been developed. Its foundation is a base schema (ForestGML-Base) containing general as well as inventory-specific base classes for stand, sample plot, and single tree inventories or other forest relevant data. On this basis, user-specific schemata are derived (e.g., for stand inventory in North Rhine-Westphalia or Rhineland-Palatinate). Thus, ForestGML is built on four levels:

1. Foundation schema (GML-3.2.1)
2. General, user and inventory independent base schema (ForestGML-Base)
3. Inventory-specific base schema (ForestGML-Base)
4. Concrete, user- and inventory-specific application schemata (e.g., ForestGML-StandInventory-NRW)

Examples for user-specific schemata already developed are stand inventory schemata for different German forestry administrations and a private forest manager, schemata for the NRW-wide sample plot inventory (Landeswaldinventur) and the German-wide sample plot inventory (Bundeswaldinventur), and a schema for growth succession simulation (Siehoff et al., 2011) data.

There are various motivations for developing ForestGML:

- A consistent data exchange format facilitates communication among users,
- A consistent data management minimizes "friction loss" caused by converting between different data formats,
- Based on a common data schema, data processing algorithms can be unified allowing for synergy effects and reuse,
- Using a consistent data format for the different inventory types (stand, sample plot, and single tree inventory) simplifies reuse and combination of components of the Virtual Forest, and
- A consistent base schema for different users (different forestry administrations) allows for the reuse of basic system structures.

Like GML, ForestGML schemata are defined as XML Schemata Files (XSD). However, following the mapping to UML (Unified Modeling Language) as shown in the GML standard's Annex E and F (Open Geospatial Consortium (OGC), n.d.), the structure of ForestGML can also be described in terms of UML class diagrams. The core base classes of ForestGML-Base (from level 1 and 2) are depicted in Figure 13. Two base classes for features (AbstractForestGMLFeature) and other objects (AbstractForestGML) are derived from their GML counterparts, both extended by

Figure 13. The core base classes of ForestGML-Base

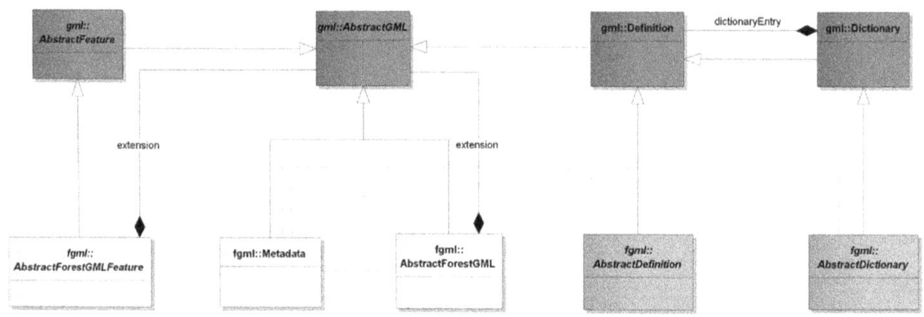

a generic extension property for arbitrary connections to other objects as well as a ForestGML metadata object. For definitions and dictionaries, the corresponding base classes are similarly derived and extended by metadata.

Figure 14 shows another excerpt from ForestGML-Base. Level 2 classes AbstractHierarchy*Feature are used to generically describe hierarchically structured features often found in forestry data. AbstractAttributeSets are used to define property combinations like a tree species code combined with an average height and an area percentage. The features they shall describe reference such data containers. An example for level 3 are stand-inventory-specific base classes. AbstractStand describes a stand feature comprising strata, which describe a stand's vertical structure. An AbstractStratum in turn comprises AbstractStandAttributeSets describing the tree species found in a stratum. Using these base classes, concrete (level 4) application schemata can be built for different users.

ForestGML also allows to model, control and document processes (Figure 15). Specializations of the base class AbstractProcessControl model, e.g., forest inventory or reorganization processes. They allow to define the process's state (planned, active, or finished), a reference and a finalization date, or information regarding associated geo data packages (e.g., containing remote sensing and thematic data) needed in offline processes. More specific information can be added in derived classes. Any AbstractHierarchyFeature can be associated with such a process specification. Moreover, a process control object can be associated with Metadata objects.

When combined with the aforementioned versioning approach, any change is associated with a Metadata object, as well. This allows to store information on the user ("Who changed it?") and the motivation ("Why was it changed?"). Thus, Metadata objects are used to model process steps (e.g., preparation, execution and quality assurance of a forest inventory process) by being associated with the corresponding process control object on the one side and with all the changes representing a process step on the other side.

Figure 14. ForestGML-Base classes for hierarchically structured features, attribute sets, and stand inventory base classes

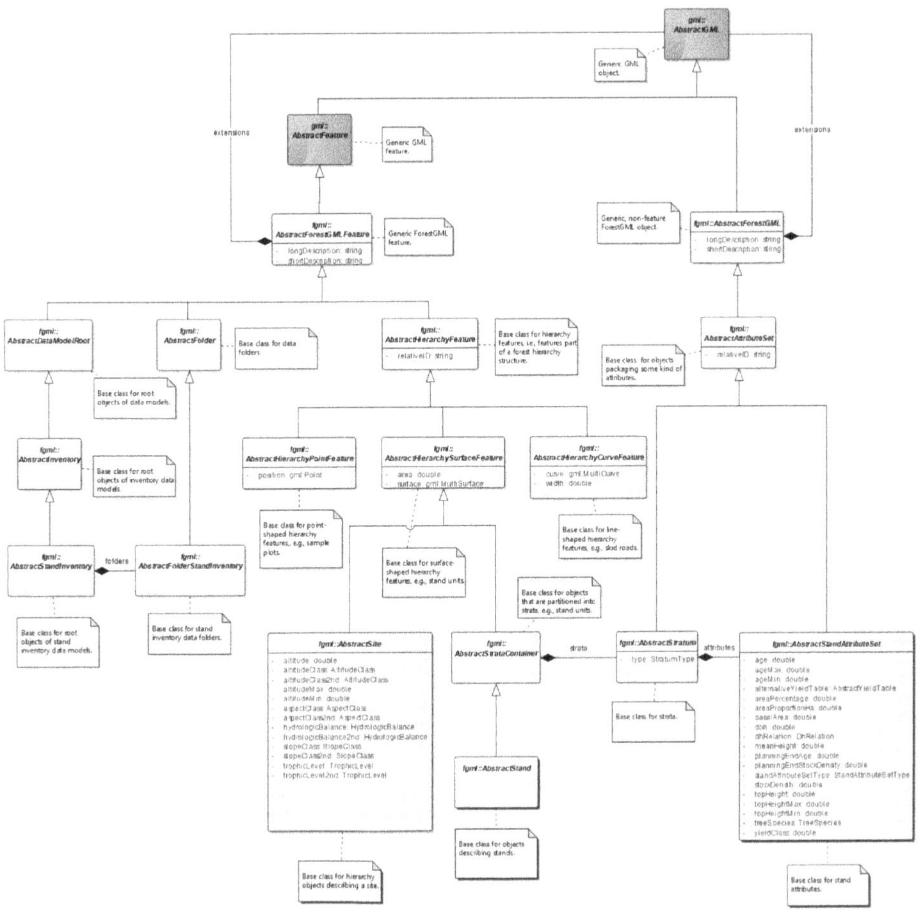

APPLICATIONS

First of all, highly detailed forest models can be used in realistic landscape visualizations. In this section, we introduce some more sophisticated applications benefiting from such models.

VFR (visual flight rules) flight training requires realistic landscape models, especially in areas close to airports. In this context, current commercial flight simulators use polygonal boxes to represent forests. This approach only provides a forest's size as an obstacle but does not allow realism comparable to a real cockpit's view. Realistic, large scale forest models can serve as an alternative. When tree data

Figure 15. ForestGML-Base classes process control

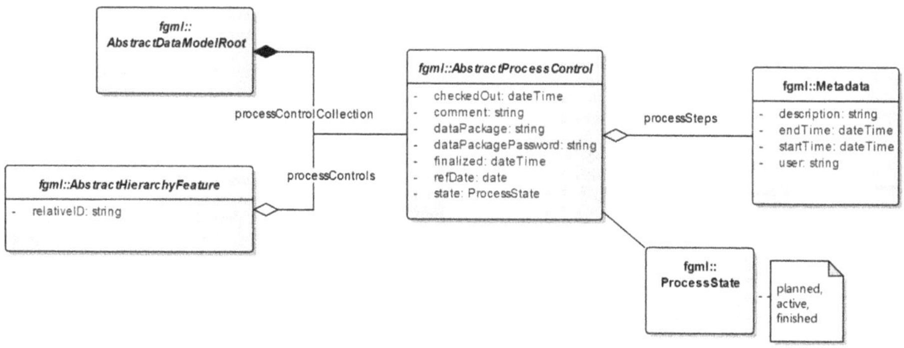

is streamed from a central database as mentioned above, nearly endless landscapes become possible. Figure 16 shows an example of a flight simulator using a highly detailed forest model as a landscape.

As another example, forest machine simulators (Figure 1) require even more realistic models of their environment. Here, the forest model is no longer limited to visualization. An important part in the training of harvester drivers is to learn

Figure 16. A drone simulator based on the described forest model

the physical limits of the forest machine. A harvester has quite strict limitations on the maximum size of a tree that can be felled. The first limitation is the size and maximum opening of the aggregate that grabs the tree. This requires trees with a realistic DBH. The second limitation is the maximum weight that the crane can handle at a certain distance and a certain gradient. When the weight is too high, the harvester will flip over. The weight of a tree can be calculated directly from its volume. There are several approaches in literature specifying the volume of a tree based on its height and DBH. Therefore, the introduced forest model can serve as a training scenario for a forest machine simulator (Rossmann & Jung, 2010).

While the two previous simulators are used for operator training, the next focus on forest development. A tree growth simulator like SILVA (Pretzsch, Biber, & Dursky, 2002) predicts, how a forest will change over the next decades. The owner benefits from such tools by finding an ideal cultivation scenario, e.g., to maximize profit while also ensuring sustainability. The presented forest models build an ideal basis for such simulations (Rossmann, Schluse, Waspe, & Moshammer, 2011). For that purpose, SILVA was integrated into the presented nD forest simulation and information system.

Besides tree growth simulation, succession simulation is another beneficial tool for forest management and planning. GraS/ForS is such a tool, designed to predict the changes in the number, species and size of plants on open areas in the forest (Siehoff, 2011). It uses a raster-based representation of the environment and calculates changes for individual cells over time. The approach goes beyond simple cellular automatons by also considering distant cells. This is needed to model seeds of a tree that can be distributed over more than 100m and thus spreading multiple cells. Figure 17 shows a schematic view of the GraS/ForS model. It uses the individual tree data from our forest model and enriches it with newly grown vegetation. Using the automatically derived tree data, the effort for initializing and parametrizing the GraS/ForS model can significantly be reduced.

Besides simulation, the same forest model can also be used by forest surveyor. Furthermore, in the project Virtual Forest, algorithms were developed estimating other parameters of the forest unit based on the individual tree model.

The last example is a forest navigation system (Rossmann, Krahwinkler, & Buecken, 2009). While GPS delivers poor localization results when used under the canopy, it is possible to use the detailed forest model as a navigation map and to recognize the current area of the forest in this map. This position lookup is similar to using a city map. For example, when seeing a large church in front of you, a city map can be used to determine your position. Tree groups in the forest have a sufficiently different layout in order to recognize them in a precise map. In this application, the GPS position is used as a coarse estimate of the real position. It specifies the area of the forest model that has to be searched for the tree group. A

Figure 17. Schematic drawing of the GraS/ForS model (Rossmann, Buecken, Lennartz, & Hudjetz, 2017)

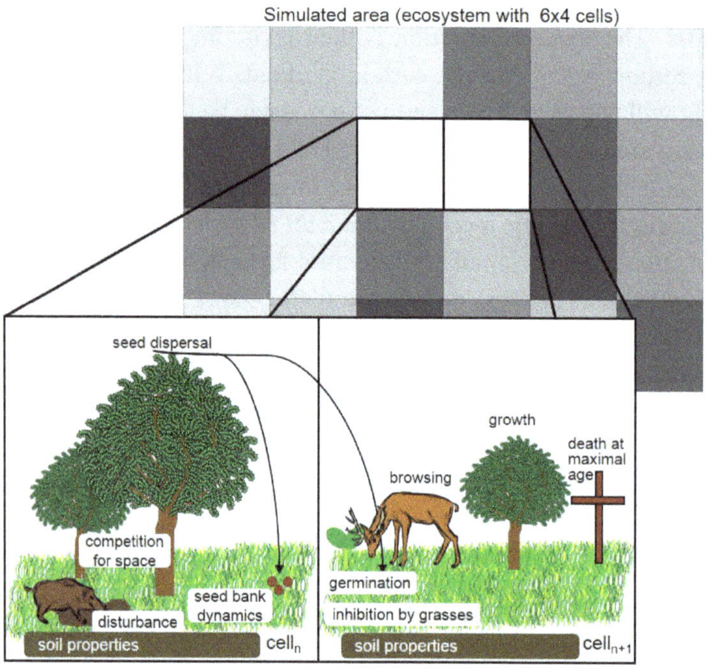

local map of the surrounding tree group is captured with a terrestrial laser scanner or a stereoscopic camera. A particle simulation is used to locate this group. It has been shown that the average positioning error with this method is only about 50cm. Before building this navigation system in reality, it was developed and tested using Virtual Testbed techniques (Rossmann et al., 2011), based on the very same forest model (Figure 18).

These are only five examples of applications that already benefit from the innovative database-driven, highly detailed forest model.

CONCLUSION

This work provides a comprehensive overview of current developments regarding data processing, modeling, and management in a large scale nD forest simulation and information system. It focuses on the integration and cooperation of two of its main components – data management and semantic world modeling based on remote sensing data – and presents various (simulation) applications already benefiting from this approach.

Figure 18. A Virtual Testbed with a wood harvester

Trees are extracted from heterogeneous remote sensing data of large areas and attributed with position, height, crown volume, type and also diameter to derive realistic forest models. This data is managed by a data management approach combining local real-time caching with a central communication hub and data versioning. In this context, an advanced multi-time-dimensional versioning approach is introduced, adding even more flexibility. ForestGML, a GML-based data schema family for forestal data, is introduced, as well. This modeling language, which already found its way into commercial software applications, builds the basis for the system's flexibility towards different users and inventory approaches. Finally, an ever-growing number of applications proves the system's practicability and the benefit of the comprehensive forest model.

ACKNOWLEDGMENT

This project is co-financed by the European Union and the federal state of North Rhine-Westphalia, European Regional Development Fund (ERDF). Europe – Investing in our future.

REFERENCES

Beetz, J., van Berlo, L., de Laat, R., & van den Helm, P. (2010). bimserver.org - An Open Source IFC Model Server. In *Proceedings of the CIB W78 2010: 27th International Conference* (pp. 16–18). Cairo, Egypt: Academic Press.

Borgatti, C., Felicori, M., Mauri, M. A., Calori, L., Guidazzoli, A., Pescarin, S., … Valentini, L. (2004). Databases and virtual environments: a good match for communicating complex cultural sites. In ACM SIGGRAPH 2004 Educators program (p. 30). ACM.

Bücken, A., & Rossmann, J. (2013). From the Volumetric Algorithm for Single-Tree Delineation Towards a Fully-Automated Process for the Generation of "Virtual Forests." *Progress and New Trends in 3D Geoinformation Sciences, Lecture Notes in Geoinformation and Cartography*, 79–99.

Buecken, A., & Rossmann, J. (2011). An Efficient, Fractal-Based, Bi-Linear Algorithm for the Interpolation of Inhomogeneously Distributed Geo-Data. In *RSPSoc Annual Conference "Earth Oberservation in a Changing World."* Academic Press.

Damer, B., Rasmussen, D., Neilson, M., Newman, P., Norkus, R., Clancey, W., … Cochrane, T. (2004). Data-Driven Virtual Environment Assembly and Operation. In *Virtual Ironbird Workshop* (p. 1).

DataFit Product Features. (n.d.). Retrieved December 1, 2017, from http://www.oakdaleengr.com/datafit.htm

Diedershagen, O., Koch, B., Weinacker, H., & Schütt, C. (2003). Combining LIDAR and GIS Data for the Extraction of Forest Inventory Parameters. In J. Hyyppä, E. Naesset, H. Olsson, T. Granqvist Phalén, & H. Reese (Eds.), *Proceedings of the Scandlaser scientific workshop on airborne laser scanning of forests* (pp. 157–165). Umeå, Sweden: Academic Press.

Doboš, J., & Steed, A. (2012). Revision Control Framework for 3D Assets. In Eurographics 2012 - Posters (p. 3). Cagliari, Italy: Academic Press.

Domínguez-Martín, B. (2014). *Methods to process low-level CAD plans and creative Building Information Models (BIM)*. University of Jaén.

Eisenmann, H., Fuchs, J., De Wilde, D., & Basso, V. (2012). ESA Virtual Spacecraft Design. In *5th International Workshop on Systems and Concurrent Engineering for Space Applications*. Academic Press.

Elmasri, R., & Navathe, S. B. (2010). *Database Systems: Models, Languages, Design, And Application Programming* (6th ed.). Prentice Hall International.

Fang, M., Yan, X., Wenhui, Y., & Sen, C. (2012). The Storage and Management of Distributed Massive 3D Models based on G/S Mode. In Lecture Notes in Information Technology (Vol. 10). Academic Press.

Forte, M., & Kurillo, G. (2010). Cyberarchaeology: Experimenting with teleimmersive archaeology. In *2010 16th International Conference on Virtual Systems and Multimedia, VSMM 2010* (pp. 155–162). Academic Press. 10.1109/VSMM.2010.5665989

Garcia, R., Suárez, J. C., & Patenaude, G. (2007). Delineation of individual tree crowns for LiDAR tree and stand parameter estimation in Scottish woodlands. *The European Information Society - Leading the Way with Geo-Information, Lecture Notes in Geoinformation and Cartography*, 55–86.

Guan, T., Ren, B., & Zhong, D. (2012). The Method of Unity3D-Based 3D Dynamic Interactive Query of High Arch Dam Construction Information. *Applied Mechanics and Materials, 256–259*, 2918–2922. doi:10.4028/www.scientific.net/AMM.256-259.2918

Guarnieri, A., Pirotti, F., & Vettore, A. (2010). Cultural heritage interactive 3D models on the web: An approach using open source and free software. *Journal of Cultural Heritage, 11*(3), 350–353. doi:10.1016/j.culher.2009.11.011

Haist, J., & Coors, V. (2005). The W3DS-Interface of Cityserver3D. In Kolbe & Gröger (Eds.), *European Spatial Data Research (EuroSDR) u.a.: Next Generation 3D City Models. Workshop Papers : Participant's Edition* (pp. 63–67). Bonn: Academic Press.

Hoerster, S., & Menzel, K. (2015). BIM based classification of building performance data for advanced analysis. In *Proceedings of International Conference CISBAT 2015 Future Buildings and Districts Sustainability from Nano to Urban Scale* (pp. 993–998). Academic Press.

Hoppen, M., Rossmann, J., Schluse, M., & Waspe, R. (2010). A New Method For Interfacing 3D Simulation Systems And Object-Oriented Geo Data Sources. In T. Kolbe, G. König, & C. Nagel (Eds.), *The 5th International Conference on 3D Geoinformation ISPRS 2010* (pp. 51–56). Shaker Verlag.

Hoppen, M., Schluse, M., Rossmann, J., & Averdung, C. (2015). A new nD Temporal Geodata Management Approach using GML. In Rückemann & Doytsher (Eds.), *GEOProcessing 2015 - The Seventh International Conference on Advanced Geographic Information Systems, Applications, and Services* (pp. 110–116). Lisbon, Portugal: IARIA (International Academy, Research, and Industry Association).

Hoppen, M., Schluse, M., Rossmann, J., & Weitzig, B. (2012). Database-Driven Distributed 3D Simulation. In *Proceedings of the 2012 Winter Simulation Conference* (pp. 1–12). Academic Press.

Hyyppä, J., & Inkinen, M. (1999). Detecting and estimating attributes for single trees using laser scanner. *The Photogrammetric Journal of Finland*, *16*(2), 27–42.

Iliescu, D., Ciocan, I., & Mateias, I. (2014). Assisted management of product data: A PDM application proposal. *Proceedings of the 18th International Conference on System Theory, Control and Computing*. 10.1109/ICSTCC.2014.6982403

Kamiura, M., Oisol, H., Tajima, K., & Tanaka, K. (1997). Spatial views and LOD-based access control in VRML-object databases. In T. Masuda, Y. Masunaga, & M. Tsukamoto (Eds.), Worldwide Computing and Its Applications (Vol. 1274, pp. 210–225). Springer Berlin / Heidelberg.

Kang, H., & Lee, G. (2009). *Development of an object-relational IFC server*. ICCEM/ICCPM.

Lobur, M., Matviykiv, O., Kernytskyy, A., & Dobosz, R. (2011). Presentation of the heat simulation model with use of relational data model. *2011 11th International Conference The Experience of Designing and Application of CAD Systems in Microelectronics CADSM*, 228–229.

Mahdjoub, M., Monticolo, D., Gomes, S., & Sagot, J. (2010). A collaborative design for usability approach supported by virtual reality and a multi-agent system embedded in a PLM environment. *Computer Aided Design*, *42*(5), 402–413. doi:10.1016/j.cad.2009.02.009

Malaikrisanachalee, S., & Vathananukij, H. (2011). Integration of java-based BIM with spatial database. *International Journal of Civil Engineering*, *9*(1), 17–22.

Manoharan, T., Taylor, H., & Gardiner, P. (2002). A collaborative analysis tool for visualisation and interaction with spatial data. In *Proceedings of the seventh international conference on 3D Web technology* (pp. 75–83). ACM. 10.1145/504502.504515

Martina, A., & Bottino, A. (2012). Using Virtual Environments as a Visual Interface for Accessing Cultural Database Contents. In *International Conference of Information Science and Computer Applications (ICISCA 2012)* (pp. 1–6). Bali, Indonesia: Academic Press.

Nour, M. (2016). Using Bounding Volumes for BIM based electronic code checking for Buildings in Egypt. *American Journal of Engineering Research*, *5*(4), 91–98.

Open Geospatial Consortium (OGC). (n.d.). *Open Geospatial Consortium*. Retrieved February 11, 2015, from http://www.opengeospatial.org/

Pacheco, D., & Wierenga, S. (2014). Spatializing experience: a framework for the geolocalization, visualization and exploration of historical data using VR/AR technologies. *Proceedings of the 2014 Virtual Reality International Conference*. 10.1145/2617841.2617842

Pretzsch, H., Biber, P., & Dursky, J. (2002). The Single Tree-Based Stand Simulator SILVA: Construction, Application and Evaluation. *Forest Ecology and Management*, *162*(1), 3–21. doi:10.1016/S0378-1127(02)00047-6

Reitberger, J. (2010). *3D-Segmentierung von Einzelbäumen und Baumartenklassifikation aus Datenflugzeuggetragener Full Waveform Laserscanner*. TU Munich.

Richards-Rissetto, H., Remondino, F., Agugiaro, G., Von Schwerin, J., Robertsson, J., & Girardi, G. (2012). Kinect and 3D GIS in archaeology. In *Proceedings of the 2012 18th International Conference on Virtual Systems and Multimedia, VSMM 2012: Virtual Systems in the Information Society* (pp. 331–337). Academic Press. 10.1109/VSMM.2012.6365942

Roberts, M., Ducheneaut, N., & Smith, T. F. (2010). The "3d Wiki": Blending virtual worlds and Web architecture for remote collaboration. In *2010 IEEE International Conference on Multimedia and Expo, ICME 2010* (pp. 1166–1171). IEEE. 10.1109/ICME.2010.5582579

Rossmann, J., Buecken, A., Lennartz, G., & Hudjetz, S. (2017). Remote Sensing Based Forests Management In Conjunction With A Succession Simulation. In *Proceedings of the SGEM Vienna Green 2017*. Vienna, Austria: Academic Press.

Roßmann, J., Hoppen, M., & Bücken, A. (2014). GML-Based Data Management and Semantic World Modelling for a 4D Forest Simulation and Information System. *International Journal of 3-D Information Modeling*, *3*(3), 50–67. 10.4018/ij3dim.2014070104

Rossmann, J., & Jung, T. (2010). Interactive Dynamics-Based Simulation of Work Machines. *VRCA (Virtual Reality Center Aachen) Progress Report*, *4*, 77–79.

Rossmann, J., Krahwinkler, P. M., & Buecken, A. (2009). Mapping and navigation of mobile robots in natural environments. In *Advances in Robotics Research - Theory, Implementation, Application (German Workshop on Robotics 2009, TU Braunschweig)* (pp. 43–52). Braunschweig: Springer.

Rossmann, J., Schluse, M., Waspe, R., & Moshammer, R. (2011). Simulation in the Woods: From Remote Sensing based Data Acquisition and Processing to Various Simulation Applications. In S. Jain, R. R. Creasey, J. Himmelspach, K. P. White, & M. Fu (Eds.), *Proceedings of the 2011 Winter Simulation Conference* (pp. 984–996). Academic Press.

Schmalstieg, D., Schall, G., Wagner, D., Barakonyi, I., Reitmayr, G., Newman, J., & Ledermann, F. (2007). Managing complex augmented reality models. *IEEE Computer Graphics and Applications*, *27*(4), 48–57. doi:10.1109/MCG.2007.85 PMID:17713234

Scully, T., Doboš, J., Sturm, T., & Jung, Y. (2015). 3drepo. io: building the next generation Web3D repository with AngularJS and X3DOM. *Proceedings of the 20th International Conference on 3D Web Technology*. 10.1145/2775292.2775312

Sendler, U. (2009). *Das PLM-Kompendium: Referenzbuch des Produkt-Lebenszyklus-Managements*. Berlin: Springer; doi:10.1007/978-3-540-87898-8

Shahabi, C., Banaei-Kashani, F., Khoshgozaran, A., Nocera, L., & Xing, S. X. S. (2010). GeoDec: A Framework to Visualize and Query Geospatial Data for Decision-Making. *IEEE MultiMedia*, *17*(3), 14–23. doi:10.1109/MMUL.2010.5692179

Shiratuddin, M. F., & Thabet, W. (2011). Utilizing a 3D game engine to develop a virtual design review system. *Journal of Information Technology in Construction, 16*, 39–68.

Siehoff, S. (2011). *Succession of semi-natural grasslands: spatially-explicit, mechanistic simulation considering various forms of land use*. Aachen: Publikationsserver der RWTH Aachen University.

Siehoff, S., Lennartz, G., Heilburg, I. C., Roß-Nickoll, M., Ratte, H. T., & Preuss, T. G. (2011). Process-based modeling of grassland dynamics built on ecological indicator values for land use. *Ecological Modelling*, *222*(23), 3854–3868. doi:10.1016/j.ecolmodel.2011.10.003

Stadler, A., Nagel, C., König, G., & Kolbe, T. H. (2009). Making interoperability persistent : A 3D geo database based on CityGML. *3D Geo-. Information Sciences*, 175–192. doi:10.1007/978-3-540-87395-2_11

Takemura, H., Kitamura, Y., Ohya, J., & Kishino, F. (1993). Distributed Processing Architecture for Virtual Space Teleconferencing. In *Proc. of ICAT* (*Vol. 93*, pp. 27–32). Academic Press.

Tarandi, V. (2011). The BIM Collaboration Hub: A model server based on IFC and PLCS for Virtual Enterprise Collaboration. In *Proceedings of the CIB W78-W102 2011: International Conference* (pp. 26–28). Sophia Antipolis, France: Academic Press.

Vakaloudis, A., & Theodoulidis, B. (1998). Spatiotemporal Database Connection to VRML. *Proceedings of the 9th UK Electronic Imaging & the Visual Arts Conference, EVA '98.*

Van Maren, G., Germs, R., & Jansen, F. (1998). Integrating 3D-GIS and Virtual Reality Design and implementation of the Karma VI system. *Proceedings of the Spatial Information Research Centre's 10th Colloquium,* 16–19.

Verein Deutscher Ingenieure (VDI). (2002). *VDI 2219 - Information technology in product development Introduction and economics of EDM/PDM Systems (Issue German/English).* Düsseldorf: Verein Deutscher Ingenieure.

Von Schweber, E. (1998). *SQL3D - Escape from VRML Island.* Retrieved February 9, 2015, from http://www.infomaniacs.com/SQL3D/SQL3D-Escape-From-VRML-Island.htm

Walczak, K. (2012). Dynamic Database Modeling of 3D Multimedia Content. In W. Cellary & K. Walczak (Eds.), *Interactive 3D Multimedia Content* (pp. 55–102). London: Springer London; doi:10.1007/978-1-4471-2497-9_4

Wang, S., Mao, Z., Zeng, C., Gong, H., Li, S., & Chen, B. (2010). A new method of virtual reality based on Unity3D. In *2010 18th International Conference on Geoinformatics* (pp. 1–5). Academic Press. 10.1109/GEOINFORMATICS.2010.5567608

Wang, Z., Cai, H., & Bu, F. (2014). Nonlinear Revision Control for Web-Based 3D Scene Editor. In *Virtual Reality and Visualization (ICVRV), 2014 International Conference on* (pp. 73–80). Academic Press.

Zhao, Y., Yan, C., Zhou, X., Zhu, Q., Wang, S., & Guo, W. (2013). The research and development of 3D urban geographic information system with Unity3D. In *Geoinformatics (GEOINFORMATICS), 2013 21st International Conference on* (pp. 1–4). Academic Press. 10.1109/Geoinformatics.2013.6626094

Chapter 9
Investigating Effects of Psychophysical Metrics on Fidelity in 3D Space Visualization

Noha Saleeb
Middlesex University, UK

ABSTRACT

Previous research tests and experiments have provided evidence for the disparity between human perception of space in the physical environment and the 3D virtual environment. This could have dire effects on the decision-making process throughout the whole construction lifecycle of an asset due to non-precision of perceived spaces. Results have shown an infidelity in displaying the actual dimensions of the space in the 3D virtual environment, and previous research by the author has identified the magnitude of this disparity. However, there has been inconclusive reasoning behind the causes for this disparity. This chapter aims to investigate and highlight different psychophysical factors that might cause this difference in perception, and compare these factors with previously investigated research.

INTRODUCTION

Digital space representation in the construction industry is essential to demonstrate prospective buildings to clients, and visualise spaces in way that is as truthful to reality as possible. Such endeavours by designers and architects include 2D CAD designs, 3D parametric representations of spaces and objects using 3D geometric

DOI: 10.4018/978-1-5225-5625-1.ch009

models. In parallel, GIS (Geographic Information Systems) allow non-graphical attributes to be linked to geometric representation through grids or matrices. The current method of depicting visualisation in the construction industry is using 3D Building Information Models (BIMs), which can be linked or federated together or with GIS, with rich non-graphical information attached inside them or to external databases (Isikdag, 2011). The visualisation can either be a solitary model or inside a virtual environment / world e.g. Second Life.

According to Parsons (1995), with these visualisation tools, both quantitative and qualitative information can be represented about spaces. Quantitative information expresses spatial relationships among people and objects e.g. length, height, size etc., in an absolute or numeric manner, while qualitative information provides a "sense of place", e.g. architectural style of building, sounds, urban characteristics (Pereira et al., 2013). However human perception of 3D models' virtual space sizes, represented by this quantitative information, has been evidenced to differ from human perception of the same space in reality that this information represents, as explained subsequently. Typically, 3D visualisations and simulations are chosen by designers to communicate with their clients – illustrating space design ideas, functionalities and sizes. However, if those digital visualisations do not portray size and dimensions of a space truthfully, this gives the client a false perception of what the space would actually look like once built. This might result in wrong decisions at design phase based on incorrect information, which would only be realised after construction is complete, rendering it impossible or expensive to change, causing both usability and financial losses. For example during planning stage, local authority might reject new development permissions due to the 'uncertainty' factor in the appearance or design. However, with realistic and accurate 3D animation and visualisation, the uncertainty can be eliminated to gain approval and building permits more easily (Triveldi, 2013).

Other applications / advantages of fidelity in representing real-life spaces is in 3D flight simulations and driving simulations for crash avoidance training or to assess the effect of distracting tasks on situation awareness. If a training pilot or driver is trying to develop sensitivity in taking split-second actions/reactions based on proximity, environment or dimensions/altitudes of spaces in the simulation, then this can be detrimental if this does not represent real-life dimensions when he actually is flying or driving; as delayed or premature life- threatening decisions can be made (Tian et al. 2012). Even an everyday application such as the Global Positioning System (GPS) device has shown problems where users misinterpret distances shown on it to those in reality and take turns on the road which are either too soon or farther than required. This is extremely inconvenient especially for those who have hearing impairment and cannot rely on the audible instructions with the GPS device (Greenberg & Blommer, 2011).

Considering this, methods to visualise 3D spaces and simulations should be enhanced to depict reality accurately. As Salamin et al. (2006) explain, these include: 3rd person view (i.e. watching an animation on screen) where the user can see his avatar moving relative to the space or imagine watching someone else moving, but does not feel embedded inside it himself; or using 1st person view (by wearing a virtual reality Head Mounted Display-HMD) where the user feels immersed inside the environment and completely surrounded by it.

While previous work by the author (summarised in the next chapter) provides evidence for the magnitude of this infidelity in space perception between virtual and real-life spaces visualisation, in all 3 spatial dimensions, a very important aspect needs to be investigated. This would be the reasons that might be causing this infidelity, and whether these are constant per user, or even changing with conditions and time for the same user, which is even more pronounced. These reasons could be related to the technology and back-end architectural programming of the software tools creating the 3D virtual scene, hence rendering this infidelity constant across users. However there could be psychophysical aspects (detailed subsequently), which differ from user to user, hence might cause the rendering and infidelity to be changeable. Some aspects could be affected both technologically and psychophysically e.g. the affect of stereo-vision associated with 1st person view which could hinder evaluating distances, but could be partially compensated by the 3rd person perspective that increases the field of view. The scope of this chapter is to discuss the possible reasons for difference in perception of space between 3D virtual and real life spaces.

PERCENTAGE DIFFERENCE IN PERCEPTION BETWEEN VIRTUAL AND PHYSICAL SPACE DIMENSIONS

In the gaming industry, according to Billger et al. (2004), “the objective is not producing correct simulations of reality, but visualisations that look good.” This however is unsuitable for conveying depictions of building spaces accurately to clients in the construction industry. Hence the author conducted a research to determine the exact percentage differences between virtual and physical perception of each space dimension, to propose solutions for more realistic 3D visualisations of spaces for clients (Saleeb, 2015). The research scope included using only Sketch Up software, with V-Ray photorealistic rendering plugin to visualise models with utmost quality and realism. A randomised, age gender and discipline diverse, sample of 18 students participated from Middlesex University, UK. Exclusion criteria for selection were visual disorders, epilepsy, tendency for motion sickness, claustrophobia or sensitivity to flashing lights. The participants partook in two experiments for each of 3 different sized rooms, detailed later. One experiment was performed in the real-life room, and the other in its virtual replica by looking at the screen in 3rd person view (1st person view and immersive virtual reality are outside of scope of this paper). Each experiment was divided into two parts, with the participant static then moving in the room, to test previous research findings, that movement enhances perception of space size.

As explained by Saleeb (2015), the following control and extraneous variables were kept constant so as not to affect the results: time of day, experiments' procedure, researcher facilitating the experiments, same room colours and content. Only one independent variable was changed/tested i.e. width or depth or height of the room, with keeping the other 2 variables constant. The real-life experiments were conducted in three rooms: Room B as the control experiment and the other 2 rooms chosen specifically to resemble one of Room B's dimensions and to be either half or double the other dimension as per figure 1. This was to fix all dimensions except one, which would be the independent variable, to compare results of rooms together against.

The rooms were emptied except for one visual cue, a cubical plain desk, which was placed in room centre to aid participants with assessing the width, depth and height of the rooms, by using relative instead of absolute sizes, thus avoiding discrepancies in humans' ability to measure using metric scales, as explained previously. The rooms were then modelled/virtually replicated using Sketch Up, as a representative of 3D model authoring software used in the construction industry and BIM projects. All dimensions, openings, colours, textures, materials, fittings were replicated exactly including the visual cue desk and its exact position in the room.

In the real-life static experiment, individual participants first analysed the space while seated, only moving around head and body, then while walking around the room and feeling the space to answer a questionnaire. These steps were repeated for all three rooms, which were experienced randomly to eliminate the effect of order on the results. For the static virtual experiments, participants observed two images representing width/height and depth/height, for which they were asked to evaluate

Figure 1. Room Size Ratios

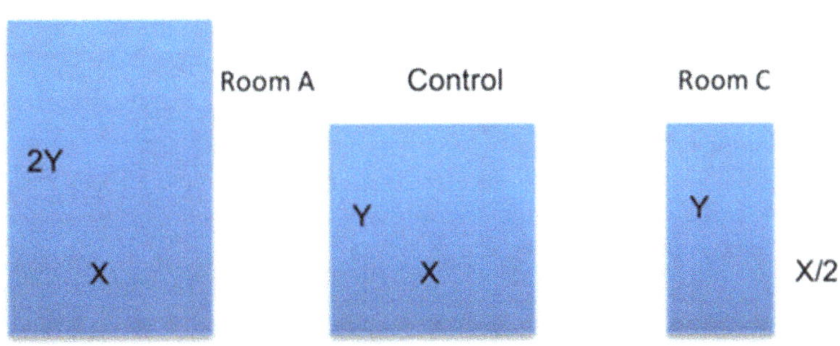

the dimension sizes (figure 2). For the moving condition, a simulation was displayed, on screen, of movement inside the room; camera height 1.7m. Those techniques were experimented with and without wearing the HMD and then compared to the experiment of static and moving observer in the physical world. To overcome bias in measurement from performing a physical followed by virtual experiment in the same room, or vice versa, these were separated by 4 weeks so participants would have little recollection of their previous answers. For each experiment participants answered questions about number of desks they perceived could fit, side by side or on top of each other, in each of the respective width, depth, height and whole area of each room.

Another important factor was brightness/light contrast of the rooms and its effect on size perception. According to Egusa (1983), perceived depth increases with increased brightness differences. Hence for all experiments participants assessed environment brightness using a numbered scale (1 for dark, 2 for shadow, 3 for medium light, 4 for bright and 5 for too much light).

Discussion of Results

Results revealed perception of space decreased from real to virtual representation, in width, depth, overall volume and slightly in height. A summary of the amount of fidelity in perception of width, depth and overall volume, from real to virtual experiments in the static and moving conditions in all 3 rooms, is illustrated in figure 3 and 4. Numbers indicated that by increasing the depth by almost twice in room

Figure 2. Left - Room B depth/height virtual image for static experiment; right – Room B width/height virtual image for static experiment.

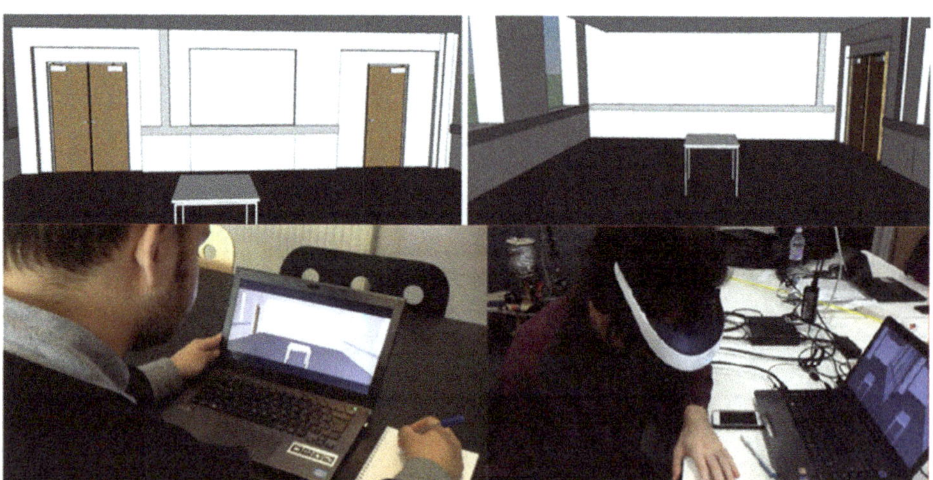

Figure 3. Summary Percentage Reduction in Space Perception fidelity from Real to Virtual Spaces in all dimensions

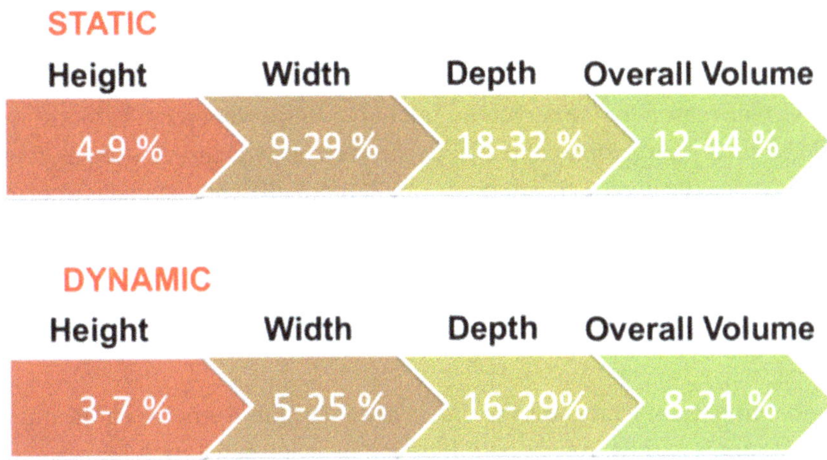

Figure 4. Percentage Perception fidelity from Real to Virtual Spaces in all Dimensions

		Static	Moving	
Width Dimension	Room A	69%	75%	Double Depth
	Room B	91%	95%	
	Room C	88%	93%	Half Width
Depth Dimension	Room A	68%	71%	Double Depth
	Room B	82%	84%	
	Room C	78%	82%	Half Width
Height Dimension	Room A	91%	93%	Double Depth
	Room B	96%	97%	
	Room C	96%	96%	Half Width
Overall Volume	Room A	56%	81%	Double Depth
	Room B	88%	92%	
	Room C	71%	79%	Half Width

Legend	
(white)	Control Room
(light)	Least affected
(dark)	Most affected

A, there was a significant increase in the difference of space perception between Real and Virtual worlds – less fidelity, hence the virtual was perceived as being much smaller than the real.

However, a less significant increase in the difference of space perception between Real and Virtual environments happened when width was decreased by almost half in room C. One reason for the decreased fidelity in space perception in rooms A and C, could be their more rectangular shape than room B, which was used as control; hence increasing the depth sensation, making it more difficult to perceive correctly virtually as depth is the dimension further from the eye. However, since room C was much smaller than A, this might have facilitated perceiving its dimensions more correctly, compensating for the depth issue and making the difference in perception less. This could indicate that different shapes affect human perception of real and virtual spaces.

Another important factor was the difference in brightness between the physical and virtual rooms and its impact on difference in space perception between them. Results revealed that the average brightness perception was extremely similar for all physical and virtual views of each room (range between 2.89-3.54). Hence the influence of this variable was considered low on the difference in perception between real and virtual spaces, although this is one of the factors discussed again consequently, since the virtual results with no HMD were slightly less.

Statistical Analysis

According to Trochim (2006), there are three criteria to meet before providing evidence for a causal relationship.

1. Temporal Precedence: to show that the cause happened before the effect. This was achieved in this research by changing the tested medium before measuring its effect on participants' perception of space size.
2. Covariation of the Cause and Effect: to show that whenever X is present, Y is also present, and whenever X is absent, Y is too. This was achieved by repeating all experiments for 18 participants, exploring similar results for all.
3. No Plausible Alternative Explanations: to rule out alternative explanations using control groups. In this research, the real-life experiments in each room are control for the virtual experiments in the respective rooms, static, moving, with and without HMD experiments. Also as aforementioned, extraneous variables were controlled to rule out their effect on the results.

In this research 2 types of statistical methods, descriptive and inferential, were used to validate the data. First, a possibility of a relationship between the identified values of decreased perception of dimensions in the virtual experiments was investigated by calculating R2 (Coefficient of Determination) between the following sets of data obtained from the observations

- Percentage decrease in perception from real to virtual experiment in depth relative to width.
- Percentage decrease in perception from real to virtual experiment in height relative to width.
- Percentage decrease in perception from real to virtual experiment in height relative to depth.

Results showed that the R2 value was 0.84 for depth relative to width, 0.66 and 0.71 for height relative to width and depth respectively. As can be seen, the percentage of goodness of fit for depth relative to width experiments is relatively high which might indicate a relationship between the causes for the decrease of perception in both dimensions, which might be attributed to the virtual environment. Even though the other two R2 results are more moderate in value, they might also indicate a similar relationship. It is worth noting that the 2 results involving height had lower R2 values, which could resonate with previous results that the height dimension was the least diminished in perception when seen virtually, i.e. the effect on it was not as significant as depth and width.

The second statistical analysis method used was ANOVA (Analysis of Variance) to compare the variation in the results of the set of participants' observations (perception of dimension size) in the three distinct conditions: real- life experiment, virtual experiment with no HMD, virtual experiment with HMD. The goal was to find whether the results are significantly different or whether they can be attributed to sampling/experimental error or randomness (i.e. unable to refute the null hypothesis). The number of observations was 18 (used to calculate the first degree of freedom). Three condition groups were used (real, virtual no HMD, virtual with HMD- used to calculate the second degree of freedom). ANOVA was calculated for each of the following results:

- No. of perceived desks in the width dimension.
- No. of perceived desks in the depth dimension.
- No. of perceived desks in the height dimension.
- No. of perceived desks overall in the area of the room.

Results showed that in all 4 cases above, the resultant F ratio was higher than the F critical value 3.68, with a p value <0.05, hence refuting the null hypothesis with evidence of a significant difference between the results, which might be attributed to change in condition between the groups investigated, i.e. presence of the virtual environment as opposed to physical environment that might have affected size of the perceived space dimension and how many desks it can hold.

PSYCHOPHYSICAL METRICS AFFECTING FIDELITY OF PERCEPTION OF VIRTUAL 3D SPACES

Results in the previous section provided evidence regarding the range of discrepancy experienced by users in perceiving 3D virtual spaces versus reality, with the depth dimension being the most affected, followed by the width then height. However the differences experienced between users (while lying within an identifiable range) suggest that there could be different reasons causing this, which relate to the individual psychology of the user not just technical issues related to the software tools creating the 3D environment. This chapter will discuss both types of causes – technical and psychophysical. Within the context of this chapter Psychophysics will be considered as the science of investigating the relationship between physical stimuli (a user's experience) and the psychological sensations and perceptions they produce (Bruce et al., 2003).

Possible Physical and Technical Issues Affecting 3D Virtual Perception

Effect of 3D View: 1st Person View / 3rd Person View / CAVE Setup

Methods to visualise 3D spaces and simulations include: 3rd person view (i.e. watching an animation on screen) where the user can see his avatar moving relative to the space or imagine watching someone else moving, but does not feel embedded inside it himself; or using 1st person view (by wearing a virtual reality Head Mounted Display-HMD) where the user feels immersed inside the environment and completely surrounded by it. Lack of stereo-vision associated with that could hinder evaluating distances, but could be partially compensated by the 3rd person perspective that increases the field of view (Salamin et al. 2006). A cave automatic virtual environment (CAVE) is an immersive virtual reality environment where projectors are directed to between three and six of the walls of a room-sized cube. While the user feels immersed in the environment, the positioning of the walls, the resolution of projection can affect the perception of the user. Furthermore, to create an image that is not distorted or out of place, the displays and sensors must be calibrated. The calibration process depends on the motion capture technology being used and calibration of electromagnetic sensors is more complex. Unless this is done correctly, distortion of the view can occur leading to infidelity on perception with varying degrees (Ottosson, 2002). In response to the realisation that the perceived distance estimation in an immersive virtual reality system is generally underestimated to the actual distance, approaches by Ng et al. (2017) have been examined to provide users with better dimensional perception. One method used in head-mounted displays is to interact by walking with visual feedback, but it is not suitable for a CAVE-like system with confined spaces for walking. Hence they used a verbal corrective feedback mechanism, which showed that estimation accuracy improved after eight feedback trials although some estimations become overestimated. One possible explanation is the need of more verbal feedback trials. Further research for improvement in depth perception was recommended.

Depth of Field

This is also referred to as focus range or effective focus range, and it is the distance between the nearest and farthest objects in a scene that appear acceptably sharp in an image. Even though a lens can accurately focus at only one distance at a time, the reduction in sharpness is gradual on each side of the focused distance, so that within the DOF, the non-clarity is unnoticeable under usual observing conditions (Ray, 2002).

Hillaire et al. (2008) introduced a model of dynamic visual blur for 1st person view of 3D virtual environment gaming spaces based on two types of blur effect: (1) a Depth-of-Field blur (DOF blur) which simulates the blurring of objects located in front or back of the focus point of the eyes, and (2) a peripheral blur which simulates the blurring of objects located at the periphery of the field of vision. Results showed that activation of visual blur effects did not impede the performance of users during multiple sessions of a multiplayer game. Furthermore, nearly half of the participants preferred when the blur effects were activated, as it increased the realism of the visual feedback. However there is no current sufficient evidence to substantiate whether the effect of DOF or blur imposed by this aspect affects the fidelity of perception of the 3D spaces, and if so to what magnitude compared to real life space perception.

Room Sizes and Shapes

The results shown in the experiments conducted by the author (presented above) indicate that there could be a difference in perception of space due to the difference in space sizes and shapes e.g. the more elongated or larger the space is, the more the error in evaluation of its true dimensions compared to reality due to enhanced depth. However previous research has also indicated a preference for human reactions towards certain geometries, e.g. the easier users can process an object, the more positive their aesthetic response, which can explain people preference for symmetric shapes, as they contain less information than asymmetric shapes (Garner, 1974), hence might be easier to evaluate in terms of size. In addition to symmetry, there is now evidence that angular shapes are less pleasing than round circles (Bar & Neta, 2008). Furthermore, people with low expertise in the fields of design prefer curved over angular shapes when they are simple (circles and hexagons), while experts show curved versus sharp preference bias for the more complex polygons. All the previous aspects related to geometry and sizes require further investigation as they not only affect the users vision but also their emotional reaction which could affect their evaluation of the space (Shemesh et al., 2015).

Software: Photorealistic Capabilities

There are numerous 3D model authoring and presentation software in the market, each with different back-end architecture for producing simulations of 3D virtual spaces. The differences in the simulation engines can affect the way 3D virtual spaces appear for the user. Hence fidelity in space perception must be investigated using different software packages. The authors' experiments utilised Sketch Up software to visualise the 3D spaces. Other software packages include Revit, Archicad etc. Furthermore photorealistic rendering capabilities of a software can impact the clarity of the 3D view, its brightness, luminosity, textures etc. which can impact evaluation of the space dimensions.

Brightness / Lightness / Glare /Colour Hues / Quality and Fidelity of Textures

Physical attributes of the space e.g. finishing material, its textures, colours and brightness of the room need to be investigated for their effect on 3D space perception. According to Fairchild (2005), brightness is an "Attribute of a visual sensation according to which an area appears to emit more or less light." Lightness is "The brightness of an area judged relative to the brightness of a similarly illuminated area that appears to be white or highly transmitting". Bright sources of light often appear to be surrounded by a halo of light, sometimes accompanied by radial streaks of light and a veiling luminance over the retinal image. This perceptual effect is called glare. It is most visible when there is a significant ratio in luminance between the glare source and the area around it, typically lowering detail visibility in that area. Differentiation between all 3 terms, brightness, lightness and glare, is essential to investigate the effects of on space perception of users.

Screen Characteristics

Viewing screens used for 3D virtual spaces can have many different attributes e.g. size, resolution, flat versus curved, brightness etc. Geng (2013) classifies 3D display technologies into three broad categories: (1) multiview 3D display, (2) volumetric 3D display, and (3) digital hologram display. For multiview 3D display technologies, there are occlusion-based technologies (parallax barrier, time-sequential aperture, moving slit, and cylindrical parallax barrier), refraction-based (lenticular sheet, multiprojector, prism, and integral imaging), reflection-based, diffraction-based, illumination-based, and projection-based 3D display mechanisms. For volumetric 3D display technologies, there is static screen (solid-state upconversion, gas medium, voxel array, layered LCD stack, and crystal cube), swept screen (rotating LED array, cathode ray sphere, varifocal mirror, rotating helix, and rotating flat screen), passive screens (no emitter) and active screens (with emitters on the screen). For digital hologram 3D displays, there are holographic display systems e.g. developed by MIT, Zebra Imaging, QinetiQ, SeeReal, IMEC, fog screens, graphic waterfalls, and virtual reality techniques, such as Vermeer from Microsoft.

These different types of screens with different attributes can affect four major physical depth cues the human brain uses to gain true 3D sensation: Accommodation, Convergence, Motion Parallax and Binocular Disparity (all explained subsequently).

Camera Viewing Height

The viewpoint of the user inside a 3D virtual space is affected by the camera height adopted by the user inside the authoring or modelling software. This can cause distortion of the space and hence alter the perception of its dimensions and needs to be investigated for its effect on virtual perception and assessment of space dimensions.

Prior Experience in 1st Person Gaming or Art Perspectives

Participant background disciplines, experience and age groups could play an important factor in perceiving 3D virtual space dimensions accurately. Architects, designers, engineers, virtual game creators, artists have more experience with 3D virtual spaces. In addition generations of 3D game users are more acquainted with interpreting visual cues in 3D virtual environments. For example Reber et al. (2004) discuss the expertise influence, claiming that training in arts gives meaning to complex structures, which results in an additional increase in processing ease. Furthermore cultural differences have been shown to have an effect in spatial perception. Saulton et al. (2017) provide evidence for that through an experiment in which participants had to judge whether a rectangular room was larger or smaller than a square room of reference. The room rectangularity was varied (depth to width aspect ratio) and the viewpoint (middle of the short wall vs. long wall) from which the room was viewed. South Koreans were significantly less biased by room rectangularity and viewpoint than their German counterparts. Hence diversifying experience, age groups, discipline and even cultural background are important in testing their effect on the magnitude of infidelity in 3D space perception between virtual and real.

Possible Psychophysical Stimuli Affecting 3D Virtual Perception

Convergence and Accommodation

The Accommodation-Convergence reflex is a reflex action of the eye, in response to focusing on a near object, then looking away at a distant object (and vice versa), involving coordinated changes in vergence (the simultaneous movement of the pupils of the eyes towards or away from one another during focusing), lens shape and pupil size. Accommodation-Convergence reflex is dependent on cranial nerve II, superior centres and cranial nerve III (efferent limb of reflex). The change in the shape of the lens is controlled by the ciliary muscles inside the eye. These alter the focal distance of the eye, causing nearer or farther images to come into focus on the retina; this process is known as accommodation (Watson & Breedlove, 2012). The reflex, involves three responses; pupil accommodation, lens accommodation, and convergence. A near object (for example, a computer screen) appears large in the field of vision, and the eye receives light from wide angles. When moving focus from a distant to a near object, the eyes converges.

As expected not only would these attributes differ from one person to another, and one space to another, but on comparison between real and virtual spaces, the confinement of the eye pupils while looking at a 3D virtual image on a screen (in 3rd person view) or within an HMD (in 1st person view) would definitely vary from looking at real life spaces with different distances that require different accommodation of the pupils and shape of the eye lens for focus. This might be a cause for a discrepancy between space perceptions of the virtual versus real 3D space.

Stereoscopic Depth and Binocular Disparity

Stereopsis refers to the perception of depth and 3D structures based on visual information received by the two eyes of users with normally developed binocular vision (ability to see a 3D image of the surroundings) (Howard & Rogers, 1995).

Because humans' eyes are positioned at different lateral areas on the head, binocular vision can result in two slightly different images projected to the retinas of the eyes hence giving different feedback and perception from user to user. These disparities are mainly in the relative horizontal position of objects in the two images. These are known as " binocular disparities". These disparities are processed in the visual cortex of the brain to produce depth perception, which due to the above reasons can cause infidelity in 3D virtual space perception as follows. While binocular disparities are naturally present when seeing a real 3D scene with two eyes, they can also be simulated by artificially presenting two different images separately to each eye using a method called stereoscopy. The perception of depth in such cases is referred to as "stereoscopic depth" (Guan & Banks, 2016). This can also be faulty with unpredictable magnitude based on the disparity presented to the user, hence causing variations in fidelity of perceiving 3D virtual spaces as seen earlier. This is one aspect, which needs to be further investigated in further research.

Motion Parallax

The perception of depth and 3D space is possible with information visible from one eye alone, such as differences in object size and "motion parallax" (differences in the image of an object over time with observer movement (Howard, 2012). However the impression of depth in these cases is often not as vivid or correct as that obtained from binocular disparities (Barry, 2009). Vishwanath (2014) indicates that the effect of "real" separation in depth is related to the accuracy with which depth is obtained, and that if the user has a conscious knowledge of this accuracy this can affect the planning of their motor action and eventual perception of space.

In three experiments conducted by Tozawa and Oyama (2006), they examined the effect of motion parallax and two perspective cues on perception of size and distance. The subject was asked to report the size and distance of a comparison stimulus relative to a standard stimulus. Two perspective cues were given by the relative heights of the two stimuli and the absolute height of the standard stimulus below the horizon. Motion parallax was expressed by both the ratio and the difference in angular velocities between the two stimuli. Results revealed that motion parallax affected size and distance estimation more than the 2 perspective cues.

Visual Focal Point

A very important factor is investigating the effect of visual focal points, e.g. setting the user vision to a focal depth at infinity, which can happen in virtual worlds maybe causing eyes to tire while looking for cues in the space, hence affecting perception of space size. This can be investigated using filmed views of the real world as control.

This effect allows controlling the angles of perspective on the model, giving the ability to add a realistic sense of scale, or to create dramatic close-up shots of the design. A shorter focal length represents a wider field of view, which gives a more exaggerated perspective effect. This could affect the perception of space and assessment of its dimensions by the user, which needs further investigation (Jakefowler, 2014).

Biological Metrics

The following factors are all biological metrics that can change with different external stimuli that the user is subjected to during experiencing a 3D visual image or scene. Measuring them during exposure to a 3D space visualisation is important in assessing whether they have an impact or are impacted by perception of the 3D space and navigation within it.

- Eye Gaze (intensity)
- Retina / Eye movement (frequency and direction while wearing a Head mounted display or viewing in 3rd person view).
- Heart and respiration rate / stress levels

Haptics and Audio Stimuli

Haptics is the science of applying touch (tactile) sensation and control to interaction with computer applications. Haptic communication recreates the sense of touch by applying forces, vibrations, or motions to the user (Freyberger & Färber, 2006). This mechanical stimulation can be used to help create virtual objects in a computer simulation, to control such virtual objects, and to enhance the remote control of machines and devices. Haptic devices may incorporate tactile sensors that measure forces exerted by the user on the interface. It is unclear what the effect of applying haptics simulations would be in improving or decreasing the fidelity of perception of 3D virtual spaces and hence is a factor that requires further investigation.

CONCLUSION AND FUTURE RESEARCH

This chapter has investigated the possible physical, technical and psychophysical factors that might have an impact on the fidelity of 3D virtual space perception and the recorded discrepancies between the assessment of dimensions in the virtual and real environment. Future experiments are required to investigate the previously mentioned factors to determine the magnitude of their impact on 3D visual perception.

Furthermore, consideration should be given to diversified age groups, gender, cultural background and educational background within all of the factors. Identifying the type of factors affecting perception fidelity will help in identifying solutions to overcome this. For example, if technical issues are involved (hence causing similar discrepancies among users), then collaboration with software and hardware designers/programmers is essential to find ways to alter or customise back end programming solutions and technical specifications to allow more realistic perception. If psychophysical factors are involved (hence causing individual disparities), then maybe best practice solutions can be recommended to allow better user 3D space perception and enhance fidelity of the visualisation in comparison with reality. This should have a direct impact on the construction life cycle and BIM applications related to client decision making processes, which would eliminate errors in future during or after construction, and the need for waste of resources (e.g. time, money and energy) to rectify these errors.

REFERENCES

Bar, M., & Neta, M. (2008). The proactive brain: Using rudimentary information to make predictive judgments. *Journal of Consumer Behaviour*, *7*(4-5), 319–330. doi:10.1002/cb.254

Barry, S. R. (2009). *Fixing my gaze: A scientist's journey into seeing in three dimensions*. Basic Books.

Billger, M., Heldal, I., Stahre, B., & Renstrom, K. (2004). Perception of color and space in virtual reality: a comparison between a real room and virtual reality models. In *Electronic Imaging 2004* (pp. 90–98). International Society for Optics and Photonics. doi:10.1117/12.526986

Bruce, V., Green, P. R., & Georgeson, M. A. (2003). *Visual perception: Physiology, psychology, & ecology*. Psychology Press.

Egusa, H. (1983). Effects of Brightness, Hue, and Saturation on perceived depth between adjacent regions in the visual field. *Perception*, *12*(2), 167–175. doi:10.1068/p120167 PMID:6657423

Freyberger, F. K. B., & Färber, B. (2006). Compliance Discrimination Of Deformable Objects By Squeezing With One And Two Fingers. Proceedings of Eurohaptics 2006, 271–276.

Garner, W. R. (1974). *The processing of information and structure*. Lawrence Erlbaum Associates Potomac.

Geng, J. (2013). Three-dimensional display technologies. *Advances in Optics and Photonics*, *5*(4), 456–535. doi:10.1364/AOP.5.000456 PMID:25530827

Greenberg, J. A., & Blommer, M. (2011). *Physical fidelity of driving simulators*. Boca Raton, FL: CRC Press. doi:10.1201/b10836-8

Guan, P., & Banks, M. S. (2016). Stereoscopic depth constancy. *Philosophical Transactions of the Royal. Society B, 371*(1697), 20150253.

Hillaire, S., Lécuyer, A., Cozot, R., & Casiez, G. (2008). Depth-of-field blur effects for first-person navigation in virtual environments. *IEEE Computer Graphics and Applications*, *28*(6), 47–55. doi:10.1109/MCG.2008.113 PMID:19004684

Howard, I. P. (2012). Perceiving in depth, volume 1: basic mechanisms. New York: Oxford University Press. doi:10.1093/acprof:oso/9780199764143.001.0001

Howard, I. P., & Rogers, B. J. (1995). *Binocular vision and stereopsis*. New York: Oxford University Press.

Isikdag, U., Zlatanova, S., & Underwood, J. (2011). *An opportunity analysis on the future role of BIMs in urban data management* (pp. 25–36). Urban and Regional Data Management–UDMS Annual. doi:10.1201/b11647-5

Jakefowler (2014). *Camera Settings – Make Your Renderings Look More Realis*tic. Retrieved October 1, 2017, From Https://Www.Autodesk.Com/Products/Fusion-360/Blog/Camera-Settings-Make-Your-Renderings-Look-More-Realistic/

Ng, A. K., Chan, L. K., & Lau, H. Y. (2017). Corrective Feedback For Depth Perception In Cave-Like Systems. In Virtual Reality (VR), 2017 IEEE (pp. 293-294). IEEE. doi:10.1109/VR.2017.7892292

Ottosson, S. (2002). Virtual Reality In The Product Development Process. *Journal of Engineering Design*, *13*(2), 159–172. doi:10.1080/09544820210129823

Parsons, E. (1995). GIS visualisation tools for qualitative spatial information. *Innovations in GIS*, *2*, 201–210.

Pereira, G. C., Rocha, M. C. F., & Florentino, P. V. (2013). Spatial Representation: City and Digital Spaces. In Computational Science and Its Applications– ICCSA 2013 (pp. 524-537). Springer Berlin Heidelberg.

Ray, S. F. (2002). *Applied photographic optics: Lenses and optical systems for photography, film, video, electronic and digital imaging*. Focal Press.

Reber, R., Schwarz, N., & And Winkielman, P. (2004). Processing Fluency And Aesthetic Pleasure: Is Beauty in the Perceiver's Processing Experience? *Personality and Social Psychology Review*, *8*(4), 364–382. doi:10.120715327957pspr0804_3 PMID:15582859

Salamin, P., Thalmann, D., & Vexo, F. (2006). The benefits of third-person perspective in virtual and augmented reality? *Proceedings of the ACM symposium on Virtual reality software and technology*, 27-30. 10.1145/1180495.1180502

Saleeb, N. (2015). Effects of Differences between Virtual and Physical Perception of Space on Building Information Modelling. *Proceedings of the International Conference on Building Information Modelling in Design, Construction & Operations*, 21-32. 10.2495/BIM150031

Saulton, A., Bülthoff, H. H., de la Rosa, S., & Dodds, T. J. (2017). Cultural differences in room size perception. *PLoS One*, *12*(4), e0176115. doi:10.1371/journal.pone.0176115 PMID:28426729

Shemesh, A., Bar, M., & Grobman, Y. J. (2015). Space and human perception: exploring our reaction to different geometries of space. *Emerging Experience in Past, Present and Future of Digital Architecture. The 20th International Conference of the Association for Computer-Aided Architectural Design Research in Asia CAADRIA.*

Tian, F., Chai, W., Wang, C., & Sun, X. (2012). *Design and Implementation of Flight Visual Simulation System.* arXiv preprint arXiv:1212.0365

Tozawa, J., & Oyama, T. (2006). Effects of motion parallax and perspective cues on perceived size and distance. *Perception*, *35*(8), 1007–1023. doi:10.1068/p5251 PMID:17076062

Trivedi, G. (2013). *The Importance of 3D Animation and Visualization in Construction and Manufacturing*. Retrieved August 20, 2017, from http://www.truecadd.com/news/the-importance-of-3d-animation-and-visualization-in-construction-and-manufacturing.

Trochim, W. (2006). *Single Group Threats*. Retrieved March 20, 2017, from http://www.socialresearchmethods.net/kb/intsing.php

Vishwanath, D. (2014). Toward a new theory of stereopsis. *Psychological Review*, *121*(2), 151–178. doi:10.1037/a0035233 PMID:24730596

Watson, N. V., & Breedlove, S. M. (2012). *The mind's machine: foundations of brain and behavior*. Sinauer Associates.

Chapter 10
Streamlining a Design, Manufacture, and Fitting Workflow Within a UK Fit-Out SME:
A BIM Implementation Case Study

Jason Underwood
University of Salford, UK

Andrew Fleming
University of Salford, UK

Mark Shelbourn
University of Salford, UK

James Heywood
Links FF&E Ltd, UK

Ian Roberts
Links FF&E Ltd, UK

ABSTRACT

A 30-month project is presented that is enabled through a knowledge transfer partnership government-funded initiative between the University of Salford and Links FF&E – a design, manufacture, and fit-out SME in the UK. The project is aiming to implement BIM as a catalyst for a lean transformation to streamline processes and operations through the adoption of a case study methodology on a design for manufacture and assembly (DfMA) BIM implementation at Links FF&E. The findings highlight that the challenges for SMEs adopting disruptive technology could be mitigated with a business case that considers the changes on organizational processes and workflows by embedding technologies within the company with the focus on eliminating waste in the processes and adding value.

DOI: 10.4018/978-1-5225-5625-1.ch010

1. INTRODUCTION

Small and Medium Enterprises (SME) are predominant in most economy structures. In the United Kingdom (UK) construction industry SMEs represent 90% of the whole industry (Statistics, 2015). To respond to the competitive pressures from low cost international nations, increasing concerns with health and safety, and the sustainability agenda, the UK Government is encouraging innovation for SMEs (Adegoke, Gerard, & Andrew, 2007; Wolstenholme et al., 2009). While the focus tends to be more on product innovation than process innovation, several studies show that the use of business approaches, such as process improvements and knowledge management, can incrementally reduce costs and increase competiveness for SMEs (Hoffman, Parejo, Bessant, & Perren, 1998; McAdam, Moffett, Hazlett, & Shevlin, 2010).

Building Information Modelling (BIM) is one of the promising approaches that has emerged to improve processes and efficiencies in the construction industry (Eastman, Teicholz, Sacks, & Liston, 2011). In 2011, the UK Government launched its Construction Strategy, which mandated that all centrally procured projects should be utilising BIM from April 2016; driven by deriving full value from public sector construction and the failings to exploit the potential for public procurement of construction and infrastructure projects to drive growth.

This paper presents a BIM implementation at Links FF&E; a UK-based SME that offers the design, manufacture, supply and installation of quality fittings and furnishings for student accommodation. The aim of the project is to ensure that the company has the expertise and capability needed to operate in a BIM environment and to comply with the Level 2 BIM mandate. The project is being delivered through a Knowledge Transfer Partnership (KTP) between the University of Salford and Links FF&E. The KTP is a program partly funded by InnovateUK (a UK government-funded initiative) with the objective of supporting businesses that want to incrementally improve their performance and competitiveness with innovative solutions by accessing and transferring the knowledge and expertise of academia. Enabled through the project, Links FF&E expect that the implementation of BIM will streamline their processes and operations, thereby facilitating the transformation of the organisation to becoming BIM-enabled via the development of a business-wide BIM strategy, and ultimately improving their overall business performance.

2. LITERATURE REVIEW

Proposed Improvements in the UK Construction Industry

In proposing radical improvements in the UK construction industry, Egan (1998) stated that construction should learn from manufacturing and services industries and rethink the way of delivering projects to achieve better performance, better products and continuous improvement. Egan (1998) advocated the use of standardised components and processes, the implementation of performance measurements, the application of Lean thinking in construction, and the use of technology as a tool to support cultural and processes improvements.

In 2011, the UK Government Construction Strategy reinforced the need for supply chain integration in order to accomplish reliability and better value for money, encouraging the use of BIM to allow the full potential of technological improvements in the construction industry, reducing coordination errors and transaction costs. In addition, there is a vision for the use of BIM to allow design to feed directly into machines, connecting design and manufacture, and discarding unnecessary intermediaries (Cabinet Office & BIS, 2011).

As one of the most propitious developments in the industry, BIM involves the application of processes, people, technologies, and tools to generate and manage information about a built environment asset during its whole life (Eastman et al., 2011; Lee, Sacks, & Eastman, 2006). Moreover, BIM increases efficiency and productivity by utilising digital technology to design one or more accurate virtual models that represent an asset prior to its actual construction. This supports the interaction of the different stakeholders around the models, which include geometrical and non-geometrical data of the built asset to enable better analysis and control of the design, construction and operations compared to manual processes (Sanchez, Hampson, & Vaux, 2016).

BIM adoption and awareness is growing globally, but particularly in the UK construction sector following the Government's commitment. According to the NBS National BIM Report (2017), 62% of organisations are currently using BIM, compared to just 13% in 2010; an average 8% increase year-on-year. Furthermore, there has been the significant influence of the "push-pull" Government Strategy for BIM that mandated the adoption of BIM on all public centrally procured projects from April 4th 2016, and encouraged a 'feeling' that BIM is a new standard for project information management, which is facilitating with transforming the construction industry landscape (Cabinet Office & BIS, 2011; NBS, 2015). Moreover, the majority of UK construction industry enterprises are SMEs and therefore the adoption of BIM within SMEs is vitally important for the transformation of the whole UK construction sector.

BIM in SMEs

Small and medium companies are usually defined as companies with less than 250 employees. In the UK, 99.9% of the enterprises are SMEs, turning over 47% of the country's economy turnover. Although there is a significant increase in the uptake of BIM within the UK in recent years, the SME uptake of BIM is still slow. The international survey, Smart Market Report (2014), stated that 34% of the large contractors have more than five years' experience with BIM, which contrasts against just 16% of small firms. However, it is not clear how "large/small" firms were categorised in the survey. In the UK, the NBS National BIM survey (2014) compared the use of BIM in organisations through the number of employees,; it reported that, whilst 61% of the companies with more than six employees were aware and using BIM, only 35% were aware and using it in organisations with one-to-five employees. Although one can argue about the representativeness of the survey (due to the small number of respondents), it provides confirmation that small organisations are lagging behind their larger counterparts in their engagement with BIM.

While the evidence regarding the slow adoption of BIM by SMEs in the UK is limited, the challenges and risks related to innovation within SMEs is well documented (Lam et al., 2017). Implementing innovation in SMEs is a complex and nonlinear process (Hosseini et al., 2016), with issues around the lack of/scarce resources (both financial and human), a lack of skills and capabilities, and the lack of systematic measurements, which can result in implementation failure and frustration for SME managers (McAdam et al., 2010). Moreover, SMEs tend to be slow on the adoption of disruptive technologies, preferring well tested technologies that pose lower risks for the investment by the organisation in change (Poirier, Staub-French, & Forgues, 2015).

Although SMEs are more agile in incorporating changes in comparison with larger organisations (Mellon & Kouider, 2016), SMEs tend to innovate to survive in business (Carrier, 1994). For example, when working collaboratively with other companies in projects that adopt innovative technologies, SMEs usually adopt the approach of learning by doing (the job), thereby missing a strategic vision for implementing innovation (Poirier et al., 2015). However, Harty (2015) argues that the benefits for SMEs are higher when not just embracing the technology, but by embedding technologies within the business processes and workflows, which poses strategic changes for the whole organisation. Such an approach to technology implementation is aligned with Lean Principles.

Synergies of Lean and BIM

BIM projects involve higher levels of collaboration between stakeholders and integrated processes. Smith and Tardif (2009) argue that the most effective BIM implementation strategy must be aligned to the business strategy, based on a review of the organisation's internal and external business processes and workflows. Moreover, it demands changes in existing processes and procedures for design and construction. While BIM represents a technology change, it is also a people and process change. It is acknowledged that technology alone will not make any significant change to business improvement (Love, Matthews, Simpson, Hill, & Olatunji, 2014). Furthermore, technology should fit the organisational infrastructure and reinforce the business process, with emphasis on management and organisational changes that are supported by the implementation of digital technologies, and not the converse, in order for the company to succeed in realising the full benefits of the implemented technology (Koskela & Kazi, 2003; Rafael Sacks, Koskela, Dave, & Owen, 2010). The concept of implementing BIM, with emphasis on streamlining company processes and workflows, are aligned with Lean Thinking, and recommended by such government and industry initiatives as those detailed by Simon (1944), Latham (1994), Egan (1998), Wolstenholme (2009), and Farmer (2016).

Lean Principles are derived from studies into the car manufacturer Toyota, which have been adapted by several authors and applied over time to other manufacturing sectors (Liker, 2004), the construction industry (Arbulu & Zabelle, 2006; Ballard, Kim, Jang, & Liu, 2007) and service industries (Womack, 1996). The core principle of Lean is to deliver a product or service, while maximising value (from the perspective of the customer) and minimising waste (Womack, 1996). Lean is process-oriented and considers the use of technology only if it serves people and processes (Liker, 2004). While Lean also offers sets of tools and techniques, it is more than simply a set of tools. It is a philosophy that is shared throughout a value stream (Ballard et al., 2007; Diekmann, Krewedl, Balonick, Stewart, & Won, 2004).

Lean Construction and BIM are independent concepts, each of which can be applied without the other. However, Sacks, Dave, Koskela & Owen (2009) argue that there are synergies between them, while the full potential for improvement in construction projects has been achieved by the adoption of both concepts together (Fakhimi et al., 2016). Sack et al. (2010) argue that any BIM implementation project should ensure that the process changes adopted aim to make the organisational process leaner, stating that BIM enables, or is the catalyst for, the organisational Lean transformation.

To enable the vision of the UK Construction Strategy 2011, in linking design through to fabrication with the use of BIM, a closer analysis of the process from design through to construction is required. Kieran (2003) suggests a review of the architectural design process, and the use of technology to manage information about the building through the whole supply chain. Moreover, it explains how complex manufacturing projects, such as airplanes, ships and cars, have changed over the years, focusing on the process of design and manufacturing parts or components that are further assembled together. The construction industry can learn from these industries in streamlining processes of design and construction, which in turn, can reduce the time and cost of projects, while creating value for the client.

Digital Design and Manufacturing

The process of design and manufacturing involves three major components: digital interactive design tools using a computer aided design (CAD) system, a computer aided manufacturing (CAM) system to specify how the digital design model is actually manufactured, and a computer numerical control (CNC) machine to fabrication (Schodek, 2005). Combining Design for Manufacture (DFM) and Design for Assembly (DFA) concepts, Design for Manufacture and Assembly (DfMA) is a product design tool that aids manufacture and assembly. Assembly can be categorised as fitting or assembly; fitting often requires power tools and skilled work and is considered a secondary manufacturing activity that allows product functionality, while assembling involves the manipulation of finished parts, transforming it into a meaningful object (Redford, 1994). DfMA is often neglected by designers who prioritise activities that are apparent to the client, like conceptual and functional design (Corsini & Moultrie, 2017). However, the application of DfMA can shorten the time to the conclusion of the finished product and consequently reduce the overall product cost (Boothroyd, Dewhurst, & Knight, 2002; Kuo, Huang, & Zhang, 2001)

Collaboration between the design and delivery team is essential for the use of DfMA (O'Rourke, 2013) and BIM supports this co-ordination to facilitate DfMA. 3D BIM models for fabrication need to be populated with data sets and attributes for manufacturing (Fereday & Potter, 2013), requiring parameters for design and close co-ordination (Irizarry, Karan, & Jalaei, 2013). Once designed, 3D models can be archived as BIM libraries for further reproduction and fabrication. Furthermore, BIM libraries allow customisation with control of the parameters for fabrication.

Being relatively mature in terms of their BIM capabilities, major contractors in the UK are expanding the use of DfMA in construction. Such organisations have programs to increase off-site fabrication by up to 70% on all projects. These programs incentivise innovation and emphasise that, in a competitive industry, those organisations that can deliver innovative, integrated solutions most cost effectively are more successful (O'Rourke, 2013). However, one consideration to be made is the challenges identified for pre-fabrication relating to the cost of projects. On low cost projects, there is less space for product and service innovation, while high end projects can be highly customised, using more reliable and state-of-the-art technology, which can see the creation of higher quality products (Thomas & Thomas, 2012).

Although it is possible that the use of BIM could enable the use of digital fabrication tools, such as CNC machines, to improve the construction performance and the quality of products, there is a need to integrate the design and manufacturing processes to manage the production system and the supply chain, which is majorly composed of SMEs (Gann, 1996). Lean principles and tools adopted by other industries, have proved to increase the effectiveness and reliability of projects, and the integration of the supply chain around the design to manufacturing process (Ariaratnam & Rojas, 2009). Furthermore, collaborative BIM projects could provide mechanisms to control processes and communication around the supply chain (Irizarry et al., 2013). Therefore, the adoption of Lean together with BIM has the potential to reduce the gap between design through to manufacture. However, there is a need to address the slow uptake of BIM by SMEs to achieve integration across the whole supply chain.

3. RESEARCH METHODOLOGY

The research presented is currently being undertaken through a Knowledge Transfer Partnership (KTP) project between the University of Salford and Links FF&E. The overarching aim of the KTP is to ensure the company has the expertise needed to operate in a BIM environment from design through to assembly, while filling the gaps between design to manufacture, streamlining processes, reducing duplication of information, ensuring control, and increasing projects' efficiencies.

Dealing with real-world issues, the research adopts a pragmatic philosophic position, considering that a proposition is valid if it works satisfactorily, dealing with practical consequences of accepting the proposition, and rejecting unpractical ideas (Mounce, 1992). It is an action research initiative with the purpose of studying a system (Links FF&E culture and process), diagnosing the current situation, exploring alternatives, proposing changes towards a desirable direction, and measuring the results (French, 1999). Due to the nature of the project, the research focuses on a single case study and applies qualitative and quantitative methods. Data collection was conducted through the combination of interviews, documents and observations during the time of collaboration with Links employees.

The 30-month project comprises five key stages: 1. establishing and consolidating best practice knowledge in BIM; 2. conducting a detailed review and analysis of the organisation's current situation; 3. developing a BIM-based collaborative strategy; 4. piloting an implementation of the BIM-based collaborative strategy for DfMA (which is currently in progress, following on from the previous three completed stages); and 5. conducting a project review and evaluation, to further disseminate the results. Table 1 outlines the project stages' main objectives and outputs.

4. CASE STUDY ORGANISATION

The case study company, Links FF&E, operates with three core areas, namely design, project management and support services; manufacturing provided by a sister company based in Lithuania; and fitting and installation, with teams based

Table 1. BIM implementation approach at Links

Project Stage	Outputs
Stage 1: Establish and consolidate best practice knowledge in BIM	State-of-the-art best practice knowledge in BIM and for collaborative DfMA report and presentation (Output 1A)
Stage 2: Detailed review and analysis of the organisation's current situation	Detailed and validated current process maps and information flows (Output 2A) IT systems, file formats, information exchanges review and recommendations (Output 2B)
Stage 3: Develop BIM-based collaborative strategy	Improvement gains analysis (Output 3A) BIM-enabled processes and practices mapped and documented (Output 3B) IT systems and information requirements documented (Output 3C) Training plan formulated (Output 3D) Organisational BIM implementation strategy (Output 3E) DfMA BIM implementation strategy plan formulated (Output 3F)
Stage 4: Pilot implementation of BIM-based collaborative strategy for DfMA	Pilot implementation DfMA project identified (Output 4A) IT system(s) selected, procured and integrated (Output 4B) Component libraries developed and implemented (Output 4C) Training plan implemented rolled out in Links UK&Lithuania (Output 4D) New processes and practices embedded (Output 4E)
Stage 5: Project review, evaluation, and dissemination	Implementation project impact assessment (Output 5A) Project review and evaluation (Output 5B) Academic and industry dissemination (Output 5C)

on sites across the UK. This KTP project concentrates on the design service and its relationship with the manufacturing and fitting.

In 2014, in response to market demands, Links started offering full design services for bedrooms, kitchens and common areas of student accommodation. Initially, all drawings were outsourced; therefore the department is currently developing its capabilities and trying to address the lack of standardised processes, lack of internal capabilities and lack of applied technology. The company is addressing this challenge through the implementation of BIM with the support of the University of Salford through the KTP project. Moreover, as the industry begins to embark on the challenge set down by the Government Construction Strategy/BIM Task Group, this project also aims to demonstrate an SME's organisational transformation in meeting this challenge.

The core challenge posed for the KTP is in changing current work practices, and processes, and adopting a culture that a BIM approach brings in terms of Links FF&E at an organisational level, and in enabling even more effective collaborative engagement with their supply chain in a move towards whole-life thinking. BIM requires shifting from an innate traditional 'fragmented' culture, currently embedded within the industry, towards operating in a collaborative working environment, whereby project teams use standardised protocols and agreed standards, methods, and procedures, to ensure the same form and quality of information is managed and produced, enabling it to be used and reused without change or interpretation. This shift towards 'true collaboration' requires people, systems, processes and practices to be collaboratively aligned, whereby project partners share a common 'project-focused' goal.

5. PROJECT STAGES DESCRIPTION AND FINDINGS

To date, the project is in its final stages. It has completed the first three key stages and is currently finalising the fourth stage of the project to develop BIM libraries and embed the new processes and procedures through a pilot implementation prior to moving into the final stage of conducting a review and evaluation of the outcomes of the project. The results of the project are discussed below.

Project Stage 1: Establish and Consolidate Best Practice Knowledge in BIM

The main objective of Stage 1 was to establish and consolidate best practice knowledge in BIM and collaborative processes for collaborative DfMA. The initial stage of the implementation aimed to benchmark the best practice of BIM in the UK. Based on a state-of-the-art literature review, it discussed and found that a brief overview of BIM implementation, an outline of the UK Government BIM Strategy, an overview of BIM protocols and data formats, relevant results from recent surveys about BIM, and the main concepts of BIM for DfMA, showed the current trends in all of these areas. Next, based on primary data collected from semi-structured interviews with industry key players, a report established the state-of-the-art BIM implementations in the UK covering the key aspects that shape BIM implementation, namely: main drivers, steps for implementation, and challenges related to people and SMEs. A content analysis of the data collected from both the literature review and interviews led to a discussion of the challenges related to protocols, the development of component libraries, and the opportunities of BIM for SMEs. These different areas are discussed further below.

Based on the experiences of 'BIM mature' companies in the UK, the research undertaken in the literature review and interviews found that Links' complex business processes, which involves design, manufacturing and fitting, could potentially benefit from BIM, especially regarding information exchange, integration with the supply chain and DfMA (Eastman et al., 2011). However, in order to implement BIM successfully, a change management plan was recommended to deal with risks related to the resistance of people to change and the amount of investment in training and technology (Arayici et al., 2009).

Results from the interviews indicated that dealing with the resistance of people to change is one of the main challenges experienced in implementing BIM irrespective of the company size or activity. The way in which the KTP team overcame this challenge was to demonstrate the value of BIM to the organisation through awareness and training sessions across the organisation, in order to thereby gain 'buy-in' from those affected by the change. For example, regular lunchtime seminars were held to increase the awareness of BIM, working in a more collaborative environment, as well as progress of the research project across the organisation. During these seminars, presentations were made to update the organisation on the results of the project, information was presented regarding BIM implementation progress and updates in the UK and on process improvement, and other relevant aspects to ensure that the whole of the organisation is engaged in the transformation taking place through the project and the adoption of BIM, and working more collaboratively with BIM. To date, the seminars have proved to have had a positive impact on employee's understanding and engagement with BIM and also on the project, thus creating a more positive mindset for the success of the BIM implementation. Further formal evaluation is planned to be rolled out during Stage 5 in order to establish their effectiveness in increasing the awareness and engagement of the organisation with BIM.

Project Stage 2: Detailed Review and Analysis of the Organisation's Current Situation

The main objective of Stage 2 was to produce a detailed review and analysis of the organisation's current situation. The paradigm of BIM comprises people, process, information and enabling technologies. Following the approach proposed by Smith & Tardif (2009), which involved reviewing the organisation's business processes and workflows, and exploiting the enabling technologies, Stage 2 aimed to explore Links FF&E's processes and understand their business through the mapping of their current business processes, digital technologies, systems and infrastructure, file formats and information exchange. Therefore, this involved identifying, through an improvement gain analyses, the areas where BIM processes and technologies could support the organisation to achieve its strategic objectives.

Links FF&E processes, which had previously been documented for attaining their Quality Management System ISO 9001, were found to be out of date and the company was operating with processes that have not been previously documented. Therefore, through the KTP project, the organisation began mapping their processes, which while an extremely time consuming exercise, brought several benefits of standardising the organisation's processes and operation (Indulska, Green, Recker, & Rosemann, 2009).

The technique adopted to map their current processes was a series of interviews with each department of the organisation in order to capture the information about their current activities and workflows. Based on the interviews' data, a series of process maps were designed and subsequently validated with the interviewees and amended where required.

Several different notations can be used map business processes. The key to choosing the notation for adoption is to consider who is going to use the process maps produced (Harmon, 2003). For this project, the Business Process Modelling Notation (BPMN) was adopted as it is considered to be user friendly and produces diagrams that are easy for business managers to understand and analyse their business processes (Harmon, 2003). Moreover, the BPMN notation was recommended for use on BIM projects by the National BIM Standards United States (Eastman, Jeong, Sacks, & Kaner, 2010).

The resulting "As-Is" process maps represented how Links FF&E currently operate and immediately, some discrepancies of process and procedures adopted from different employers in the same function were highlighted. In addition, the discussion around the processes was important to promote process standardisation, which is aligned with the Lean principle stated by Liker (2004): "*standardised tasks are the foundation for continuous improvement and employee empowerment*".

Furthermore, a workshop was organised to identify areas for improvement, involving the key stakeholders of the project (head of departments, academic team and company directors), and clarified the vision as to how the company could increase efficiency with the same resource through the re-engineering of their existing processes. The workshop promoted an interesting debate between employees from different business functions of the company; promoting a better understanding of Links business processes from end-to-end. Therefore, the company agreed that actions needed to be taken in order to reduce waste and increase profitability by determining the processes that could benefit from the use of digital technologies for automation or to reduce cycle times.

The relevant process for the BIM implementation is the design to manufacture. Figure 1 depicts a summary of the current Design process including interfaces with Customer and Manufacturer. In the current process, following the client signing the contract, the *Design Preparation* starts, when the designer's ideas for the space are sketched by hand on paper. Next, the *Concept Design* takes place, when specification sheets are produced and 2D layouts and elevations are drawn using AutoCAD software, and 3D visualizations (CGIs) in 3DMax. At this stage the project is then presented to the client. Once the client approves the design, the design outputs are sent to their manufacturing sister company in *Project Planning & Procurement.* During the *Furniture Detail Design,* the manufacturer develops detailed furniture drawings in 2D AutoCAD, which are then sent back to Links FF&E for approval. Next, Links sign off the drawings and the manufacturer sister company produces the *3D Models* of the furniture including all manufacturer information to be input into their CNC machines. Finally, all the project furniture is manufactured and sent to site for *Fitting/Installation.* During the process, all information is shared in pdf, i.e. non-editable files.

Process Analysis

Bicheno & Holweg (2009) propose a set of tools to apply Lean Principles to services, redefining the types of waste initially described in the Toyota Production System to support service operations to eliminate waste in their value stream in order to produce more value for the customer. Based on these principles, Links' design processes were analysed to identify the areas of waste in their current processes as follows:

Over-production/duplication of information: This relates to the copying of the same information in multiple file formats, which are not interoperable, with a lack of the "single version of the truth" in design projects. Duplication of information is considered a waste, which can cause errors and an excess of inventory.

Motion/waiting: This relates to unnecessary information movement between departments due to the lack of skills and technology. When information is moved from one department to another, if not well planned, the process can be delayed by the availability of the next department, and further waiting occurs.

Over processing/defects: This relates to unclear communications and the lack of systematic procedures to capture client requirements, which can cause misunderstandings, and projects that can be developed that do not conform to a client's requirements, ultimately causing re-work.

Skills: This relates to the lack of knowledge transfer throughout the organisation. For example, Links FF&E has specialists in product development, manufacturing process, and fitting; however, those skills and knowledge are not transferred to the design department.

A final consideration is the lack of a clear design freeze moment in the project. The client must have a clear understanding of when the design needs to freeze as the root cause of many problems during procurement and installation is the client not making decisions during design on time and the provision of inadequate drawings (Bildsten & Guan, 2011). To enable manufacturing, the client has to acknowledge that the design has to freeze earlier for the benefit of all concerned (Gibb & Isack, 2003). However, considering the Lean Project Delivery System, it is also important to consider several design alternatives and find the last responsible moment to freeze the design (Ballard et al., 2007).

Findings from this stage of the project highlighted that mappings of the business process and workflows serve to clarify the steps that are necessary for the organisation to produce its products. Moreover, it clarifies the relationship between stakeholders, helping to establish roles and integrate the company departments towards the final company objective, which is to deliver the product to the client, while maximising its value. By reviewing the process mappings, it is possible to identify inefficiencies in the current processes. Furthermore, shifting the discussion from managing people to managing process, enables employees to engage in proposing process improvements that can both benefit the organisation and their clients along with empowering the employees.

Project Stage 3: Develop BIM-Based Collaborative Strategy

During the third stage, the strategy to implement BIM-based collaborative DfMA was developed by establishing the areas of improvement gain followed by the process, procedures, systems, practices and people capabilities to facilitate these

Figure 1.

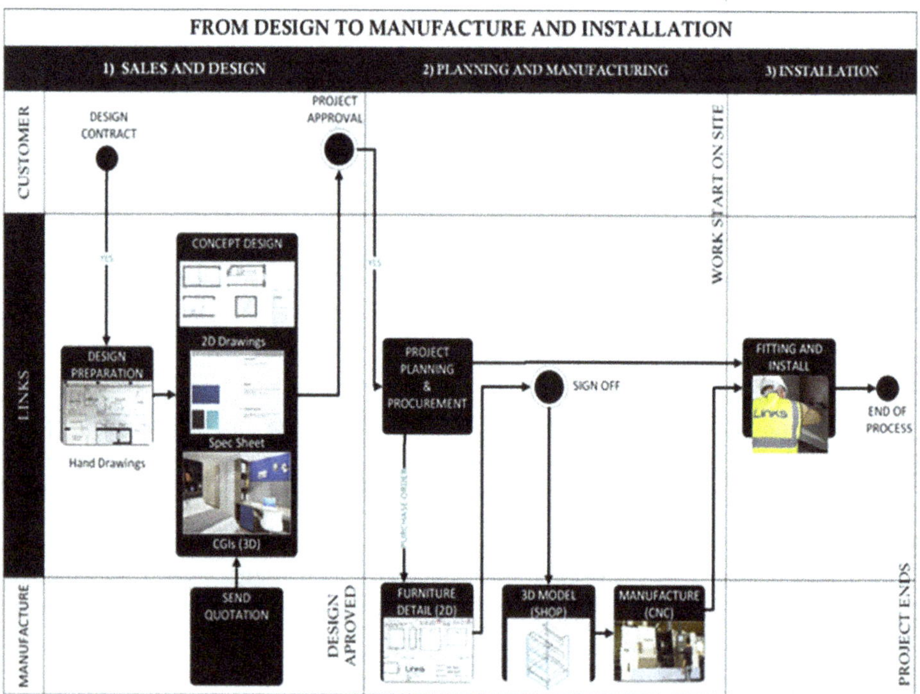

improvements, including an improvement gains analysis and the development of a training plan. The state-of-the-art review and the organisation's vision/business strategy was brought together through focus group meetings with the key stakeholder group in order to establish and review the areas of potential improvement gain of a BIM-enabled approach across the business, together with identifying the potential risks. Based on the vision of how to improve the current design process, a "To-be" target process was established.

Target Process

The development of a long term BIM strategy for DfMA, enabled by appropriate technology and consideration of the interoperability of the different software utilised in design into manufacture, would eliminate the current information duplication. In addition, part of the BIM implementation is concerned with the development of standard component object libraries with parametric models that will be used in concept design and sent direct to manufacture after design approval (Kolarevic, 2005). As outlined below, it is expected that the use of such component-based design will speed up the design process, reduce errors, and increase manufacture efficiency (Gann, 1996).

Based on the data captured in the first stages and the technology selected for the organisation, Table 2 compares the current process with the target process in relation to the estimated time savings. The left-side of the table has the 23 steps that Links FF&E currently takes from design to manufacture. The right-side of the table has the target process, and marked in red are actions that incorporate the suggested changes with BIM. Comparing the two tables the initial steps (1 and 2), which are related to design preparation, remain the same, but in step 3 the development of a 3D BIM model enables plans and elevations to be extracted (step 4) and to use the model as a base to produce CGIs (step 5). This avoids duplication of the information and unnecessary motion, thereby saving time, while improving information consistency. The use of BIM can also speed up the handover to the estimators, as all the furniture quantities can be extracted automatically from the model. Finally, at the bottom of the table are the activities performed by the manufacturer's sister company. In the target process, it is proposed to incorporate 3D component object libraries at the design stages that are compliant with the manufacturer's requirements; therefore, the BIM model could potentially diminish or eliminate the necessity of re-design for manufacturing purposes and thereby, enable a seamless flow of information through to the CNC machines (Figure 2). Furthermore, the full BIM process can save time and increase efficiencies in the overall design process. However, for the

Table 2.

Title: Design to Manufacture (As-Is Process) Date: 20/11/14

Step	Flow	Time (min)	Operation ●	Move ➡	Delay D	Store ▼	Inspect ■
1 Request Suppliers Quotation and Samples	●	2 days					
2 Spec Sheet Development	●	1 day					
3 Send information to CAD/3D Visualizers	➡	0,5 days					
4 Wait until CAD/3D Visualizers can allocate design	D	up to 7 days					
5 Prepare Layout and Elevations	●	2 days					
6 Prepare CGIs	●	5 days					
7 Prepare Mood Boards	●	2 days					
8 Recieve and Review CGIs and 2D Drawings	■	0,5 days					
9 Finalize Spec Sheet	●	0,5 days					
10 Amend CGI	●	1 day					
11 Handover to estimators	➡	2 days					
12 Wait until estimators allocate design	D	up to 3 days					
13 Send drawings for quotation	➡	1 hour					
14 Wait until Nordic can allocate quotation request	D	up to 4 days					
15 Receive and review Cost plan	■	1 hour					
16 Prepare final presentation	●	0,5 days					
Total		**17 days 2 h**	**14d**	**2.5d 1 h**	**14d**		**0.5 d 1 h**
Process at Nordic (manufacture)							
17 Send Quotation	●	1 day					
18 Receive Purchase orders	●	1 hour					
19 Produce furniture drawings for approval	●	2 days					
20 Send drawings for approval	➡	1 hour					
21 Review and sign Drawings (Links)	■	0,5 day					
22 Wait for drawings sign off	D	up to 7 days					
23 Design for manufacture	●	3 days					
Total		**6 days 1 h**	**6d 1h**	**1h**	**7d**		**0,5d**

Title: Design to Manufacture - (Target Process) Date: 06/05/15

Step	Flow	Time (min)	Operation ●	Move ➡	Delay D	Store ▼	Inspect ■
1 Request Suppliers Quotation and Samples	●	2 days					
2 Spec Sheet Development	●	1 day					
3 Using BIM Libraries, design model in 3D	●	3 days					
4 Export and format Layout and Elevations	●	1 day					
5 Export and include light and finishes to final CGIs	●	3 days					
6 Prepare Mood Boards	●	2 days					
7 Review CGIs and 2D Drawings	■	0,5 days					
8 Finalize Spec Sheet (paralel to CGI developement)	●	0,5 days					
9 Handover to estimators with quantities	➡	1 days					
10 Wait until estimators allocate design	D	up to 3 days					
11 Send drawings for quotation	➡	1 hour					
12 Wait until Nordic can allocate quotation request	D	up to 4 days					
13 Receive and review Cost plan	■	1 hour					
14 Prepare final presentation	●	0,5 days					
Total		**14.5 days 2 h**	**13d**	**1d 1h**	**7d**		**0.5d 1h**
Process at Nordic (manufacture)							
15 Send Quotation	●	1 day					
16 Receive Purchase orders	●	1 hour					
17 Receive and Review Design for manufacture	●	2 days					
18 Send drawings for approval (Links)	➡	1 hour					
19 Review and sign Furniture Detail Drawings	■	0,5 day					
20 Wait for drawings sign off	D	up to 7 days					
Total		**3.5 days 1 h**	**3d 1h**	**1h**	**7d**		**0.5d**

development of the 3D component object libraries, there is the need to build up designers' knowledge about the manufacturing drawings requirements, which can increase the collaboration between design and manufacture.

At this stage, the organisation also decided on the digital technologies requirements for software selection and procurement by considering the workflow improvements proposed above. In addition, a training plan has been formulated, including technical skills (software), design to manufacture skills to improve the integration of these areas of the organisation, and change management skills to support the transition to the next stage of the project.

Figure 2.

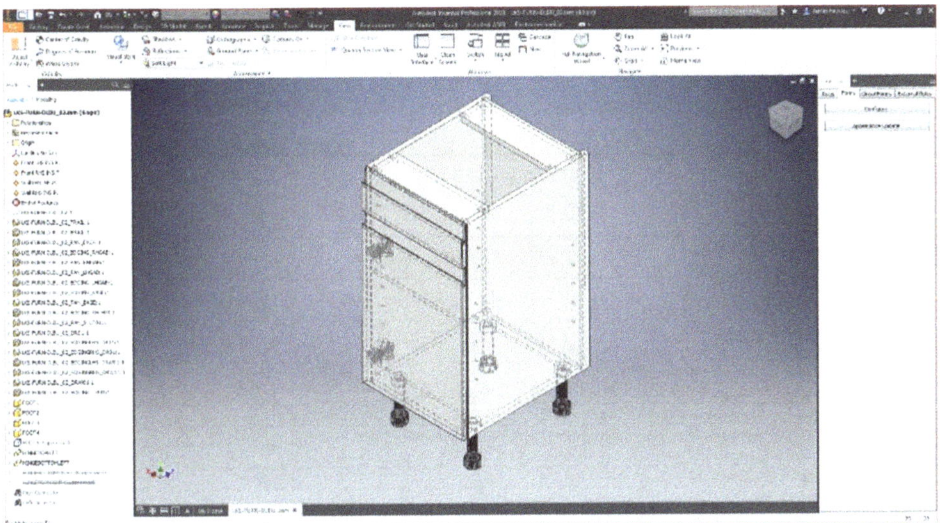

Digital Technologies

In addition to the shift from 2D drawings to 3D models using BIM-enabled digital technologies, the choice for Links considered the integration between a computer-aided design (CAD) system, a computer-aided manufacturing software (CAM), and a computer numerically controlled machine (CNC) for production (Schodek, 2005). Links FF&E currently utilises several pieces of software in their workflow. The aim of the new BIM-enabled technology is to provide information integration, thus avoiding error prone duplication of information. It was important to identify a software solution that could integrate technical drawings, 3D renderings for visualisation, shop drawings and extract information from the drawings for different purposes. The employees also considered the availability of training and support in the UK.

The approach to technology selection was to compare proposed digital technologies/systems against a list of requirements produced in collaboration with the employees involved in the project. The team considered ImosCAD, which is software specifically for furniture design and manufacturing, utilised by the manufacturer sister company; Solidworks, which is software currently in use by company furniture designers; and the chosen system was Autodesk Inventor, part of a software package called Factory Design Suite that includes AutoCAD, 3DMax and Navisworks. The list of requirements for the design software is listed in Table 3. Autodesk Revit and Graphisoft ArchiCAD, two of the most used BIM authoring tools, were reviewed, but due to the nature of the organisation's work, i.e. furniture design, neither of these

Table 3.

IT REQUERIMENTS DESIGN

	Facet	Weight	Autodesk Factory Design Suite (AutoCAD, Inventor, 3DMax, Showcase, Naviswork)	Solidworks + 3DMax	ImosCAD + 3D Max
Design (layouts, elevations, specification sheets)	Ability to input data with accurate dimensions		5	5	3
	Ability to use customize libraries		5	5	3
	Ease of setting up standards and templates		4	5	5
	Speed to design a simple room		5	3	5
	Revision control management		5	4	1
	Accuracy of relation of plans & elevations		5	5	5
	Ease of use furniture libraries		5	5	5
	Ease of development of 3D Models for CGIs (export-import models)		5	4	4
	Facility to find compatible 3d models from furniture suppliers		5	3	5
	Ease of development of shop drawing from the layouts & elevations		5	5	4
	Flexibility to design bespoke shapes and bespoke furniture for common rooms		5	5	3
	Metric measurement system		5	5	5
	Speed to get design visualization for project development purpose		5	4	5
CGI	**Ability to produce Photorealistic images**				
	Ease to input bespoke materials for CGI				
	Facility to find object libraries for interiors				
	Ability to save backups that can be opened in older versions of the same software				
	Ease to detail furniture for photorealistic images (bold edges, etc.)				
Shop Drawings	Ease of input design drawings into production		4	4	5
	Interoperability with CAD/CAM system of Manufacturer (including Nordic - Imos)		5	4	5
Internal Information exchange	**Interoperability between layout, CGI and production systems**		**5**	**5**	**4**
	Ability to connect objects with data base of prices		**5**	**5**	**4**
	Ability to link objects with data base of manuals		**5**	**5**	**5**
	Ability to extract schedule of quantities of furniture		**5**	**5**	**5**
	Ability to produce estimating from the 3D models		**5**	**5**	**5**
External Information exchange	**Interoperability with architects/contractors systems**		**4**	**3**	**3**
	Ability to extract COBie data (spreadsheet)		**2**	**3**	**2**
	Ability to export IFC		**3**	**4**	**1**
	Interoperability with coordination software (e.g. Solibri and Navisworks)		**5**	**4**	**3**
Cost	Cost of Licence		4	3	4
	Cost of Training		4	4	4
	Year maintenance cost		4	4	5
Total			124	116	108

Score

1-Doesn't meet criteria
2-Below Average
3-Average (e.g. requires advanced setting to meet criteria)
4- Good
5-Excellent (100% meet criteria)

software tools would support the level of detail and customisation that could facilitate the design to manufacturing process integration. The decision for the selected digital technologies/systems was aligned with the BIM strategy of adopting technology that could streamline business processes. The systematic approach to collaborative decision making also reinforced the awareness of BIM around the organisation's employees, thereby facilitating the change process from the current "as is" process, to the BIM enabled target process.

Project Stage 4: Pilot Implementation of BIM-Based Collaborative Strategy for DfMA

The next stage of the project was designed to implement the revised policies and procedures on a project and to determine whether they were appropriate and make changes if necessary. The identification of a suitable project was easy as Links has the partner in Lithuania.

Initially, the aim of the project was to send a completed kitchen unit from the 3D object library developments to Lithuania. The research project had already determined the software to be used, i.e. Inventor and Solidworks (during Stage 3). It was determined that the CAM software in Lithuania could take the 3D information and produce the unit; however there were some observations that were subsequently found to be an issue. While the cutting machine software knew there was a hole for a dowel, it did not know how deep to drill that hole. The same issue occurred for any surface details including slots, chamfers and countersinks.

However, the main issue found from the pilot emanated from Lithuania needing the 3D information from the kitchen unit object to go through their IMOS CAD software in order that material optimising and monitoring, and automated invoicing and checking can be achieved. The lessons learnt from this part of the pilot have enabled the finalisation of a design sign off procedure between Lithuania and Links in the UK. The sign off procedure has improved efficiency when communicating design, and decreased the possibility of inaccurate production through incorrectly interpreting layout drawings, whilst mitigating the risk of manufacturing from superseded drawings with drawing numbers and file names using BIM status and suitability coding. The DfMA process for Links with Lithuania still requires them to work with 2D information, which can easily be extracted from the 3D component library objects. Furthermore, through the development of their 3D component library objects, Links have established the capability for a 3D DfMA process for operating with other manufacturers, whereby the 3D object model can be imported directly by their CAM software system. The pilot project findings are summarised in Figure 3.

The pilot project is showing signs of achieving the goal of a Lean DfMA process; using 3D CAD Models to improve efficiency in communication and decreasing instances of duplication of information, developing processes to minimise "Motion/ Waiting" and establish clear design freeze. The research team are working closely with the software developers (IMOS CAD) to improve the process in striving to solve many of the interoperability issues. This work is currently ongoing.

Figure 3. Pilot project timeline and main findings

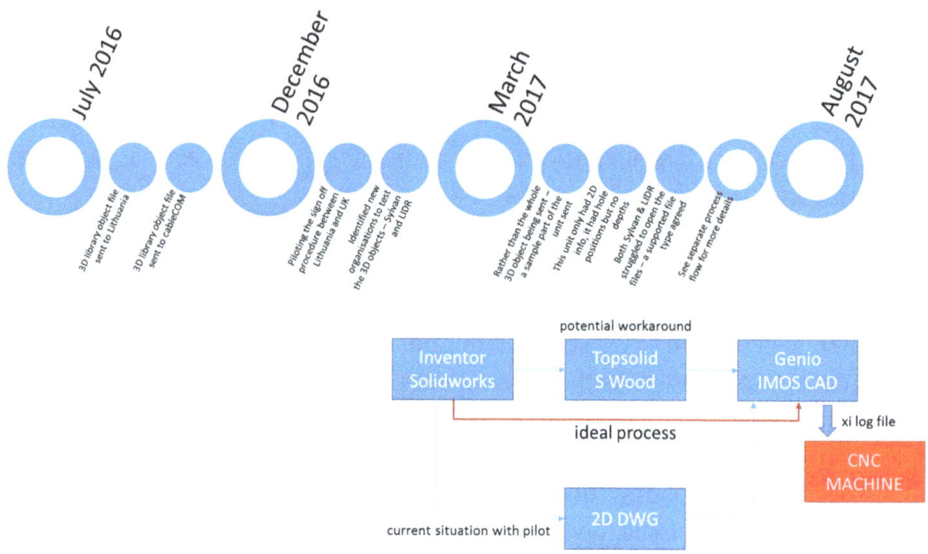

Project Stage 5: Review, Evaluation, and Dissemination

The implementation of BIM at Links is currently moving into the final stage of the project, namely the review, evaluation and dissemination. The component object library was developed using the chosen software package. These libraries are currently being used in parallel with the deployment of a BIM pilot project that is putting into practice the new processes and procedures that enables Links to operate in a Level 2 BIM environment. Links FF&E's component object library, which consists of 40 typical furniture components, including shop drawings attached to each furniture item and the capability of exporting the data into BIM exchangeable file formats is continuing to be tested and verified in the pilot project. Figure 2 summarises the results of the pilot study.

Internal training is being scheduled to outline the new BIM enabled policies and procedures required to work on a Level 2 BIM project, highlighting the reason why this is being undertaken.

The change management process is gradual and follows a process of cultural change that has occurred since the project began. The change from 2D project design to the use of 3D BIM models requires time and effort from employees to develop their capabilities. Furthermore, the design to manufacture and installation knowledge has to be matured alongside the project. Moreover, the adoption and adherence to BS1192 2007 through a Common Data Environment (CDE) aligned to PAS1192:2 needs to be embedded within the culture of the organisation. All these changes are currently ongoing and expect to be finalised at the end of this project.

6. CONCLUSION

The launch of the UK Government Construction Strategy in 2011 has witnessed the building of momentum within the construction industry with a significant increase in the awareness and adoption of BIM following the mandate for the use of collaborative BIM on all centrally procured public projects by 2016. In addition, the large majority of enterprises in the UK construction industry are SMEs; therefore SMEs are vitally important in the whole UK sector's approach to BIM and in transforming the industry. Moreover, the use of business approaches, such as process improvements and knowledge management, can incrementally reduce costs and increase competiveness for SMEs.

This paper has presented the findings to date of a 30 month KTP project in support of beginning to implement BIM within a design, manufacture and fitting SME based in the UK. The project is being delivered through five key stages. The paper has presented the findings of four of the stages of the project. Stages four and five are currently still progressing. The first stage was focused on establishing and consolidating best practice knowledge in BIM prior to stage 2, which mapped the current business processes, and various associated waste in the process was identified through their analysis. Stage 3 then established and reviewed the areas of potential improvement gain of a BIM-enabled approach across the business together with the potential risks. Stage 4 is currently implementing the developed BIM-based collaborative strategy for DfMA through an identified pilot project. The final stage will conduct a project review and evaluation together with further dissemination of the results.

In conclusion, the findings from the work to date, suggest that the proposed BIM workflow from design through to manufacture and fitting within Links could reduce cycle times in design from concept to shop drawings, saving time and increasing profits. BIM can address issues commonly found in the design to manufacture and fitting, supporting a better integration between the organisation's business functions, increasing predictability and reducing the overproduction of drawings.

However, for the company to incorporate the streamlined BIM workflow there is a need to manage the change and get their employees engaged with BIM. As highlighted by the results of stage 1 of the project, there are challenges in managing the changes that are common for SMEs. Although very enthusiastic with the potential improvements proposed by the project, issues in dealing with the resistance of people to change, allocation of financial resources and the time and effort of people, have impacted on the BIM implementation project so far. Culturally, changes have been embedded and the organisation is progressing towards its strategic objectives and continuous improvement. However, to conclude the project successfully, the implementation of BIM should focus on change management, process standardisation, training and metrics, thereby creating the right environment for continuous improvement in a learning organisation.

REFERENCES

Adegoke, O., Gerard, B., & Andrew, M. (2007). Innovation types and performance in growing UK SMEs. *International Journal of Operations & Production Management, 27*(7), 735–753. doi:10.1108/01443570710756974

Arayici, Y., Coates, S. P., Koskela, L. J., Kagioglou, M., Usher, C., & O'Reilly, K. (2009). *BIM implementation for an architectural practice*. CRC Press.

Arbulu, R., & Zabelle, T. (2006). *Implementing Lean in construction: How to succeed.* Paper presented at the Intl. Group for Lean Construction, Santiago, Chile.

Ariaratnam, S. T., & Rojas, E. M. (2009). Assessing the Environmental Impacts of Lean Supply System: A Case Study of Rebar Supply in High-Rise Condominium Construction Projects. In *Construction Research Congress 2009* (pp. 1009-1018). Reston, VA: American Society of Civil Engineers.

Ballard, G., Kim, Y. W., Jang, J. W., & Liu, M. (2007). *Road Map for Lean Implementation at the Project Level* (Vol. Research Report 234-11, pp. 426). Construction Industry Institute, Bureau of Engineering Research, The University of Texas at Austin.

Bicheno, J., & Holweg, M. (2009). The Lean toolbox: the essential guide to Lean transformation (4th ed.). Buckingham, UK: PICSIE.

Bildsten, L., & Guan, W. (2011). *The Study of a Kitchen Assembly Process in Industrial Housing*. Paper presented at the Nordic Conference of Construction Economics and Organization, Copenhagen, Denmark.

Boothroyd, G., Dewhurst, P., & Knight, W. A. (2002). Product design for manufacture and assembly (2nd ed.). New York: Marcel Dekker.

Cabinet Office. (2011). *E. a. R. G., & BIS, C. S. U.* London, UK: Government Construction Strategy.

Carrier, C. (1994). Intrapreneurship in large firms and SMEs: A comparative study. *International Small Business Journal*, *12*(3), 54–62. doi:10.1177/0266242694123005

Construction, M. H. (2014). *The Business Value of BIM for Construction in Major Global markets*. Retrieved from https://synchroltd.com/newsletters/Business%20Value%20Of%20BIM%20In%20Global%20Markets%202014.pdf

Corsini, L., & Moultrie, J. (2017) An exploratory study into the impact of new digital design and manufacturing tools on the design process. *DS 87-2 Proceedings of the 21st International Conference on Engineering Design (ICED 17) Vol 2: Design Processes, Design Organisation and Management.*

Diekmann, J. E., Krewedl, M., Balonick, J., Stewart, T., & Won, S. (2004). *Application of lean manufacturing principles to construction*. Austin, TX: The Construction Industry Institute The University of Texas at Austin.

Eastman, C., Jeong, Y.-S., Sacks, R., & Kaner, I. (2010). Exchange Model and Exchange Object Concepts for Implementation of National BIM Standards. *ASCE Journal of Computing in Civil EngineeringASCE Journal of Computing in Civil Engineering*, *24*(1), 25–34. doi:10.1061/(ASCE)0887-3801(2010)24:1(25)

Eastman, C., Teicholz, P., Sacks, R., & Liston, K. (2011). BIM handbook: a guide to building information modeling for owners, managers designers, engineers, and contractors. Hoboken, NJ: John Wiley Sons Inc.

Egan, J. L. S. (1998). *Rethinking construction: the report of the Construction Task Force to the Deputy Prime Minister, John Prescott, on the scope for improving the quality and efficiency of UK construction*. London: Department of the Environment, Transport and the Regions.

Fakhimi, A. H., Majrouhi, S. J., & Azhar, S. (2016). How can Lean, IPD and BIM Work Together? ISARC. In *Proceedings of the International Symposium on Automation and Robotics in Construction*. Vilnius: Vilnius Gediminas Technical University, Department of Construction Economics & Property.

Farmer, M. (2016). *The Farmer Review of the UK Construction Labour Model*. Construction Leadership Council.

Fereday, S., & Potter, M. (2013). From lonely BIM to Design for Manufacture and Assembly (DfMA): A learning curve for one SME. *The Structural Engineer*, *91*(11).

French, W. L. (1999). Organization development: behavioral science interventions for organization improvement. Upper Saddle River, NJ: Prentice Hall.

Gann, D. M. (1996). Construction as a manufacturing process? Similarities and differences between industrialized housing and car production in Japan. *Construction Management and Economics*, *14*(5), 437–450. doi:10.1080/014461996373304

Gibb, A. G. F., & Isack, F. (2003). Re-engineering through pre-assembly: client expectations and drivers. *Building Research & Information, 31*(2), 146-160. doi:10.1080/09613210302000

Harmon, P. (2003). *Business process change: a manager's guide to improving, redesigning, and automating processes.* San Francisco, CA: Morgan Kaufmann.

Harty, J. (2015). Getting to Grips with BIM A Guide for Small and Medium-Sized Architecture, Engineering and Construction Firms. In G. Paterson (Ed.), Getting to grips with BIM. Taylor and Francis.

Hoffman, K., Parejo, M., Bessant, J., & Perren, L. (1998). Small firms, R&D, technology and innovation in the UK: A literature review. *Technovation, 18*(1), 39–55. doi:10.1016/S0166-4972(97)00102-8

Hosseini, M. R., Namzadi, M. O., Rameezdeen, R., Banihashemi, S., & Chileshe, N. (2016) Barriers to BIM adoption: perceptions from Australian Small and Medium Enterprises (SMEs). In *Proceedings of the 40th AUBEA Conference 'Radical Innovation in the Built Environment'*. Central Queensland University.

Indulska, M., Green, P., Recker, J., & Rosemann, M. (2009). *Business process modeling: Perceived benefits.* Paper presented at the International Conference on Conceptual Modeling.

Irizarry, J., Karan, E. P., & Jalaei, F. (2013). Integrating BIM and GIS to improve the visual monitoring of construction supply chain management. *Automation in Construction*, *31*, 241–254. doi:10.1016/j.autcon.2012.12.005

Kieran, S. (2003). Refabricating architecture. New York: McGraw-Hill.

Kolarevic, B. (2005). *Architecture in the digital age: design and manufacturing*. New York: Taylor & Francis.

Koskela, L. J., & Kazi, A. S. (2003). *Information technology in construction: How to realise the benefits?* IGI Publishing. doi:10.4018/978-1-59140-104-9.ch004

Kuo, T.-C., Huang, S. H., & Zhang, H.-C. (2001). Design for manufacture and design for ' X': Concepts, applications, and perspectives. *Computers & Industrial Engineering*, *41*(3), 241–260. doi:10.1016/S0360-8352(01)00045-6

Lam, T. T., Mahdjoubi, L., & Mason, J. (2017). A framework to assist in the analysis of risks and rewards of adopting BIM for SMEs in the UK. *Journal of Civil Engineering and Management*, *23*(6), 740–752. doi:10.3846/13923730.2017.1281840

Lee, G., Sacks, R., & Eastman, C. M. (2006). Specifying parametric building object behavior (BOB) for a building information modeling system. *Automation in Construction*, *15*(6), 758–776. doi:10.1016/j.autcon.2005.09.009

Liker, J. K. (2004). *The Toyota way: 14 management principles from the world's greatest manufacturer*. New York: McGraw-Hill.

Love, P. E. D., Matthews, J., Simpson, I., Hill, A., & Olatunji, O. A. (2014). A benefits realization management building information modeling framework for asset owners. *Automation in Construction*, *37*(0), 1–10. doi:10.1016/j.autcon.2013.09.007

McAdam, R., Moffett, S., Hazlett, S. A., & Shevlin, M. (2010). Developing a model of innovation implementation for UK SMEs: A path analysis and explanatory case analysis. *International Small Business Journal*, *28*(3), 195–214. doi:10.1177/0266242609360610

Mellon, S., & Kouider, T. (2016). SMEs and Level 2 BIM, the way forward. In *Proceedings of the 6th International Congress of Architectural Technology - Healthy Buildings: Innovation, Design & Technology*. University of Alicante.

Mounce, H. O. (1992). Pragmatism from Peirce to Davidson. *Philosophy (London, England)*, *67*(260), 260–262. doi:10.1017/S003181910003967X

NBS. (2014). *NBS National BIM Report 2014*. NBS.

NBS. (2015). *NBS National BIM Report 2015*. NBS.

NBS. (2017). National BIM Report 2017. Retrieved from https://www.thenbs.com/-/media/uk/files/pdf/nbs-national-bim-report-2017.pdf'?la=en

O'Rourke, L. (2013). *The future of DfMA is the future of construction* [Press release]. Retrieved from https://www.laingorourke.com/~/media/LOR/Files/LOR%20Engineering%20Excellence%20Journal%202013.pdf

Poirier, E., Staub-French, S., & Forgues, D. (2015). Embedded contexts of innovation: BIM adoption and implementation for a specialty contracting SME. *Construction Innovation*, *15*(1), 42–65. doi:10.1108/CI-01-2014-0013

Redford, A. H. (1994). *Design for assembly: principles and practice. London.* London: McGraw-Hill.

Sacks, R., Dave, B., Koskela, L. J., & Owen, R. L. (2009). *Analysis framework for the interaction between lean construction and building information modelling*. Academic Press.

Sacks, R., Koskela, L., Dave, B. A., & Owen, R. (2010). Interaction of Lean and Building Information Modeling in Construction. *Journal of Construction Engineering and Management*, *136*(9), 968–980. doi:10.1061/(ASCE)CO.1943-7862.0000203

Sacks, R., Radosavljevic, M., & Barak, R. (2010). Requirements for building information modeling based lean production management systems for construction. *Automation in Construction*, *19*(5), 641–655. doi:10.1016/j.autcon.2010.02.010

Sanchez, A. e., Hampson, K. e., & Vaux, S. e. (2016). *Delivering value with BIM: a whole-of-life approach*. Academic Press.

Schodek, D. L. (2005). Digital design and manufacturing: CAD/CAM applications in architecture and design. Hoboken, NJ: John Wiley.

Smith, D. K., & Tardif, M. (2009). *Building information modeling: a strategic implementation guide for architects, engineers, constructors, and real estate asset managers*. Hoboken, NJ: John Wiley & Sons. doi:10.1002/9780470432846

Statistics, O. N. (2015). *Construction Statistics - No. 16, 2015 Edition*. Retrieved from http://www.ons.gov.uk/ons/rel/construction/construction-statistics/no--15--2014-edition/index.html

Thomas, L., & Thomas, B. (2012). Evolution of large- scale industrialisation and service innovation in Japanese prefabrication industry. *Construction Innovation*, *12*(2), 156–178. doi:10.1108/14714171211215921

Wolstenholme, A., Austin, S. A., Bairstow, M., Blumenthal, A., Lorimer, J., McGuckin, S., . . . Davies, R. (2009). *Never waste a good crisis: A review of progress since Rethinking Construction and thoughts for our future*. Retrieved from https://dspace.lboro.ac.uk/2134/6040

Womack, J. (1996). *Lean thinking*. Simon & Schuster.

Compilation of References

Abbott, E. A. (1884). *Flatland: A romance of many dimensions*. London: Seeley.

Abdul Ghafar, M. (2016). *Professional Collaborative Cultural Model*. Universiti Putra Malaysia. Retrieved from https://upm.academia.edu/MaszuraAbdulGhafar

Abdul Ghafar, M., Ibrahim, R., & Shari, Z. (2014). Embedding cultural knowledge in Building Information Modeling (BIM) for fabrication efficicncy to reduce industrialized construction waste. In *15th International Conference on Computing in Civil and Building Engineering (ICCCBE)*. Orlando, FL: ASCE. 10.1061/9780784413616.025

Achille, C., Fassi, F., Lombardini, N., Gaudio, & Galbusera. (2011). Survey of the archaeological site of Nemi. A training experience. In *Proceedings of XXIIIrd International CIPA Symposium*. Prague, Czech Republic: Academic Press.

Achille, C., Lombardini, N., & Valentini, M. (2012). Aree archeologiche e sistemi di protezione: modelli di valutazione di compatibilità, in 'La conservazione del patrimonio architettonico all'aperto. Superfici, strutture, finiture e contesti'. In *Proceedings of the XXVIII Congress Science and Cultural Heritage* (pp. 269-277). Bressanone, Italy: Academic Press.

Adamson, I. T. (1995). *A general topology workbook*. Birkhäuser.

Adegoke, O., Gerard, B., & Andrew, M. (2007). Innovation types and performance in growing UK SMEs. *International Journal of Operations & Production Management*, *27*(7), 735–753. doi:10.1108/01443570710756974

AIA California Council. (2007). *Integrated Project Delivery – A Working Definition*. Retrieved from http://www.aia.org/aiaucmp/groups/aia/documents/pdf/aias076695.pdf

AIA California Council. (2008). *Integrated project delivery frequently asked questions*. Retrieved March 12, 2017, from http://aiacc.org/wpcontent/uploads/2010/07/AIACC_1108FAQ.pdf

Aibinu, A., & Papadonikolaki, E. (2016). BIM implementation and project coordination in Design-Build procurement. In *Proceedings of the 32nd Annual ARCOM Conference* (pp. 15-24). Manchester, UK: Association of Researchers in Construction Management.

Aldous, J. M., & Wilson, R. J. (2003). *Graphs and applications: an introductory approach*. Springer Science & Business Media.

Alexander, K. (Ed.). (2013). *Facilities management: theory and practice*. Routledge.

Alexandrov, P. (1937). Diskrete Räume. *Matematicheskii Sbornik, 44*(2), 501–518.

American Institute of Architects. (2007). *Integrated Project Delivery: A Guide*. New York: American Institute of Architects.

Ammari, K. E., & Hammad, A. (2014). Collaborative BIM-based markerless mixed reality framework for facilities maintenance. *Computing in Civil and Building Engineering - Proceedings of the 2014 International Conference on Computing in Civil and Building Engineering*, 657–664. 10.1061/9780784413616.082

Anh, N., Vinh, P. T., & Duy, H. (2012). A study on 4d gis spatio-temporal data model. *Proceedings of the fourth int. conf. on knowledge and systems engineering (KSE) 2012*, 34-38. doi: 10.1109/KSE.2012.1

ANIMA. (n.d.). *Legislative Decree 18 2016 April, n. 50*. Retrieved from http://en.anima.it/contenuti/10654/appalti-pubblici

Arayici, Y., Onyenobi, T. C., & Egbu, C. O. (2012). Building information modelling (BIM) for facilities management (FM): The MediaCity case study approach. *International Journal of 3D Information Modelling, 1*(1), 55-73.

Arayici, Y., Coates, S. P., Koskela, L. J., Kagioglou, M., Usher, C., & O'Reilly, K. (2009). *BIM implementation for an architectural practice*. CRC Press.

Arbulu, R., & Zabelle, T. (2006). *Implementing Lean in construction: How to succeed.* Paper presented at the Intl. Group for Lean Construction, Santiago, Chile.

Ariaratnam, S. T., & Rojas, E. M. (2009). Assessing the Environmental Impacts of Lean Supply System: A Case Study of Rebar Supply in High-Rise Condominium Construction Projects. In *Construction Research Congress 2009* (pp. 1009-1018). Reston, VA: American Society of Civil Engineers.

ARUP. (2016). *The Circular Economy in the Built Environment*. Retrieved from http://publications.arup.com/publications/c/circular_economy_in_the_built_environment

Association of Project Management. (2012). *APM body of knowledge* (6th ed.). Buckinghamshire, UK: Association of Project Management.

Atkin, B., & Brooks, A. (2009). *Total Facilities Management, Facilities* (3rd ed.; Vol. 12). Oxford, UK: Wiley-Blackwell.

Austin, S., Baldwin, A., Li, B., & Waskett, P. (2000). Analytical design planning technique (ADePT): A dependency structure matrix tool to schedule the building design process. *Construction Management and Economics*, *18*(2), 173–182. doi:10.1080/014461900370807

Autodesk Revit. (2012-2013). *Auto Desk*. Retrieved from http://www.autodesk.it

Azhar, S. (2011). Building Information Modeling (BIM): Trends, Benefits, Risks, and Challenges for the AEC Industry. *Leadership and Management in Engineering*, *11*(3), 241–252. doi:10.1061/(ASCE)LM.1943-5630.0000127

Ballard, G., Kim, Y. W., Jang, J. W., & Liu, M. (2007). *Road Map for Lean Implementation at the Project Level* (Vol. Research Report 234-11, pp. 426). Construction Industry Institute, Bureau of Engineering Research, The University of Texas at Austin.

Ballard, G., Jung, W., Kim, Y., & Han, S. (2012). Understanding of target value design for integrated project delivery with the context of game theory. In *Proceedings of Construction Research Congress 2012*, (pp. 556-563). West Lafayette, IN: American Society of Civil Engineers

Ballesty, S., Morris, J., Ding, L., Drogemuller, R., Mitchell, J., Schevers, H., . . . Spink, G. (2006). An Integrated Collaborative Approach for FM – Sydney Opera House FM Exemplar, Clients Driving Innovation: Moving Ideas into Practice. Gold Coast, Australia: Academic Press.

Bar, M., & Neta, M. (2008). The proactive brain: Using rudimentary information to make predictive judgments. *Journal of Consumer Behaviour*, *7*(4-5), 319–330. doi:10.1002/cb.254

Barrett, P., & Baldry, D. (2009). *Facilities management: Towards best practice*. John Wiley & Sons.

Barry, S. R. (2009). *Fixing my gaze: A scientist's journey into seeing in three dimensions*. Basic Books.

Barton, D., & Court, D. (2012). Making Advanced Analytics Work For You. *Harvard Business Review*, *90*(10), 78–83. PMID:23074867

Batra, S. (2014). Big data analytics and its reflections on DIKW hierarchy. *Review of Management*, *4*(1), 5–17.

Becerik-Gerber, B., Jazizadeh, F., Li, N., & Calis, G. (2012). Application Areas and Data Requirements for BIM-Enabled Facilities Management. *Journal of Construction Engineering and Management*, *138*(May), 431–442. doi:10.1061/(ASCE)CO.1943-7862.0000433

Beck, K. (2012, Spring). The state of Wisconsin: BIM – digital FM handover pilot projects. *Journal of Building Information Modeling*, 22-23.

Beetz, J., van Berlo, L., de Laat, R., & van den Helm, P. (2010). bimserver.org - An Open Source IFC Model Server. In *Proceedings of the CIB W78 2010: 27th International Conference* (pp. 16–18). Cairo, Egypt: Academic Press.

Bercerik-Gerber, B., & Kent, D. C. (2010). Understanding construction industry experience and attitudes toward integrated project delivery. *Journal of Construction Engineering and Management*, *136*(8), 815–825. doi:10.1061/(ASCE)CO.1943-7862.0000188

Berinato, S. (2016). Visualizations that really work. *Harvard Business Review*.

Bernstein, P. G., & Pittman, J. H. (2004). *Barriers to the adoption of building information modeling in the building industry*. Autodesk building solutions.

Best, R., & De Valence, G. (Eds.). (1999). *Building in value: predesign issues*. Sydney, Australia: Routledge.

Bicheno, J., & Holweg, M. (2009). The Lean toolbox: the essential guide to Lean transformation (4th ed.). Buckingham, UK: PICSIE.

Biddle, B., Curci, F. X., Haslach, T. F., Marchant, G. E., Askland, A., & Gaudet, L. (2012). The expanding role and importance of standards in the information and communications technology industry. *Jurimetrics*, 177–208.

Bilal, M., Oyedele, L. O., Qadir, J., Munir, K., Ajayi, S. O., Akinade, O. O., & Pasha, M. (2016). Big Data in the construction industry: A review of present status, opportunities, and future trends. *Advanced Engineering Informatics*, *30*(3), 500–521. doi:10.1016/j.aei.2016.07.001

Bildsten, L., & Guan, W. (2011). *The Study of a Kitchen Assembly Process in Industrial Housing*. Paper presented at the Nordic Conference of Construction Economics and Organization, Copenhagen, Denmark.

Billger, M., Heldal, I., Stahre, B., & Renstrom, K. (2004). Perception of color and space in virtual reality: a comparison between a real room and virtual reality models. In *Electronic Imaging 2004* (pp. 90–98). International Society for Optics and Photonics. doi:10.1117/12.526986

BIM Industry Working Group. (2011). *A report for the government construction client group Building Information Modelling (BIM) working party strategy paper*. Retrieved from http://www.bimtaskgroup.org/wp-content/uploads/2012/03/BIS-BIM-strategy-Report.pdf

BIM Task Group. (n.d.). *British Standards Institution, Department of Business, Energy & Industrial Strategy, 2016. BIM Level 2 Standards*. Retrieved from: http://bim-level2.org/en/standards/

BIS. (2011). *A report for the Government Construction Client Group Building Information Modelling (BIM)*. Working Party Strategy Paper. Retrieved from: http://www.bimtaskgroup.org/wp-content/uploads/2012/03/BIS-BIM-strategy-Report.pdf

Biscaya, S. (2013). *Coordination and Management of Information for Construction Design Projects – A Framework for Portugal*. Manchester, UK: University of Salford. Retrieved from http://usir.salford.ac.uk/29281/

Bockstael, D., & Issa, M. (2016). A methodology for contractor clash detection using Building Information Modelling on commercial construction projects. *Journal of Information Technology in Construction*, *21*, 233–249.

Boothroyd, G., Dewhurst, P., & Knight, W. A. (2002). Product design for manufacture and assembly (2nd ed.). New York: Marcel Dekker.

Borgatti, C., Felicori, M., Mauri, M. A., Calori, L., Guidazzoli, A., Pescarin, S., … Valentini, L. (2004). Databases and virtual environments: a good match for communicating complex cultural sites. In ACM SIGGRAPH 2004 Educators program (p. 30). ACM.

Borrmann, A., & Rank, E. (2009a). Topological analysis of 3D building models using a spatial query language. *Advanced Engineering Informatics*, *23*(4), 370–385. doi:10.1016/j.aei.2009.06.001

Borrmann, A., & Rank, E. (2009b). Specification and implementation of directional operators in a 3D spatial query language for building information models. *Advanced Engineering Informatics*, *23*(1), 32–44. doi:10.1016/j.aei.2008.06.005

Bouazza, T., & Greenwood, D. (2017). Revolutionizing the AEC Industry through the use of BIM and KM. In *Proceedings of The Ninth International Conference on Construction in the 21st Century (CITC-9)*. Dubai, UAE: Academic Press.

Bouchlaghem, D., Shang, H., Whyte, J., & Ganah, A. (2005). Visualisation in architecture, engineering and construction (AEC). *Automation in Construction*, *14*(3), 287–295. doi:10.1016/j.autcon.2004.08.012

Bradley, P. E., & Paul, N. (2013). *Dimension of Alexandrov topologies*. Retrieved August 25, 2017, from http://arxiv.org/abs/1305.1815

Bradley, P. (2015). Supporting data analytics for smart cities: An overview of data models and topology. In A. Gammerman, V. Vovk, & H. Papadopoulos (Eds.), *Statistical Learning and Data Sciences SLDS 2015, LNCS 9047* (pp. 406–413). Springer. doi:10.1007/978-3-319-17091-6_35

Bradley, P. E., & Paul, N. (2010). Using the relational model to capture topological information of spaces. *The Computer Journal*, *53*(1), 69–89. doi:10.1093/comjnl/bxn054

Bradley, P. E., & Paul, N. (2014). Comparing G-maps with other topological data structures. *GeoInformatica*, *18*(3), 595–620. doi:10.100710707-013-0191-1

Brilakis, I. K., & Soibelman, L. (2008). Shape-based retrieval of construction site photographs. *Journal of Computing in Civil Engineering*, *22*(1), 14–20. doi:10.1061/(ASCE)0887-3801(2008)22:1(14)

British Standards Institution. (2013). *PAS1192-2:2013 Specification for information management for the capital/delivery phase of construction projects using building information modelling*. London, UK: British Standards Institution.

Brodlie, K. (1995). Scientific visualization - Past, present and future. *Nuclear Instrument and Methods in Physics Reserach A*, 104–111.

Brooke, J. (1996). SUS-A quick and dirty usability scale. *Usability Evaluation in Industry*, *189*(194), 4-7.

Bruce, V., Green, P. R., & Georgeson, M. A. (2003). *Visual perception: Physiology, psychology, & ecology*. Psychology Press.

Bryde, D., Broquetas, M., & Volm, J. M. (2013). The project benefits of Building Information Modelling (BIM). *International Journal of Project Management*, *31*(7), 971–980. doi:10.1016/j.ijproman.2012.12.001

BSI. (2007). *Facility Management - Part 1: Terms and definitions: BS EN 15221-1:2006* (Vol. 44). London: BSI Standards Publication.

BSI. (2014). Asset management. Overview, principles and terminology. *BS ISO*, *55000*, 2014.

Bücken, A., & Rossmann, J. (2013). From the Volumetric Algorithm for Single-Tree Delineation Towards a Fully-Automated Process for the Generation of "Virtual Forests." *Progress and New Trends in 3D Geoinformation Sciences, Lecture Notes in Geoinformation and Cartography*, 79–99.

Buecken, A., & Rossmann, J. (2011). An Efficient, Fractal-Based, Bi-Linear Algorithm for the Interpolation of Inhomogeneously Distributed Geo-Data. In *RSPSoc Annual Conference "Earth Oberservation in a Changing World."* Academic Press.

Building Smart. (2012). *Common BIM Requirements*. Retrieved from http://www.en.buildingsmart.kotisivukone.com/3

Buildingsmart. (2008). *Industry Foundation Classess (IFC) data model*. Retrieved from http://www.buildingsmart.org/standard/ifc

buildingSMART. (2013a). *IfcExtrudedAreaSolid specification*. Retrieved from: http://www.buildingsmarttech.org/ifc/IFC4/final/html/schema/ifcgeometricmodelresource/lexical/ifcextrudedareasolid.htm

buildingSMART. (2013b). *IfcTriangulatedFaceSet specification*. Retrieved from: http://www.buildingsmarttech.org/ifc/IFC4/final/html/schema/ifcgeometricmodelresource/lexical/ifctriangulatedfaceset.htm

buildingSMART. (2017). *Summary of IFC Releases*. Retrieved from: http://www.buildingsmart-tech.org/specifications/ifc-releases/summary

Bungar, S. (2012). *Facilities Management-UK Overview*. London: Facilities Management.

Business Watch. (2005). ICT and Electronic Business in the construction Industry, IT adoption and e-business activity in 2005. *The European E-Business Market Watch. European Commission, Enterprise and Industry Directorate General.*

Cabinet Office. (2011). *E. a. R. G., & BIS, C. S. U.* London, UK: Government Construction Strategy.

Cabinet Office. (2011). *Government Construction Strategy*. Retrieved from https://www.gov.uk/government/uploads/system/uploads/attachment_data/file/61152/Government-Construction-Strategy_0.pdf

Cabinet Office. (2011). *Government Construction Strategy*. Retrieved November 5, 2017 from: https://www.gov.uk/government/uploads/system/uploads/attachment_data/file/61152/Government-Construction-Strategy_0.pdf

Cabinet Office. (2012). *The Government Soft Landings Policy*. Retrieved from BIM task group website: http://www.bimtaskgroup.org/gsl-policy-2

Cairns, G. (2003). Seeking a facilities management philosophy for the changing workplace. *Facilities*, *21*(5/6), 95–105. doi:10.1108/02632770310476705

Caldas, C. H., Soibelman, L., & Han, J. (2002). Automated classification of construction project documents. *Journal of Computing in Civil Engineering*, *16*(4), 234–243. doi:10.1061/(ASCE)0887-3801(2002)16:4(234)

Camastra, C., & Innocenti, A. (2015). *Dal dato storico-materico alla modellazione: il metodo BIM per i Beni Culturali nella Chiesa di Santa Maria delle Grazie a Cesate* (Unpublished thesis dissertation). Politecnico di Milano, Italy.

Carbonara, G. (2012). An Italian contribution to architectural restoration. *Frontiers of Architectural Research*, *1*(1), 2–9. doi:10.1016/j.foar.2012.02.007

Card, S. K., Mackinlay, J. D., & Shneiderman, B. (Eds.). (1999). *Readings in information visualization: using vision to think*. Morgan Kaufmann.

Carrier, C. (1994). Intrapreneurship in large firms and SMEs: A comparative study. *International Small Business Journal*, *12*(3), 54–62. doi:10.1177/0266242694123005

Cerovsek, T. (2011). A review and outlook for a "Building Information Model" (BIM): A multi-standpoint framework for technological development. *Advanced Engineering Informatics, Information mining and retrieval in design, 25*, 224–244.

Chachere, J. M., Kunz, J., & Levitt, R. (2009). The Role of Reduced Latency in Integrated Concurrent Engineering. *CIFE Working Paper#WP116.*

Chang, C. Y., Pan, W., & Howard, R. (2015). *Will the application of Building Information Modelling increase acceptance of integrated delivery systems in China?* London, UK: University College London.

Charalambous, G., Thorpe, T., Demian, P., Yeomans, S. G., Doughty, N., & Peters, C. (2013). Collaborative BIM in the Cloud and the Communication tools to support it. *Proceedings of the 30th CIB W78 International Conference on Applications of IT in the AEC industry*, 58 - 67.

CIC - Construction Industry Council. (2014). *Built Environment 2050: A Report on our Digital Future*. BIM 2050. Available at: https://www.google.co.uk/url?sa=t&rct=j&q=&esrc=s&source=web&cd=1&cad=rja&uact=8&ved=0ahUKEwjbyszd8pDYAhWpLcAKHckiBJgQFggpMAA&url=http%3A%2F%2Fcic.org.uk%2Fdownload.php%3Ff%3Dbe2050-cic-bim2050-2014-1.pdf&usg=AOvVaw3FocdkM1OqsWG9LLuznY2M

Codd, E. F. (1979). Extending the database relational model to capture more meaning. *ACM Transactions on Database Systems*, *4*(4), 397–434. doi:10.1145/320107.320109

Codd, E. F. (1990). *The relational model for database management: version 2*. Boston: AddisonWesley.

Codice dei Beni Culturali e del Paesaggio. (2014). *Gazzetta Ufficiale n°45*.

Codinhoto, R., Kiviniemi, A., Kemmer, S., & Da Rocha, C. G. (2011). *BIM Implementation: Manchester Town Hall Complex*. SCRI Report Series 6. Available at: http://live.scri.salford.ac.uk/wp-content/uploads/2011/12/MCC_Final_Research_Report.pdf

Codinhoto, R., Kiviniemi, A., Kemmer, S., & da Rocha, C. G. (2013a). BIM-FM implementation: an exploratory invesigation. *International Journal of 3-D Information Modeling, 2*(2), 1-15.

Codinhoto, R., Kiviniemi, A., Kemmer, S., & Gravina da Rocha, C. (2013). BIM-FM Implementation: An Exploratory Invesigation. *International Journal of 3-D Information Modeling, 2*(2), 1–15.

Codinhoto, R., Tzortzopoulos, P., Rooke, J., Kagioglou, M., & Koskela, L. (2008). *Facilitators and barriers to the integration of healthcare service and building design*. Academic Press.

Codinhoto, R., Kiviniemi, A., Kemmer, S., Martin, E., Donato, V., & Tonso, L. (2013b). *BIM FM at the Manchester Town Hall Complex: A Case Study Investigation. Research Report*. The University of Salford.

Cohen, J. (2010). *Integrated Project Delivery: Case studies*. Sacramento, CA: AIA California Council.

Comiskey, D., McKane, M., Jaffrey, A., Wilson, P., & Mordue, S. (2017). An analysis of data sharing platforms in multidisciplinary education. *Architectural Engineering and Design Management*, 1–18.

Comiskey, D.; McKane, M.; Eadie, R.; & Goldberg, D. E. (2016). Providing Collaborative Education with an International Dimension: An Ulster University and Pennsylvania State University Case Study. *International Journal of 3-D Information Modeling, 5*(2), 1–15.

Comlay, J. (2015). *Facilities Management (FM) Organisational Transformation: Building Information Modelling (BIM) is the Proposition* (MSc dissertation). The University of Salford.

Comlay, J., & Codinhoto, R. (2017). Facilities Management: Granularity of Information for a Digital Information Platform. *Proceedings of the International Conference on Sustainable Futures (ICSF) Bahrain 2017*.

Constructing Excellence. (2009). *Never waste a good crisis: a review of progress since rethinking construction and a thought for our future*. London, UK: Quick Brown and Fox Ltd.

Construction Industry Council. (2006). *Design quality indicators*. Retrieved May 14, 2017 from www.dqi.org.uk/DQI/Common/history.html

Construction Industry Training Board. (2016). *UK industry performance report based on the UK construction industry key performance indicators*. Watford, UK: Construction Industry Training Board.

Construction, M. H. (2014). *The Business Value of BIM for Construction in Major Global markets*. Retrieved from https://synchroltd.com/newsletters/Business%20Value%20Of%20BIM%20In%20Global%20Markets%202014.pdf

Corrocher, N. (2013). The Development of Short Message Services: Standard Organizations as Engines of Innovations. *Revue Economique*, *64*(1), 149–163. doi:10.3917/reco.641.0149

Corsini, L., & Moultrie, J. (2017) An exploratory study into the impact of new digital design and manufacturing tools on the design process. *DS 87-2 Proceedings of the 21st International Conference on Engineering Design (ICED 17) Vol 2: Design Processes, Design Organisation and Management*.

Cotts, D., Roper, K., & Payant, R. (2010). *The facility management handbook* (3rd ed.). New York: American Management Association.

Cracknell, D. (2012). *Sector BIM strategy: Improving BIM education & training*. The Higher Education Academy. Available at http://www.heacademy.ac.uk/assets/documents/disciplines/built_environment/BIM/BIM_Skills_CIC_David_Cracknell.pdf

Cragun, D., Pal, T., Vadaparampil, S. T., Baldwin, J., Hampel, H., & DeBate, R. D. (2015). Qualitative Comparative Analysis. *Journal of Mixed Methods Research*, *10*(3), 251–272. doi:10.1177/1558689815572023 PMID:27429602

CRC Construction Innovation. (2007). *Adopting BIM for Facilities Management: Solutions for Managing the Sydney Opera House*. Brisbane, Australia: Cooperative Research Center for Construction Innovation.

Crown. (2013). *Construction 2025. (URN BIS/13/955)*. London: HMSO.

Cruz, F., Chong, W. K., & Grau, D. (2015). Design for Disassembly and Deconstruction - Challenges and Opportunities. *Procedia Engineering*, *118*, 1296–1304. doi:10.1016/j.proeng.2015.08.485

Damen, T., Sebastian, R., MacDonald, M., Soetanto, D., Hartmann, T., Di Giulio, R., Bonsma, P., & Luig, K. (2015). The Application of BIM as Collaborative Design Technology for Collective Self-Organised Housing. *International Journal of 3-D Information Modeling*, *4*(1), 1–18.

Damer, B., Rasmussen, D., Neilson, M., Newman, P., Norkus, R., Clancey, W., ... Cochrane, T. (2004). Data-Driven Virtual Environment Assembly and Operation. In *Virtual Ironbird Workshop* (p. 1).

Darrington, J. (2011). Using a design-build contract for lean integrated project delivery. *Lean Construction Journal*, *7*, 85–91.

DataFit Product Features. (n.d.). Retrieved December 1, 2017, from http://www.oakdaleengr.com/datafit.htm

Daum, S., & Borrmann, A. (2014). Processing of topological BIM queries using boundary representation based methods. *Advanced Engineering Informatics*, *28*(4), 272–286. doi:10.1016/j.aei.2014.06.001

De Dartein, F. (1865). *Étude sur l'Architecture Lombarde et sur les origins de l'architecture Romano-Byzantine*. Paris, France: Dunod Éditeur.

Degasperi, M. (2015). *Il progetto della conservazione nel confronto con la metodologia BIM: problemi di modellazione e di 'linguaggio' nello studio di una porzione della Basilica di San Michele Maggiore a Pavia* (unpublished thesis). Politecnico di Milano, Italy.

Delavari, N., Said, N. S., Ibrahim, R., & Abdullah, M. (2011). HCI to Engage Design Team Members in IT-integrated Design Collaboration Process. *WSEAS Transaction Information Science and Applications*, *8*(9), 341–355.

Della Torre, S. (2016). *Beni Culturali e antisismica: il commento di Stefano Della Torre sull'annullamento della Circ. 18 del MiBACT*. Retrieved September 2, 2016, from http://www.ingenio-web.it/

Demian, P., & Fruchter, R. (2005). Measuring relevance in support of design reuse from archives of building product models. *Journal of Computing in Civil Engineering*, *19*(2), 119–136. doi:10.1061/(ASCE)0887-3801(2005)19:2(119)

Demian, P., & Walters, D. (2014). The advantages of information management through building information modelling. *Construction Management and Economics*, *32*(12), 1153–1165. doi:10.1080/01446193.2013.777754

Deshpande, A., Azhar, S., & Amireddy, S. (2014). A framework for a BIM-based knowledge management system. *Procedia Engineering*, *85*, 113–122. doi:10.1016/j.proeng.2014.10.535

Design-Build Institute of America. (2017). *What is Design-Build?* Retrieved April 21, 2017 from https://www.dbia.org/about/Pages/What-is-Design-Build.aspx

Di Benedetto, C., & Gozzo, M. (2013). *Aree Archeologiche e Bim: Verifiche di compatibilita* (Unpublished master's thesis dissertation). Politecnico di Milano, Italy.

Diedershagen, O., Koch, B., Weinacker, H., & Schütt, C. (2003). Combining LIDAR and GIS Data for the Extraction of Forest Inventory Parameters. In J. Hyyppä, E. Naesset, H. Olsson, T. Granqvist Phalén, & H. Reese (Eds.), *Proceedings of the Scandlaser scientific workshop on airborne laser scanning of forests* (pp. 157–165). Umeå, Sweden: Academic Press.

Diekmann, J. E., Krewedl, M., Balonick, J., Stewart, T., & Won, S. (2004). *Application of lean manufacturing principles to construction*. Austin, TX: The Construction Industry Institute The University of Texas at Austin.

Doboš, J., & Steed, A. (2012). Revision Control Framework for 3D Assets. In Eurographics 2012 - Posters (p. 3). Cagliari, Italy: Academic Press.

Domínguez-Martín, B. (2014). *Methods to process low-level CAD plans and creative Building Information Models (BIM)*. University of Jaén.

Dominich, S. (2008). *The modern algebra of information retrieval*. Heidelberg, Germany: Springer.

Donato, V. (2017). Towards design process validation integrating graph theory into BIM. *Architectural Engineering and Design Management, 13*(1), 22–38. doi:10.1080/17452007.2016.1208602

Dore, C., & Murphy, M. (2013). Semi-automatic modelling of building façades with shape grammars using historic building information modelling. *Proc. of the Int. Conf, on 3D ARCH 2013 - 3D Virtual Reconstruction and Visualization of Complex Architectures*, 57-64. 10.5194/isprsarchives-XL-5-W1-57-2013

Dore, C., & Murphy, M. (2012). Integration of Historic BIM (HBIM) and 3D GIS for Recording and Managing Cultural Heritage Sites. *Proc. of the 18th Int. Conf. on Virtual Systems and Multimedia (VSMM)*, 369–376.

Eadie, R., Odeyinka, H., Browne, M., McKeown, C., & Yohanis, M. (2014). Building Information Modeling adoption: An analysis of the barriers to implementation. *Journal of Engineering and Architecture*, *2*(1), 77–101.

Eastman, C., Teicholz, P., Sacks, R., & Liston, K. (2011). BIM handbook: a guide to building information modeling for owners, managers designers, engineers, and contractors. Hoboken, NJ: John Wiley Sons Inc.

Eastman, C. M. (1999). *Building product models: computer environments, supporting design and construction*. CRC Press.

Eastman, C. M., Eastman, C., Teicholz, P., & Sacks, R. (2011). *BIM Handbook: A Guide to Building Information Modeling for Owners, Managers, Designers, Engineers and Contractors*. Hoboken, NJ: John Wiley & Sons.

Eastman, C., Jeong, Y.-S., Sacks, R., & Kaner, I. (2010). Exchange Model and Exchange Object Concepts for Implementation of National BIM Standards. *ASCE Journal of Computing in Civil EngineeringASCE Journal of Computing in Civil Engineering*, *24*(1), 25–34. doi:10.1061/(ASCE)0887-3801(2010)24:1(25)

Eastman, C., Teicholz, P., Sacks, R., & Liston, K. (2008). *BIM handbook: a guide to Building Information Modeling for owners, managers, designers, engineers, and contractors*. John Wiley & Sons. doi:10.1002/9780470261309

Eastman, C., Teicholz, P., Sacks, R., & Liston, K. (2011). *BIM Handbook: A Guide to Building Information Modeling for Owners, Managers, Designers, Engineers, and Contractors* (2nd ed.). John Wiley & Sons.

Egan, J. (1998). *Rethinking construction: the report of the construction task force*. London, UK: HMSO.

Egan, J. L. S. (1998). *Rethinking construction: the report of the Construction Task Force to the Deputy Prime Minister, John Prescott, on the scope for improving the quality and efficiency of UK construction*. London: Department of the Environment, Transport and the Regions.

Egenhofer, M. J. (1991, January). Reasoning about binary topological relations. In *Advances in spatial databases* (pp. 141–160). Springer Berlin Heidelberg. doi:10.1007/3-540-54414-3_36

Egenhofer, M. J. (1994). Spatial SQL: A query and presentation language. *Knowledge and data engineering. IEEE Transactions on*, *6*(1), 86–95.

Egusa, H. (1983). Effects of Brightness, Hue, and Saturation on perceived depth between adjacent regions in the visual field. *Perception*, *12*(2), 167–175. doi:10.1068/p120167 PMID:6657423

Eisenmann, H., Fuchs, J., De Wilde, D., & Basso, V. (2012). ESA Virtual Spacecraft Design. In *5th International Workshop on Systems and Concurrent Engineering for Space Applications*. Academic Press.

Eisenmann, S., & Park, B. (2012). Building Information Modeling's impact on team performance. In *Proceedings of the 48th ASC Annual International Conference*. Normal, IL: Associated Schools of Construction.

Ellul, C., & Haklay, M. (2007). Requirements for Topology in 3D GIS. *Transactions in GIS*, *10*(2), 157–175. doi:10.1111/j.1467-9671.2006.00251.x

Ellul, C., & Haklay, M. M. (2009). *Using a B-rep structure to query 9-intersection topological relationships in 3D GIS–reviewing the approach and improving performance. In 3D geo-information sciences* (pp. 127–151). Springer Berlin Heidelberg.

Elmasri, R., & Navathe, S. B. (2010). *Database Systems: Models, Languages, Design, And Application Programming* (6th ed.). Prentice Hall International.

Emmitt, S., & Ruikar, K. (2013). *Collaborative design management*. Routledge.

Engelking, R. (1978). *Dimension theory*. Amsterdam: North-Holland.

EU BIM Task Group. (2017). *EUBIM Task Group*. Retrieved from http://www.eubim.eu

EUPPD. (2014). *D.L. n° 42 January 22nd*. EUPPD.

European Commission. (2017). *Energy Efficiency: Buildings*. Available at https://ec.europa.eu/energy/en/topics/energy-efficiency/buildings

European Procurement Directive. (2014). *Official Journal of the European Union*. Art.22, c.4.

Fai, S., Graham, K., Duckworth, T., Wood, N., & Attar, R. (2011). Building Information Modeling and heritage documentation. *Proc. of the 23rd Int. CIPA Symposium*, 43–48.

Fakhimi, A. H., Majrouhi, S. J., & Azhar, S. (2016). How can Lean, IPD and BIM Work Together? ISARC. In *Proceedings of the International Symposium on Automation and Robotics in Construction*. Vilnius: Vilnius Gediminas Technical University, Department of Construction Economics & Property.

Fallon, C. (1971). *Value analysis* (2nd ed.). Washington, DC: Wiley-Interscience.

Fang, M., Yan, X., Wenhui, Y., & Sen, C. (2012). The Storage and Management of Distributed Massive 3D Models based on G/S Mode. In Lecture Notes in Information Technology (Vol. 10). Academic Press.

Farmer, M. (2016). *The Farmer Review of the UK Construction Labour Model*. Construction Leadership Council.

Fereday, S., & Potter, M. (2013). From lonely BIM to Design for Manufacture and Assembly (DfMA): A learning curve for one SME. *The Structural Engineer*, *91*(11).

Fischer, M., & Kam, C. (2002). *PM4D final report, CIFE technical report*. CIFE. Retrieved from https://web.stanford.edu/group/4D/download/PM4D_Final_Report.pdf

Fischer, M. (2006). Formalizing Construction Knowledge for Concurrent Performance-Based Design. In *Intelligent Computing in Engineering and Architecture* (pp. 186–205). Berlin: Springer. doi:10.1007/11888598_20

Fischer, M., & Kunz, J. (2004). The Scope and Role of Information Technology in Construction. In *Proceeding-Japan Society of Civil Engineers* (pp. 1–32). DOTOKU GAKKAI. doi:10.2208/jscej.2004.763_1

Flanagan, T., Eckert, C., & Clarkson, P. J. (2007). Externalizing tacit overview knowledge: A model based approach to supporting design teams. *Artificial Intelligence for Engineering Design, Analysis and Manufacturing*, *21*(03), 227–242. doi:10.1017/S089006040700025X

Forns-Samso, F., Bogus, S. M., & Migliaccio, G. C. (2011). *Use of Building Information Modeling (BIM) in Facilities Management.* Paper presented at the 3rd International/9th Construction Speciality Conference, Ottawa, Ontario, Canada.

Forte, M., & Kurillo, G. (2010). Cyberarchaeology: Experimenting with teleimmersive archaeology. In *2010 16th International Conference on Virtual Systems and Multimedia, VSMM 2010* (pp. 155–162). Academic Press. 10.1109/VSMM.2010.5665989

French, W. L. (1999). Organization development: behavioral science interventions for organization improvement. Upper Saddle River, NJ: Prentice Hall.

Freyberger, F. K. B., & Färber, B. (2006). Compliance Discrimination Of Deformable Objects By Squeezing With One And Two Fingers. Proceedings of Eurohaptics 2006, 271–276.

Friendly, M., & Denis, D. J. (2001). *Milestones in the history of thematic cartography, statistical graphics, and data visualization.* Retrieved from http://www.math.yorku.ca/SCS/Gallery/milestone/milestone.pdf

Froese, T. (2003, April). Future Directions For IFC-Based Interoperability. *ITcon: Special Issue IFC- Product Models for the AEC Arena, 8*, 231–246.

Gann, D. M. (1996). Construction as a manufacturing process? Similarities and differences between industrialized housing and car production in Japan. *Construction Management and Economics*, *14*(5), 437–450. doi:10.1080/014461996373304

Garagnani, S., & Manferdini, A. M. (2013). Parametric accuracy: building information modeling process applied to the cultural heritage preservation. *Proc. of the Int. Conf, on 3D ARCH 2013 - 3D Virtual Reconstruction and Visualization of Complex Architectures*, 87-92. 10.5194/isprsarchives-XL-5-W1-87-2013

Garcia, A. C. B., Kunz, J., & Fischer, M. (2003). *Meeting Details : Methods to Instrument Meetings and Use Agenda Voting to Make Them More Effective.* CIFE Technical Report #147.

Garcia, R., Suárez, J. C., & Patenaude, G. (2007). Delineation of individual tree crowns for LiDAR tree and stand parameter estimation in Scottish woodlands. *The European Information Society - Leading the Way with Geo-Information, Lecture Notes in Geoinformation and Cartography*, 55–86.

Garner, W. R. (1974). *The processing of information and structure*. Lawrence Erlbaum Associates Potomac.

GCCG. (2011). Government Construction Client Group: BIM Working Party Strategy Paper. GCCG.

Gemignani, M. C. (1972). *Elementary Topology*. Courier Corporation.

Geng, J. (2013). Three-dimensional display technologies. *Advances in Optics and Photonics*, *5*(4), 456–535. doi:10.1364/AOP.5.000456 PMID:25530827

Gerrish, T., Ruikar, K., Cook, M., Johnson, M., Phillip, M., & Lowry, C. (2017). BIM application to building energy performance visualisation and management: Challenges and potential. *Energy and Building*, *144*, 218–228. doi:10.1016/j.enbuild.2017.03.032

Gibb, A. G. F., & Isack, F. (2003). Re-engineering through pre-assembly: client expectations and drivers. *Building Research & Information, 31*(2), 146-160. doi:10.1080/09613210302000

Giel, B. K., Mayo, G., & Issa, R. R. A. (2015). BIM Use and Requirements Among Building Owners. In R. R. A. Issa & S. Olbina (Eds.), Building Information Modeling - Applications and Practices. American Society of Civil Engineers (ASCE). doi:10.1061/9780784413982.ch10

Graphisoft Archicad. (n.d.). *Graphisoft*. Retrieved from https://www.graphisoft.com

Gray, C., & Hughes, W. (2001). *Building design management*. Routledge.

Greenberg, J. A., & Blommer, M. (2011). *Physical fidelity of driving simulators*. Boca Raton, FL: CRC Press. doi:10.1201/b10836-8

Grilo, A., & Jardim-Goncalves, R. (2011). Challenging electronic procurement in the AEC sector: A BIM-based integrated perspective. *Automation in Construction*, *20*(2), 107–114. doi:10.1016/j.autcon.2010.09.008

Guan, P., & Banks, M. S. (2016). Stereoscopic depth constancy. *Philosophical Transactions of the Royal. Society B, 371*(1697), 20150253.

Guan, T., Ren, B., & Zhong, D. (2012). The Method of Unity3D-Based 3D Dynamic Interactive Query of High Arch Dam Construction Information. *Applied Mechanics and Materials*, *256–259*, 2918–2922. doi:10.4028/www.scientific.net/AMM.256-259.2918

Guarnieri, A., Pirotti, F., & Vettore, A. (2010). Cultural heritage interactive 3D models on the web: An approach using open source and free software. *Journal of Cultural Heritage*, *11*(3), 350–353. doi:10.1016/j.culher.2009.11.011

Gu, N., & London, K. (2010). Understanding and facilitating BIM adoption in the AEC industry. *Automation in Construction*, *19*(8), 988–999. doi:10.1016/j.autcon.2010.09.002

Haist, J., & Coors, V. (2005). The W3DS-Interface of Cityserver3D. In Kolbe & Gröger (Eds.), *European Spatial Data Research (EuroSDR) u.a.: Next Generation 3D City Models. Workshop Papers : Participant's Edition* (pp. 63–67). Bonn: Academic Press.

Hallberg, D., & Tarandi, V. (2011). On the use of open bim and 4d visualisation in a predictive life cycle management system for construction works. *Journal of Information Technology in Construction, 16*, 445–466. Retrieved from http://www.diva-portal.org/smash/record.jsf?pid=diva2%3A478341&dswid=-8750

Hamil, S. (2012). *Building Information Modeling and Interoperability*. Retrieved from https://www.thenbs.com/topics/BIM/articles/bimAndInteroperability.asp

Harmon, P. (2003). *Business process change: a manager's guide to improving, redesigning, and automating processes*. San Francisco, CA: Morgan Kaufmann.

Harpaceas. (2017). *Harpaceas the BIM Expert*. Retrieved from http://www.harpaceas.it

Hartshorne, R. (1993). *Algebraic geometry*. New York: Springer.

Harty, J. (2015). Getting to Grips with BIM A Guide for Small and Medium-Sized Architecture, Engineering and Construction Firms. In G. Paterson (Ed.), Getting to grips with BIM. Taylor and Francis.

Hatcher, A. (2002). *Algebraic topology*. Cambridge University Press. Retrieved August 12, 2013, from http://www.math.cornell.edu/~hatcher/AT/ATpage.html

Haymaker, J., Chachere, J. M., & Senescu, R. R. (2011). Measuring and improving rationale clarity in a university building design process. *Journal of Architectural Engineering, 17*(3), 97–11. doi:10.1061/(ASCE)AE.1943-5568.0000041

Hazelton, N., Leahy, F., & Williamson, I. (1990). On the design of temporally-referenced, 3-D geographical information systems: development of four-dimensional GIS. *Proceedings of GIS/LIS'90*, 357-372.

Heesom, D., & Mahdjoubi, L. (2004). Trends of 4D CAD applications for construction planning. *Construction Management and Economics, 22*(2), 171–182. doi:10.1080/0144619042000201376

Her Majesty's Government. (2013). *Construction 2025: industrial strategy for construction – Government and industry in partnership*. London, UK: Her Majesty's Government.

Hickethier, G., Tommelein, I. D., & Lostuvali, B. (2013). Social network analysis of information flow in an IPD-project design organisation. In Proceedings of the 21st Annual Conference of the International Group for Lean Construction (IGLC). Fortaleza, Brazil: International Group for Lean Construction.

Hillaire, S., Lécuyer, A., Cozot, R., & Casiez, G. (2008). Depth-of-field blur effects for first-person navigation in virtual environments. *IEEE Computer Graphics and Applications, 28*(6), 47–55. doi:10.1109/MCG.2008.113 PMID:19004684

Ho, P., Fischer, M., & Haymaker, J. (2009). *Automated Identification of Occupant Interactions in Renovations of Occupied Buildings*. CIFE Working Paper#WP122.

Ho, P., Fischer, M., & Kam, C. (2009). *Prospective Validation of Virtual Design and Construction Methods: Framework, Application, and Implementation Guidelines*. CIFE Working Paper#WP123.

Hoerster, S., & Menzel, K. (2015). BIM based classification of building performance data for advanced analysis. In *Proceedings of International Conference CISBAT 2015 Future Buildings and Districts Sustainability from Nano to Urban Scale* (pp. 993–998). Academic Press.

Hoffman, K., Parejo, M., Bessant, J., & Perren, L. (1998). Small firms, R&D, technology and innovation in the UK: A literature review. *Technovation*, *18*(1), 39–55. doi:10.1016/S0166-4972(97)00102-8

Hofstede, G. (1997). *Cultures and Organizations: Software of the mind*. McGraw Hill International.

Hoppen, M., Rossmann, J., Schluse, M., & Waspe, R. (2010). A New Method For Interfacing 3D Simulation Systems And Object-Oriented Geo Data Sources. In T. Kolbe, G. König, & C. Nagel (Eds.), *The 5th International Conference on 3D Geoinformation ISPRS 2010* (pp. 51–56). Shaker Verlag.

Hoppen, M., Schluse, M., Rossmann, J., & Averdung, C. (2015). A new nD Temporal Geodata Management Approach using GML. In Rückemann & Doytsher (Eds.), *GEOProcessing 2015 - The Seventh International Conference on Advanced Geographic Information Systems, Applications, and Services* (pp. 110–116). Lisbon, Portugal: IARIA (International Academy, Research, and Industry Association).

Hoppen, M., Schluse, M., Rossmann, J., & Weitzig, B. (2012). Database-Driven Distributed 3D Simulation. In *Proceedings of the 2012 Winter Simulation Conference* (pp. 1–12). Academic Press.

Hosseini, M. R., Namzadi, M. O., Rameezdeen, R., Banihashemi, S., & Chileshe, N. (2016) Barriers to BIM adoption: perceptions from Australian Small and Medium Enterprises (SMEs). In *Proceedings of the 40th AUBEA Conference 'Radical Innovation in the Built Environment'*. Central Queensland University.

Howard, I. P. (2012). Perceiving in depth, volume 1: basic mechanisms. New York: Oxford University Press. doi:10.1093/acprof:oso/9780199764143.001.0001

Howard, R., & Björk, B.-C. (2008). Building information modelling – Experts' views on standardisation and industry deployment. *Advanced Engineering Informatics, Network methods in engineering, 22*, 271–280.

Howard, I. P., & Rogers, B. J. (1995). *Binocular vision and stereopsis*. New York: Oxford University Press.

Howard, R., & Bjork, B. (2008). Building information modelling – Experts' views on standardisation and industry deployment. *Advanced Engineering Informatics*, *22*(2), 271–280. doi:10.1016/j.aei.2007.03.001

Hyyppä, J., & Inkinen, M. (1999). Detecting and estimating attributes for single trees using laser scanner. *The Photogrammetric Journal of Finland*, *16*(2), 27–42.

Ibrahim, M., Krawczyk, R., & Schipporiet, G. (2004). No Title. Two Approaches to BIM: a Comparative Study. Proceedings of eCAADe 2004.

Ibrahim, R., & Paulson, B. C. (2008). Discontinuity in organizations: Identifying business environment affecting of knowledge flow in PLM. *International Journal of Product Lifecycle Management*, *3*(1), 21–36. doi:10.1504/IJPLM.2008.019972

Ibrahim, R., & Pour, F. (2010). Comparison of CAD and manual sketching tools for teaching architectural design. *Automation in Construction*, *19*(8), 978–987. doi:10.1016/j.autcon.2010.09.003

ICOMOS. (2008). *Illustrated glossary on stone deterioration patterns*. Retrieved from https://www.icomos.org/publications/monuments_and_sites/15/pdf/Monuments_and_Sites_15_ISCS_Glossary_Stone.pdf

Iliescu, D., Ciocan, I., & Mateias, I. (2014). Assisted management of product data: A PDM application proposal. *Proceedings of the 18th International Conference on System Theory, Control and Computing*. 10.1109/ICSTCC.2014.6982403

Ilozor, B., & Kelly, D. (2011). Building Information Modelling and Integrated Project Delivery in the commercial construction industry: A conceptual study. *Journal of Engineering, Project, and Production Management*, *2*(1), 23–36.

Indulska, M., Green, P., Recker, J., & Rosemann, M. (2009). *Business process modeling: Perceived benefits*. Paper presented at the International Conference on Conceptual Modeling.

Irizarry, J., Gheisari, M., Williams, G., & Walker, B. N. (2013). InfoSPOT: A mobile Augmented Reality method for accessing building information through a situation awareness approach. *Automation in Construction*, *33*, 11–23. doi:10.1016/j.autcon.2012.09.002

Irizarry, J., Karan, E. P., & Jalaei, F. (2013). Integrating BIM and GIS to improve the visual monitoring of construction supply chain management. *Automation in Construction*, *31*, 241–254. doi:10.1016/j.autcon.2012.12.005

Isikdag, U. (2015). Building Information Models: An Introduction. In Enhanced Building Information Models. In Enhanced Building Information Models (pp. 1–12). Springer International Publishing.

Isikdag, U., Zlatanova, S., & Underwood, J. (2011). *An opportunity analysis on the future role of BIMs in urban data management* (pp. 25–36). Urban and Regional Data Management–UDMS Annual. doi:10.1201/b11647-5

Jakefowler (2014). *Camera Settings – Make Your Renderings Look More Realis*tic. Retrieved October 1, 2017, From Https://Www.Autodesk.Com/Products/Fusion-360/Blog/Camera-Settings-Make-Your-Renderings-Look-More-Realistic/

Jalaei, F., & Jrade, A. (2014). Developing a spatial data framework for facility management supply chain. In *Construction Research Congress 2014* (pp. 140–149). Academic Press. 10.1061/9780784413517.176

Jänich, K. (1995). *Topology*. Springer.

Jezernik, A., & Hren, G. (2003). A solution to integrate computer-aided design (CAD) and virtual reality (VR) databases in design and manufacturing processes. *International Journal of Advanced Manufacturing Technology*, *22*(11–12), 768–774. doi:10.100700170-003-1604-3

Jin, Y., Levitt, R., Kunz, J., & Christiansen, T. R. (1995). The Virtual Design Team : A Computer Simulation Framework for Studying Organizational Aspects of Concurrent design. *Simulation*, *64*(3), 42–57. doi:10.1177/003754979506400304

Juan, D., & Zheng, Q. (2014). Cloud and Open BIM-Based Building Information Interoperability Research. *Journal of Service Science and Management*, *7*(April), 47–56. doi:10.4236/jssm.2014.72005

Jung, Y., & Gibson, G. E. (1999). Planning for computer integrated construction. *Journal of Computing in Civil Engineering*, *13*(4), 217–225. doi:10.1061/(ASCE)0887-3801(1999)13:4(217)

Jung, Y., & Joo, M. (2011). Building Information Modeling (BIM): Framework for Practical Implementation. *Automation in Construction*, *20*(2), 126–133. doi:10.1016/j.autcon.2010.09.010

Jylhä, T., & Suvanto, M. E. (2015). Impacts of poor quality of information in the facility management field. *Facilities*, *33*(5/6), 302–319. doi:10.1108/F-07-2013-0057

Kamat, V. R., & Martinez, J. C. (2001). Visualizing Simulated Construction Operations in 3D. *Journal of Computing in Civil Engineering*, *15*(4), 329–337. doi:10.1061/(ASCE)0887-3801(2001)15:4(329)

Kam, C., & Fischer, M. (2004). Capitalizing on early project decision-making opportunities to improve facility design, construction, and life-cycle performance — POP, PM4D, and decision dashboard approaches. *Automation in Construction*, *13*(1), 53–65. doi:10.1016/j.autcon.2003.08.004

Kam, C., Fischer, M., Manager, G., Oy, O. G., Karjalainen, A., Manager, C., ... Solutions, C. (2003). The product model and fourth dimension project. *ITcon*, *8*, 137–166.

Kamiura, M., Oisol, H., Tajima, K., & Tanaka, K. (1997). Spatial views and LOD-based access control in VRML-object databases. In T. Masuda, Y. Masunaga, & M. Tsukamoto (Eds.), Worldwide Computing and Its Applications (Vol. 1274, pp. 210–225). Springer Berlin / Heidelberg.

Kang, H., & Lee, G. (2009). *Development of an object-relational IFC server*. ICCEM/ICCPM.

Karasulu, Y., Pishdad-Bozorgi, P., & Moghaddam, E. H. (2013). Advancing target price and target value design process in IPD using BIM and risk-sharing approaches. In *Proceedings of the 49th ASC Annual International Conference*. San Luis Obispo, CA: Associated Schools of Construction.

Kassem, M., Iqbal, N., Kelly, G., Lockley, S., & Dawood, N. (2014). Building Information Modelling: Protocols for collaborative design processes. *Journal of Information Technology in Construction*, *19*, 126–149.

Kassem, M., Jenaban, M., Craggs, D., & Dawood, D. (2016). A tool for the assessment of project compliance with Level 2 BIM. *16th International Conference on Construction Applications of Virtual Reality*, Hong Kong.

Kassem, M., Kelly, G., Dawood, N., Serginson, M., & Lockley, S. (2015). BIM in facilities management applications: A case study of a large university complex. *Built Environment Project and Asset Management*, *5*(3), 261–277. doi:10.1108/BEPAM-02-2014-0011

Kelly, J. (2007). Making client values explicit in value management workshops. *Construction Management and Economics*, *25*(4), 435–442. doi:10.1080/01446190601071839

Khanzode, A. (2010). *An Integrated, Virtual Design and Construction and Lean (IVL) Method for Coordination of MEP*. CIFE Technical Report #TR187.

Khanzode, A., Fischer, M., Reed, D., & Ballard, G. (2006). *A Guide to Applying the Principles of Virtual Design & Construction (VDC) to the Lean Project Delivery Process*. CIFE Working Paper#093.

Khuan, C. T., Abdul-rahman, A., & Zlatanova, S. (2008). 3D Solids and Their Management. In *DBMS. In Advances in 3D geoinformation systems* (pp. 279–311). Springer Berlin Heidelberg. doi:10.1007/978-3-540-72135-2_16

Kieran, S. (2003). Refabricating architecture. New York: McGraw-Hill.

Kiviniemi, A., & Codinhoto, R. (2014). Challenges in the Implementation of BIM for FM—Case Manchester Town Hall Complex. *Computing in Civil and Building Engineering*, 665–672. Retrieved from http://ascelibrary.org/doi/abs/10.1061/9780784413616.083

Kiviniemi, A., Fischer, M., Bazjanac, V., & Paulson, B. (2004). *PREMISS - Requirements Management Interface to Building Product Models: Problem Definition and Research Issues*. CIFE Working Paper#092.

Kiviniemi, A., Tarandi, V., Karlshøj, J., Bell, H., & Karud, O. J. (2008). *Review of the development and implementation of IFC compatible BIM*. Erabuild Funding Organizations, Europe.

Klepeis, N. E., Nelson, W. C., Ott, W. R., Robinson, J. P., Tsang, A. M., Switzer, P., ... Engelmann, W. H. (2001). The National Human Activity Pattern Survey (NHAPS): A resource for assessing exposure to environmental pollutants. *Journal of Exposure Science & Environmental Epidemiology*, *11*(3), 231–252. doi:10.1038j.jea.7500165 PMID:11477521

Knotten, V., Svalestuen, F., Hansen, G. K., & Laedre, O. (2015). Design management in the building process - a review of current literature. *Procedia Economics and Finance*, *21*, 120–127. doi:10.1016/S2212-5671(15)00158-6

Knutt, E. (2016). *Low confidence on level 2 BIM rests on weak uptake of '8 pillars': BIM + survey*. Retrieved April 21, 2017 from http://www.bimplus.co.uk/news/bim-survey-lowconfid3enc4e-limi1ted-experience/

Kolarevic, B. (2005). *Architecture in the digital age: design and manufacturing*. New York: Taylor & Francis.

Konukcu, S., & Koseoglu, O. (2012). Knowledge Management Through Bim in Construction Supply Chains. In *CIB W78 2012: 29th International Conference* (pp. 17–19). Beirut: Academic Press.

Koo, B., & Fischer, M. (2000). Feasibility study of 4D CAD in commercial construction. *Journal of Construction Engineering and Management*, *126*(4), 251–260. doi:10.1061/(ASCE)0733-9364(2000)126:4(251)

Korsgaard, C. M. (1986). Aristotle and Kant on the source of value. *Ethics*, *96*(3), 486–505. doi:10.1086/292771

Kosara, R. (2007). Visualization Criticism - The Missing Link Between Information Visualization and Art. In *2007 11th International Conference Information Visualization (IV '07)* (pp. 631–636). IEEE.

Koskela, L., & Kazi, A. S. (2003). Information technology in construction: how to realize the benefits? In S. Clarke, E. Coakes, M. G. Hunter, & A. Wenn (Eds.), Socio-technical and human cognition elements of information systems (pp. 60-75). Hershey, PA: Idea Group Inc (IGI) Publishing. doi:10.4018/978-1-59140-104-9.ch004

Kunz, J., & Fischer, M. (2012). *Virtual Design and Construction: Themes, Case Studies and Implementation Suggestions*. CIFE Working Paper#097, 14.

Kunz, J. C., Christiansen, T. R., Cohen, G. P., Jin, Y., & Levitt, R. E. (1998). The virtual design team. *Communications of the ACM*, *41*(11), 84–91. doi:10.1145/287831.287844

Kuo, T.-C., Huang, S. H., & Zhang, H.-C. (2001). Design for manufacture and design for ' X': Concepts, applications, and perspectives. *Computers & Industrial Engineering*, *41*(3), 241–260. doi:10.1016/S0360-8352(01)00045-6

Laakso, M., & Kiviniemi, A. O. (2012). The IFC standard: A review of history, development, and standardization, information technology. *ITcon*, *17*(9), 134–161.

Lam, T. T., Mahdjoubi, L., & Mason, J. (2017). A framework to assist in the analysis of risks and rewards of adopting BIM for SMEs in the UK. *Journal of Civil Engineering and Management*, *23*(6), 740–752. doi:10.3846/13923730.2017.1281840

Lansdale, M., & Edmonds, E. (1992). Using memory for events in the design of personal filing systems. *International Journal of Man-Machine Studies*, *36*(1), 97–126. doi:10.1016/0020-7373(92)90054-O

Latham, M. (1994). *Constructing the team*. London, UK: HMSO.

Lee, A., Marshall Pointing, A., Aouad, G., Tah, J., Cooper, R., & Fu, C. F. (2005). *nD modeling- a driver or enabler for construction improvement?* RICS Research paper Series, University of Salford.

Lee, G., Sacks, R., & Eastman, C. M. (2006). Specifying parametric building object behavior (BOB) for a building information modeling system. *Automation in Construction*, *15*(6), 758–776. doi:10.1016/j.autcon.2005.09.009

Leite, F., Akcamete, A., Akinci, B., Atasoy, G., & Kiziltas, S. (2011). Analysis of modeling effort and impact of different levels of detail in building information models. *Automation in Construction*, *20*(5), 601–609. doi:10.1016/j.autcon.2010.11.027

Leslie, H. G. (1996). Strategy for information in the AEC industry. *INCIT 96 Proceedings: Bridging the Gap*.

Lienhardt, P. (1991). Topological models for boundary representation: A comparison with ndimensional generalized maps. *Computer Aided Design*, *23*(1), 59–82. doi:10.1016/0010-4485(91)90082-8

Lienhardt, P. (1994). N-dimensional generalized combinatorial maps and cellular quasi-manifolds. *International Journal of Computational Geometry & Applications*, *4*(03), 275–324. doi:10.1142/S0218195994000173

Liker, J. K. (2004). *The Toyota way: 14 management principles from the world's greatest manufacturer*. New York: McGraw-Hill.

Lillrank, P. (1995). The transfer of management innovations from Japan. *Organization Studies*, *16*(6), 971–989. doi:10.1177/017084069501600603

Lin, C. J. (2013). Smart Spatial Ontology: Bridging Semantic Ontology to Spatial Topology. *Computer-Aided Design and Applications*, *10*(3), 489–497. doi:10.3722/cadaps.2013.489-497

Lindblad, H. (2013). *Study of the implementation process of BIM in construction projects: analysis of the barriers limiting BIM adoption in the AEC industry* (Master's Thesis). KTH Royal Institute of Technology, Stockholm, Sweden.

Lin, K. Y., & Soibelman, L. (2009). Incorporating domain knowledge and information retrieval techniques to develop an architectural/engineering/construction online product search engine. *Journal of Computing in Civil Engineering*, *23*(4), 201–210. doi:10.1061/(ASCE)0887-3801(2009)23:4(201)

Linning, C. (Producer). (2015). *BIM for Facility Management* [PowerPoint slides]. Retrieved from https://bips.dk/files/article_files/keynote_bim_i_fm_paa_sydney_opera_house.pdf

Liston, K., Fischer, M., Kunz, J., & Dong, N. (2007). *Observations of Two MEP iRoom Coordination Meetings: An Investigation of Artifact Use in AEC Project Meetings.* CIFE Working Paper#WP106.

Liu, R., & Issa, R. R. A. (2013). Issues in BIM for Facility Management from Industry Practitioners' Perspectives. *Computing in Civil Engineering*, 411–418.

Liu, Y. (2015). Big data and predictive business analytics. *The Journal of Business Forecasting*, *33*(4), 40–42.

Lobur, M., Matviykiv, O., Kernytskyy, A., & Dobosz, R. (2011). Presentation of the heat simulation model with use of relational data model. *2011 11th International Conference The Experience of Designing and Application of CAD Systems in Microelectronics CADSM*, 228–229.

Lombardini, N. (Ed.). (2015). Un approccio interdisciplinare al progetto di conservazione. La valorizzazione dell'area archeologica dedicata al culto della Diana Nemorense. Firenze, Italy: Altralinea edizioni.

Lombardini, N., & Cantini, L. (in press). Non-standardized data in the bim process. the management of construction systems data in the cultural heritage conservation. In *Proceedings of the 3rd International Conference on Protection of Historical Constructions*. Lisbon, Portugal: Academic Press.

Looi, M. (2013). *Integrated project insurance – an inevitable step to BIM implementation.* Retrieved April 21, 2017 from: http://www.lexology.com/library/detail.aspx?g=249c6f9c-c364-4364-a3b6-eb4881e903f3

Love, P. E., Matthews, J., Simpson, I., Hill, A., & Olatunji, O. A. (2014). A benefits realization management building information modeling framework for asset owners. *Automation in Construction*, *37*, 1–10. doi:10.1016/j.autcon.2013.09.007

Lucas, P. (1999). *The trillion-node network.* MAYA Technical Report #MTR-00001.

MacLeamy, P. (2008). *BIM, BAM, BOOM! How to build greener, high-performance buildings.* Urban Land Green Magazine.

Mahdjoub, M., Monticolo, D., Gomes, S., & Sagot, J. (2010). A collaborative design for usability approach supported by virtual reality and a multi-agent system embedded in a PLM environment. *Computer Aided Design*, *42*(5), 402–413. doi:10.1016/j.cad.2009.02.009

Malaikrisanachalee, S., & Vathananukij, H. (2011). Integration of java-based BIM with spatial database. *International Journal of Civil Engineering*, *9*(1), 17–22.

Male, S., Kelly, J., Gronqvist, M., & Graham, D. (2007). Managing value as a management style for projects. *International Journal of Project Management*, *25*(2), 107–114. doi:10.1016/j.ijproman.2006.09.001

Manoharan, T., Taylor, H., & Gardiner, P. (2002). A collaborative analysis tool for visualisation and interaction with spatial data. In *Proceedings of the seventh international conference on 3D Web technology* (pp. 75–83). ACM. 10.1145/504502.504515

Martina, A., & Bottino, A. (2012). Using Virtual Environments as a Visual Interface for Accessing Cultural Database Contents. In *International Conference of Information Science and Computer Applications (ICISCA 2012)* (pp. 1–6). Bali, Indonesia: Academic Press.

Matthews, O., & Howell, G. (2005). Integrated Project Delivery: An example of relational contracting. *Lean Construction Journal*, *2*(1), 46–61.

McAdam, R., Moffett, S., Hazlett, S. A., & Shevlin, M. (2010). Developing a model of innovation implementation for UK SMEs: A path analysis and explanatory case analysis. *International Small Business Journal*, *28*(3), 195–214. doi:10.1177/0266242609360610

McCord, M. C. (1966). Singular homology and homotopy groups of finite topological spaces. *Duke Mathematical Journal*, *33*(3), 465–474. doi:10.1215/S0012-7094-66-03352-7

McCormack, K., & Lockamy, A. III. (2004). The development of a supply chain management process maturity model using the concepts of business process orientation. *Supply Chain Management*, *9*(4), 272–278. doi:10.1108/13598540410550019

McGraw-Hill Construction. (2012). *SmartMarket report - the business value of BIM in North America*. Bedford, MA: McGraw-Hill Construction.

Mckinney, K., & Fischer, M. (1998). Generating, evaluating and visualizing construction schedules with CAD tools. *Automation in Construction*, *7*(6), 433–447. doi:10.1016/S0926-5805(98)00053-3

Mêda, P., Hjelseth, E., & Sousa, H. (2017). Construction information framework—the role of classification systems. In *eWork and eBusiness in Architecture, Engineering and Construction: ECPPM 2016: Proceedings of the 11th European Conference on Product and Process Modelling (ECPPM 2016)* (p. 447). CRC Press.

Mellon, S., & Kouider, T. (2016). SMEs and Level 2 BIM, the way forward. In *Proceedings of the 6th International Congress of Architectural Technology - Healthy Buildings: Innovation, Design & Technology*. University of Alicante.

Mignard, C., & Nicolle, C. (2015). A Building Information Management (BIM) Framework and Supporting Case Study for Existing Building Operations, Maintenance and Sustainability. *Procedia Engineering*, *118*, 1104–1111. doi:10.1016/j.proeng.2015.08.450

Ministry of the public education. (1972). Carta italiana del restauro. *Circular n° 117*, April 6th, 1972.

Monteiro, A., Mêda, P., & Martins, J. P. (2014). Framework for the coordinated application of two different integrated project delivery platforms. *Automation in Construction*, *38*, 87–99. doi:10.1016/j.autcon.2013.11.003

Moore, G. E. (1998). The conception of intrinsic value. In J. Rachels (Ed.), *Philosophical Studies*. Oxford, UK: Oxford University Press.

Motamedi, A., Hammad, A., & Asen, Y. (2014). Knowledge-assisted BIM-based visual analytics for failure root cause detection in facilities management. *Automation in Construction*, *43*, 73–83. doi:10.1016/j.autcon.2014.03.012

Motawa, I., & Almarshad, A. (2013). A knowledge-based BIM system for building maintenance. *Automation in Construction*, *29*, 173–182. doi:10.1016/j.autcon.2012.09.008

Mounce, H. O. (1992). Pragmatism from Peirce to Davidson. *Philosophy (London, England)*, *67*(260), 260–262. doi:10.1017/S003181910003967X

National Building Specification. (2014). *BIM management handbook*. Newcastle, UK: National Building Specification.

National Institutes of Building Sciences. (2018). *National BIM Guide for Owners*. Retrieved from https://www.nibs.org/?nbgo

Naylor, S. R. (2017). *NHS Property and Estates: Why the estate matters for patients*. London, UK: Crown.

NBS. (2014). *NBS National BIM Report 2014*. NBS.

NBS. (2015). *NBS National BIM Report 2015*. NBS.

NBS. (2017). National BIM Report 2017. Retrieved from https://www.thenbs.com/-/media/uk/files/pdf/nbs-national-bim-report-2017.pdf'?la=en

NBS. (2017). *National BIM Report 2017*. Retrieved from https://www.thenbs.com/knowledge/nbs-national-bim-report-2017

Ng, A. K., Chan, L. K., & Lau, H. Y. (2017). Corrective Feedback For Depth Perception In Cave-Like Systems. In Virtual Reality (VR), 2017 IEEE (pp. 293-294). IEEE. doi:10.1109/VR.2017.7892292

Nicał, A. K., & Wodyński, W. A. (2016). Enhancing Facility Management through BIM 6D. In *Creative Construction Conference* (pp. 656–661). Budapest: Academic Press.

Nissen, M. E. (2006). *Harnessing Knowledge Dynamics Principled Organizational Knowing and Learning* (Vol. 2). IRM Press. doi:10.4018/978-1-59140-773-7

Normal Cultural Heritage Code, U. N. I. (2006). *UNI 11182, 01.020 91.100.15*. retrieved from http://www00.unibg.it/dati/corsi/20067/70707-UNI11182.pdf

Nour, M. (2016). Using Bounding Volumes for BIM based electronic code checking for Buildings in Egypt. *American Journal of Engineering Research*, *5*(4), 91–98.

NTC. (2008). *Norme Tecniche per le Costruzioni, D.M., January 14th*. Retrieved from http://www.ingegneriasoft.com/NTC2008_Norme_tecniche_per_le_costruzioni.htm

O'Keeffe, S. E. (2016). OpenBIM Framework for a Collaborative Historic Preservation System. *International Journal of 3-D Information Modeling, 5*(4), 1–11.

O'Rourke, L. (2013). *The future of DfMA is the future of construction* [Press release]. Retrieved from https://www.laingorourke.com/~/media/LOR/Files/LOR%20Engineering%20Excellence%20Journal%202013.pdf

OED Online. (2017). Oxford University Press. Retrieved from http://www.oed.com/view/Entry/203426?redirectedFrom=topology

Office of Government Commerce. (2007). *Management of risks: Guidance for practitioners*. Office of Government Commerce.

Open Geospatial Consortium (OGC). (n.d.). *Open Geospatial Consortium*. Retrieved February 11, 2015, from http://www.opengeospatial.org/

Ottosson, S. (2002). Virtual Reality In The Product Development Process. *Journal of Engineering Design, 13*(2), 159–172. doi:10.1080/09544820210129823

Owen, R., Amor, R., Palmer, M., Dickinson, J., Tatum, C., Kazi, A. S., ... East, B. (2010). Challenges for Integrated Design Delivery Solutions. *Architectural Engineering and Design Management*, *6*(4), 232–240. doi:10.3763/aedm.2010.IDDS1

Pacheco, D., & Wierenga, S. (2014). Spatializing experience: a framework for the geolocalization, visualization and exploration of historical data using VR/AR technologies. *Proceedings of the 2014 Virtual Reality International Conference*. 10.1145/2617841.2617842

Papadonikolaki, E., Vrijhoef, R., & Wamelink, H. (2015). BIM adoption in integrated supply chains: a multiple case study. In *Proceedings of the 31st Annual ARCOM Conference* (pp. 631-640). Lincoln, UK: Association of Researchers in Construction Management.

Park, R. J. (1999). *Value engineering: a plan for invention*. CRC Press.

Parsons, E. (1995). GIS visualisation tools for qualitative spatial information. *Innovations in GIS*, *2*, 201–210.

Paul, N. (2009). Basic topological notions and their relation to BIM. Handbook of Research on Building Information Modeling and Construction Informatics: Concepts and Technologies, 451-472.

Paul, N. (2010). Basic topological notions and their relation to BIM. In J. Underwood & U. Isikdag (Eds.), *Handbook of research on building information modeling and construction informatics: Concepts and technologies*. IGI Global. doi:10.4018/978-1-60566-928-1.ch019

Paulson, B. C., & Fondahl, J. (1980). *Towards Improved Transportation Construction Through Research*. Stanford, CA: The Construction Institute, Stanford University.

Pena, W., & Parshall, S. (2001). *Problem seeking: an architectural programming primer* (4th ed.). New York: John Wiley and Sons Inc.

Pereira, G. C., Rocha, M. C. F., & Florentino, P. V. (2013). Spatial Representation: City and Digital Spaces. In Computational Science and Its Applications– ICCSA 2013 (pp. 524-537). Springer Berlin Heidelberg.

Pirolli, P., & Card, S. (1999). Information foraging. *Psychological Review*, *106*(4), 643–675. doi:10.1037/0033-295X.106.4.643

Poirier, E., Staub-French, S., & Forgues, D. (2015). Embedded contexts of innovation: BIM adoption and implementation for a specialty contracting SME. *Construction Innovation*, *15*(1), 42–65. doi:10.1108/CI-01-2014-0013

Politecnico di Milano, B. I. M. Manager. Methods, Models and Applications. III Edition. (n.d.). Retrieved from https://www.polimi.it

Popov, V., Mikalauskas, S., Migilinskas, D., & Vainiunas, P. (2006). Complex Usage of 4D Information Modelling Concept for Building Design, Estimation, Scheduling and Determination of Effective Variant. *Technological and Economic Development of Economy*, *12*, 91–98.

Pretzsch, H., Biber, P., & Dursky, J. (2002). The Single Tree-Based Stand Simulator SILVA: Construction, Application and Evaluation. *Forest Ecology and Management*, *162*(1), 3–21. doi:10.1016/S0378-1127(02)00047-6

Project Management Institute. (2013). *The project management body of knowledge* (5th ed.). Project Management Institute Inc.

Rahimian, F. P., & Ibrahim, R. (2011). Impacts of VR 3D sketching on novice designers' spatial cognition in collaborative conceptual architectural design. *Design Studies*, *32*(3), 255–291. doi:10.1016/j.destud.2010.10.003

Raisbeck, P., Millie, R., & Maher, A. (2010). Assessing Integrated Project Delivery: a comparative analysis of IPD and alliance contracting procurement routes. In *Proceedings of the 26th Annual ARCOM Conference* (pp. 1019-1028). Leeds, UK: Association of Researchers in Construction Management.

Ramsey, M. S., & Levitt, R. E. (2005). *A computational framework for experimentation with edge organizations*. Academic Press.

Ray, S. F. (2002). *Applied photographic optics: Lenses and optical systems for photography, film, video, electronic and digital imaging*. Focal Press.

Reber, R., Schwarz, N., & And Winkielman, P. (2004). Processing Fluency And Aesthetic Pleasure: Is Beauty in the Perceiver's Processing Experience? *Personality and Social Psychology Review*, *8*(4), 364–382. doi:10.120715327957pspr0804_3 PMID:15582859

Redford, A. H. (1994). *Design for assembly: principles and practice. London*. London: McGraw-Hill.

Redmond, A., Hore, A., Alshawi, M., & West, R. (2012). Exploring how information exchanges can be enhanced through cloud BIM. *Automation in Construction*, *24*, 175–183.

Redmond, A., Hore, A., Alshawi, M., & West, R. (2012). Exploring how information exchanges can be enhanced through Cloud BIM. *Automation in Construction*, *24*, 175–183. doi:10.1016/j.autcon.2012.02.003

Reginald, A. I. (2015). Integrating BIM with BMS in Energy Performance Assessment: Case Study of a University Building in UK. *International Journal of 3-D Information Modeling*, *4*(1), 19–44.

Reinhardt, R., & Gurtner, S. (2015). Differences between early adopters of disruptive and sustaining innovations. *Journal of Business Research*, *68*(1), 137–145. doi:10.1016/j.jbusres.2014.04.007

Reitberger, J. (2010). *3D-Segmentierung von Einzelbäumen und Baumartenklassifikation aus Daten flugzeuggetragener Full Waveform Laserscanner*. TU Munich.

Rezgui, Y. (2006). Ontology-centered knowledge management using information retrieval techniques. *Journal of Computing in Civil Engineering*, *20*(4), 261–270. doi:10.1061/(ASCE)0887-3801(2006)20:4(261)

RIBA. (2017). *NBS National BIM Report 2017*. London: RIBA Enterprises Ltd.

Richards-Rissetto, H., Remondino, F., Agugiaro, G., Von Schwerin, J., Robertsson, J., & Girardi, G. (2012). Kinect and 3D GIS in archaeology. In *Proceedings of the 2012 18th International Conference on Virtual Systems and Multimedia, VSMM 2012: Virtual Systems in the Information Society* (pp. 331–337). Academic Press. 10.1109/VSMM.2012.6365942

Roberts, M., Ducheneaut, N., & Smith, T. F. (2010). The "3d Wiki": Blending virtual worlds and Web architecture for remote collaboration. In *2010 IEEE International Conference on Multimedia and Expo, ICME 2010* (pp. 1166–1171). IEEE. 10.1109/ICME.2010.5582579

Robertson, G., Czerwinski, M., Larson, K., Robbins, D. C., Thiel, D., & Van Dantzich, M. (1998). Data mountain: using spatial memory for document management. In *Proceedings of the 11th annual ACM symposium on User interface software and technology* (pp. 153-162). ACM. 10.1145/288392.288596

Rohrer, R., & Swing, E. (1997). Web-Based Information Visualization. *IEEE, 17*(4), 52–59.

Rossignac, J., & O'Connor, M. (1989). SGC: A dimension-independent model for pointsets with internal structures and incomplete boundaries. *Proceedings of the IFIP Workshop on CAD/CAM*, 145-180.

Rossmann, J., Buecken, A., Lennartz, G., & Hudjetz, S. (2017). Remote Sensing Based Forests Management In Conjunction With A Succession Simulation. In *Proceedings of the SGEM Vienna Green 2017*. Vienna, Austria: Academic Press.

Roßmann, J., Hoppen, M., & Bücken, A. (2014). GML-Based Data Management and Semantic World Modelling for a 4D Forest Simulation and Information System. *International Journal of 3-D Information Modeling, 3*(3), 50–67. 10.4018/ij3dim.2014070104

Rossmann, J., Krahwinkler, P. M., & Buecken, A. (2009). Mapping and navigation of mobile robots in natural environments. In *Advances in Robotics Research - Theory, Implementation, Application (German Workshop on Robotics 2009, TU Braunschweig)* (pp. 43–52). Braunschweig: Springer.

Rossmann, J., Schluse, M., Waspe, R., & Moshammer, R. (2011). Simulation in the Woods: From Remote Sensing based Data Acquisition and Processing to Various Simulation Applications. In S. Jain, R. R. Creasey, J. Himmelspach, K. P. White, & M. Fu (Eds.), *Proceedings of the 2011 Winter Simulation Conference* (pp. 984–996). Academic Press.

Rossmann, J., & Jung, T. (2010). Interactive Dynamics-Based Simulation of Work Machines. *VRCA (Virtual Reality Center Aachen) Progress Report, 4*, 77–79.

Rosson, M. B., & Carroll, J. M. (2001). *Usability engineering: scenario-based development of human-computer interaction*. Elsevier.

Royal Institute of British Architects. (2013). *Plan of work*. Retrieved April 14, 2017 from http://www.ribaplanofwork.com/Download.aspx

Ruikar, K. (2014). BIM and Regulation 38. Fire Sector Summit 214.

Sacks, R., Dave, B., Koskela, L. J., & Owen, R. L. (2009). *Analysis framework for the interaction between lean construction and building information modelling*. Academic Press.

Sacks, R., Kaner, I., Eastman, C.M., & Jeong, Y.-S. (2010). The Rosewood experiment — Building information modeling and interoperability for architectural precast facades. *Automation in Construction, Building information modeling and interoperability, 19*, 419–432.

Sacks, R., Koskela, L., Dave, B. A., & Owen, R. (2010). Interaction of Lean and Building Information Modeling in Construction. *Journal of Construction Engineering and Management, 136*(9), 968–980. doi:10.1061/(ASCE)CO.1943-7862.0000203

Sacks, R., Radosavljevic, M., & Barak, R. (2010). Requirements for building information modeling based lean production management systems for construction. *Automation in Construction, 19*(5), 641–655. doi:10.1016/j.autcon.2010.02.010

Sakr, M. A., & Güting, R. H. (2009). Spatiotemporal pattern queries in secondo. In SSTD 2009: Advances in spatial and temporal databases (pp. 422-426). Springer.

Salamin, P., Thalmann, D., & Vexo, F. (2006). The benefits of third-person perspective in virtual and augmented reality? *Proceedings of the ACM symposium on Virtual reality software and technology*, 27-30. 10.1145/1180495.1180502

Saleeb, N. (2015). Effects of Differences between Virtual and Physical Perception of Space on Building Information Modelling. *Proceedings of the International Conference on Building Information Modelling in Design, Construction & Operations*, 21-32. 10.2495/BIM150031

Sanchez, A. e., Hampson, K. e., & Vaux, S. e. (2016). *Delivering value with BIM: a whole-of-life approach*. Academic Press.

Saulton, A., Bülthoff, H. H., de la Rosa, S., & Dodds, T. J. (2017). Cultural differences in room size perception. *PLoS One, 12*(4), e0176115. doi:10.1371/journal.pone.0176115 PMID:28426729

Saxon, R. (2013). *Growth through BIM*. London, UK: Construction Industry Council.

Schevers, H., Mitchell, J., Akhurst, P., Marchant, D., Bull, S., McDonald, K., ... Linning, C. (2007). Towards digital facility modelling for Sydney opera house using IFC and semantic web technology. *Journal of Information Technology in Construction, 12*(24), 347–362.

Schmalstieg, D., Schall, G., Wagner, D., Barakonyi, I., Reitmayr, G., Newman, J., & Ledermann, F. (2007). Managing complex augmented reality models. *IEEE Computer Graphics and Applications, 27*(4), 48–57. doi:10.1109/MCG.2007.85 PMID:17713234

Schneider, M., Chen, T., Viswanathan, G., & Yuan, W. (2012). Cardinal directions between complex regions. *ACM Transactions on Database Systems, 37*(2), 8. doi:10.1145/2188349.2188350

Schodek, D. L. (2005). Digital design and manufacturing: CAD/CAM applications in architecture and design. Hoboken, NJ: John Wiley.

Schreyer, M., Scholar, V., Hartmann, T., Engineer, S., & Fischer, M. (2005). Supporting Project Meetings With Concurrent Interoperability in a Construction Information Workspace. *ITcon, 10*, 153–167.

Scianna, A., Gristina, S., & Paliaga, S. (2014). Experimental BIM Applications in Archaeology: A Work-Flow. *Proc. Of the 5th Int. Conf. on Digital Heritage. Progress in Cultural Heritage: Documentation, Preservation, and Protection*, 490–498. 10.1007/978-3-319-13695-0_48

Scully, T., Doboš, J., Sturm, T., & Jung, Y. (2015). 3drepo. io: building the next generation Web3D repository with AngularJS and X3DOM. *Proceedings of the 20th International Conference on 3D Web Technology*. 10.1145/2775292.2775312

Seeam, A., Zheng, T., Lu, Y., Usmani, A., & Laurenson, D. (2013). BIM Integrated Workflow Management and Monitoring System for Modular Buildings. *Journal of 3-D Information Modeling, 2*(1), 17–28.

Sendler, U. (2009). *Das PLM-Kompendium: Referenzbuch des Produkt-Lebenszyklus-Managements*. Berlin: Springer; doi:10.1007/978-3-540-87898-8

Shaaban, S., Lockley, S., & Elkadi, H. (2001). Information visualisation for the architectural practice. In *Information Visualisation, 2001. Proceedings. Fifth International Conference* (pp. 43-50). IEEE.

Shahabi, C., Banaei-Kashani, F., Khoshgozaran, A., Nocera, L., & Xing, S. X. S. (2010). GeoDec: A Framework to Visualize and Query Geospatial Data for Decision-Making. *IEEE MultiMedia, 17*(3), 14–23. doi:10.1109/MMUL.2010.5692179

Shahrin, F., & Johansen, E. (2013). Challenges in engaging the client during the capture, translation, transformation, and delivery (CTTD) of client requirements (CR) within the Building Information Modelling (BIM) environment. In *Proceedings of the 7th Nordic Conference on Construction Economics and Organisation*, (pp. 469-479). Trondheim, Norway: Academic Press.

Shemesh, A., Bar, M., & Grobman, Y. J. (2015). Space and human perception: exploring our reaction to different geometries of space. *Emerging Experience in Past, Present and Future of Digital Architecture. The 20th International Conference of the Association for Computer-Aided Architectural Design Research in Asia CAADRIA*.

Shen, S., & Codinhoto, R. (2017). *Post-Occupancy Evaluation Report: Manchester Central Library Case Study* (Master Dissertation). The University of Bath.

Shiratuddin, M. F., & Thabet, W. (2011). Utilizing a 3D game engine to develop a virtual design review system. *Journal of Information Technology in Construction, 16*, 39–68.

Shumate, M., Ibrahim, R., & Levitt, R. (2010). Dynamic Information Retrieval and Allocation Flows in Project Teams With Discontinuous Membership. *European Journal of International Management, 4*(6), 556–575. doi:10.1504/EJIM.2010.035589

SiCAR. (n.d.). *Sistema Informatico per la documentazione e la Progettazione dei Cantieri di Restauro*. Retrieved from http://www.infobc.uniud.it/progetto/attivita/ricerca/sicar-sistema-informatico-per-i-cantieri-di-restauro

Siehoff, S. (2011). *Succession of semi-natural grasslands: spatially-explicit, mechanistic simulation considering various forms of land use*. Aachen: Publikationsserver der RWTH Aachen University.

Siehoff, S., Lennartz, G., Heilburg, I. C., Roß-Nickoll, M., Ratte, H. T., & Preuss, T. G. (2011). Process-based modeling of grassland dynamics built on ecological indicator values for land use. *Ecological Modelling*, *222*(23), 3854–3868. doi:10.1016/j.ecolmodel.2011.10.003

Smith, D., & Tardiff, M. (2009). *Building Information Modeling: a strategic implementation guide for architects, engineers, constructors, and real estate asset manager*. John Wiley and Sons. doi:10.1002/9780470432846

Specialist Engineering Contractors. (2007). *Supplementary memorandum submitted by Specialist Engineering Contractors' Group*. Retrieved April 22, 2017 from http://www.publications.parliament.uk/pa/cm200708/cmselect/cmberr/127/127we57.htm

Stadler, A., Nagel, C., König, G., & Kolbe, T. H. (2009). Making interoperability persistent : A 3D geo database based on CityGML. *3D Geo-. Information Sciences*, 175–192. doi:10.1007/978-3-540-87395-2_11

Stapleton, K. A. J., Gledson, B. J., & Alwan, Z. (2014). Understanding technological interoperability through observations of data leakage in Building Information Modelling (BIM) based transactions. *Proceedings of the 32nd eCAADe Conference*, *2*, 515–524. 10.13140/RG.2.1.3882.5849

Stastbygg. (2013). *Stastbygg BIM Manual 1.2.1*. Retrieved from http://www.statsbygg.no/files/publikasjoner/manualer/StatsbyggBIM-manual-ver1-2-1eng-2013-12-17.pdf

Statistics, O. N. (2015). *Construction Statistics - No. 16, 2015 Edition*. Retrieved from http://www.ons.gov.uk/ons/rel/construction/construction-statistics/no--15--2014-edition/index.html

Steel, J., Drogemuller, R., & Toth, B. (2012). Model interoperability in building information modelling. *Software & Systems Modeling*, *11*(1), 99–109. doi:10.100710270-010-0178-4

Succar, B. (2009). Building information modelling framework: A research and delivery foundation for industry stakeholders. *Automation in Construction*, *18*(3), 357–375. doi:10.1016/j.autcon.2008.10.003

Succar, B., & Kassem, M. (2015). Macro-BIM adoption: Conceptual structures. *Automation in Construction*, *57*, 64–79. doi:10.1016/j.autcon.2015.04.018

Sveikauskas, L., Rowe, S., Mildenberger, J., Price, J., & Young, A. (2014). *Productivity growth in construction*. U.S. Bureau of Labor Statistics.

Szulanski, G. (2000). The process of knowledge transfer: A diachronic analysis of stickiness. *Organizational Behavior and Human Decision Processes*, *82*(1), 9–27. doi:10.1006/obhd.2000.2884

Takemura, H., Kitamura, Y., Ohya, J., & Kishino, F. (1993). Distributed Processing Architecture for Virtual Space Teleconferencing. In *Proc. of ICAT* (*Vol. 93*, pp. 27–32). Academic Press.

Talebi, S. (2014). Exploring advantages and challenges of adaptation and implementation of BIM in project life cycle. In Proceedings of the 2nd BIM International Conference on Challenges to Overcome (pp. 1–20). Lisbon, Portugal: BIM Forum.

Tarandi, V. (2011). The BIM Collaboration Hub: A model server based on IFC and PLCS for Virtual Enterprise Collaboration. In *Proceedings of the CIB W78-W102 2011: International Conference* (pp. 26–28). Sophia Antipolis, France: Academic Press.

Tay, L., & Ooi, J. T. L. (2001). Facilities management: A 'Jack of all trades'? *Facilities*, *19*(10), 357–363. doi:10.1108/EUM0000000005534

Teicholz, P. (Ed.). (2013). *BIM for facility managers*. John Wiley & Sons.

Teicholz, P., Goodrum, P. M., & Haas, C. T. (2001). U.S. Construction labor productivity trends. *Journal of Construction Engineering and Management*, *127*(5), 427–429. doi:10.1061/(ASCE)0733-9364(2001)127:5(427)

Thiry, M. (1997). *Value management practice*. Sylva, NC: PMI Publications.

Thomas, L., & Thomas, B. (2012). Evolution of large- scale industrialisation and service innovation in Japanese prefabrication industry. *Construction Innovation*, *12*(2), 156–178. doi:10.1108/14714171211215921

Thomson, D., Austin, S., Devine-Wright, H., & Mills, G. (2003). Managing value and quality in design. *Journal of Building Research & Information*, *31*(5), 334–345. doi:10.1080/0961321032000087981

Thomson, T. (1990). The essence of facilities management. *Facilities*, *8*(8), 8–12. doi:10.1108/EUM0000000002119

Thyssen, M. H., Emmitt, S., Bonke, S., & Christoffersen, A. K. (2008). Managing client values in construction design. *Proceedings of Joint CIB Conference W102 – Information and Knowledge Management in Building / W096 - Architectural Management "Improving Performance"*.

Tian, F., Chai, W., Wang, C., & Sun, X. (2012). *Design and Implementation of Flight Visual Simulation System*. arXiv preprint arXiv:1212.0365

Tolman, F. P. (1999). Product modeling standards for the building and construction industry: Past, present and future. *Automation in Construction*, *8*(3), 227–235. doi:10.1016/S0926-5805(98)00073-9

Tommasi, C., Achille, C., & Fassi, F. (2016). From point cloud to BIM: a modelling challenge in the cultural heritage field. The International Archives of the Photogrammetry, Remote Sensing and Spatial Information Sciences, XLI-B5, 429-436.

Tozawa, J., & Oyama, T. (2006). Effects of motion parallax and perspective cues on perceived size and distance. *Perception*, *35*(8), 1007–1023. doi:10.1068/p5251 PMID:17076062

Treasury and Cabinet Office. (2016). *Government Construction Strategy*. Retrieved from https://www.gov.uk/government/publications/government-construction-strategy-2016-2020

Trivedi, G. (2013). *The Importance of 3D Animation and Visualization in Construction and Manufacturing*. Retrieved August 20, 2017, from http://www.truecadd.com/news/the-importance-of-3d-animation-and-visualization-in-construction-and-manufacturing.

Trochim, W. (2006). *Single Group Threats*. Retrieved March 20, 2017, from http://www.socialresearchmethods.net/kb/intsing.php

Tzortzopoulos, P., Codinhoto, R., Kagioglou, M., Rooke, J. A., & Koskela, L. J. (2009). The gaps between healthcare service and building design: A state of the art review. *Ambiente Construído*, *9*(2), 47–55.

U.S. General Services Administration. (n.d.). Retrieved from https://www.gsa.gov/

Ugwu, O. O. (2005). A service-oriented framework for sustainability appraisal and knowledge management. *International Journal of IT-Construction ITCon*, *10*, 245–263.

UK Government. (2016). *PAS 2016: 2010: next generation access for new build homes guide*. Retrieved from https://www.gov.uk/government/publications/pas-2016-2010-next-generation-access-for-new-build-homes-guide

UNI 11337-4:2017. (2017). *Edilizia e opere di ingegneria civile - Gestione digitale dei processi informativi delle costruzioni - Part 4: Evoluzione e sviluppo informativo di modelli, elaborati e oggetti*. Retrieved from http://www.uni.com/index.php?option=com_content&view=article&id=5753:gestione-digitale-dei-processi-informativi-delle-costruzioni-pubblicate-le-parti-1-4-e-5-della-uni-11337&catid=170:commerciale&Itemid=2612

University of Edinburgh. (2015). *Kist o Riches*. Retrieved from http://www.tobarandualchais.co.uk

University of Helsinki. (2015). *CultureSampo*. Retrieved from http://www.kulttuurisampo.fi

University of Pisa. (n.d.). *BIM Master 2017. II Edition*. Retrieved from http://www.unibim.it/

Vakaloudis, A., & Theodoulidis, B. (1998). Spatiotemporal Database Connection to VRML. *Proceedings of the 9th UK Electronic Imaging & the Visual Arts Conference, EVA '98*.

Van Maren, G., Germs, R., & Jansen, F. (1998). Integrating 3D-GIS and Virtual Reality Design and implementation of the Karma VI system. *Proceedings of the Spatial Information Research Centre's 10th Colloquium*, 16–19.

Van Oosterom, P., & Stoter, J. (2012). *Principles of 5d modeling, full integration of 3d space, time and scale*. Geospatial World Weekly.

Veeramani, D., & Russell, J. (2000). Preparing the AEC industry for the knowledge economy. *Berkeley-Stanford Construction Engineering and Management Workshop. Defining a Research Agenda for AEC Process/Product Development in 2000 and Beyond*.

Verein Deutscher Ingenieure (VDI). (2002). *VDI 2219 - Information technology in product development Introduction and economics of EDM/PDM Systems (Issue German/English).* Düsseldorf: Verein Deutscher Ingenieure.

Vishwanath, D. (2014). Toward a new theory of stereopsis. *Psychological Review, 121*(2), 151–178. doi:10.1037/a0035233 PMID:24730596

Volk, R., Stengel, J., & Schultmann, F. (2014). Building Information Modeling (BIM) for existing buildings — Literature review and future needs. *Automation in Construction, 38*, 109–127. doi:10.1016/j.autcon.2013.10.023

Von Schweber, E. (1998). *SQL3D - Escape from VRML Island.* Retrieved February 9, 2015, from http://www.infomaniacs.com/SQL3D/SQL3D-Escape-From-VRML-Island.htm

Walczak, K. (2012). Dynamic Database Modeling of 3D Multimedia Content. In W. Cellary & K. Walczak (Eds.), *Interactive 3D Multimedia Content* (pp. 55–102). London: Springer London; doi:10.1007/978-1-4471-2497-9_4

Wamba, S. F., Gunasekaran, A., Akter, S., Ren, S., Dubey, R., & Childe, S. J. (2017). Big data analytics and firm performance: Effects of dynamic capabilities. *Journal of Business Research, 70*, 356–365. doi:10.1016/j.jbusres.2016.08.009

Wang, S., Mao, Z., Zeng, C., Gong, H., Li, S., & Chen, B. (2010). A new method of virtual reality based on Unity3D. In *2010 18th International Conference on Geoinformatics* (pp. 1–5). Academic Press. 10.1109/GEOINFORMATICS.2010.5567608

Wang, Y., Wang, X., Wang, J., Yung, P., & Jun, G. (2013). Engagement of facilities management in design stage through BIM: Framework and a case study. *Advances in Civil Engineering, 2013*(30836). 10.1155/2013/189105

Wang, Z., Cai, H., & Bu, F. (2014). Nonlinear Revision Control for Web-Based 3D Scene Editor. In *Virtual Reality and Visualization (ICVRV), 2014 International Conference on* (pp. 73–80). Academic Press.

Watson, N. V., & Breedlove, S. M. (2012). *The mind's machine: foundations of brain and behavior*. Sinauer Associates.

Wolstenholme, A., Austin, S. A., Bairstow, M., Blumenthal, A., Lorimer, J., McGuckin, S., . . . Davies, R. (2009). *Never waste a good crisis: A review of progress since Rethinking Construction and thoughts for our future*. Retrieved from https://dspace.lboro.ac.uk/2134/6040

Womack, J. (1996). *Lean thinking.* Simon & Schuster.

Woo, J.-H., Clayton, M. J., Johnson, R. E., Flores, B. E., & Ellis, C. (2004). Dynamic Knowledge Map: Reusing experts' tacit knowledge in the AEC industry. *Automation in Construction, 13*(2), 203–207. doi:10.1016/j.autcon.2003.09.003

Wu, I. C., & Hsieh, S. H. (2012). A framework for facilitating multi-dimensional information integration, management and visualization in engineering projects. *Automation in Construction*, *23*, 71–86. doi:10.1016/j.autcon.2011.12.010

xBimTeam. (2017). *XbimXplorer Download.* Xbim Toolkit Documentation. Retrieved from: http://docs.xbim.net/downloads/xbimxplorer.html

Yalcinkaya, M., & Singh, V. (2014, July). Building information modeling (BIM) for facilities management–literature review and future needs. In *IFIP International Conference on Product Lifecycle Management* (pp. 1-10). Springer. 10.1007/978-3-662-45937-9_1

Yan, H., & Demian, P. (2008). *Benefits and barriers of building information modelling*. Academic Press.

Yang, Q. Z., & Zhang, Y. (2006). Semantic interoperability in building design: Methods and tools. *Computer Aided Design*, *38*(10), 1099–1112. doi:10.1016/j.cad.2006.06.003

Zhang, S., Sulankivi, K., Kiviniemi, M., Romo, I., Eastman, C. M., & Teizer, J. (2015). BIM-based fall hazard identification and prevention in construction safety planning. *Safety Science*, *72*, 31–45. doi:10.1016/j.ssci.2014.08.001

Zhao, Y., Yan, C., Zhou, X., Zhu, Q., Wang, S., & Guo, W. (2013). The research and development of 3D urban geographic information system with Unity3D. In *Geoinformatics (GEOINFORMATICS), 2013 21st International Conference on* (pp. 1–4). Academic Press. 10.1109/Geoinformatics.2013.6626094

Zlatanova, S., Rahman, A. A., & Shi, W. (2004). Topological models and frameworks for 3D spatial objects. *Computers & Geosciences*, *30*(4), 419–428. doi:10.1016/j.cageo.2003.06.004

Zou, Y., Jones, S. W., & Kiviniemi, A. (2015). BIM and Knowledge Based Risk Management System A Conceptual Model. *Proceedings of CitA BIM Gathering 2015*, 63–68.

About the Contributors

Maszura Abdul Ghafar is a Senior Lecturer teaching Architectural Studies and Sustainable Design in the Architecture Department, Faculty of Design and Architecture, Universiti Putra Malaysia since 2011. Previously, she was a Lecturer in the Architecture Department, Kampus Sri Iskandar, University Technology MARA (UiTM), Perak for 8 years. After servicing UiTM, she worked abroad in The Republic of Ireland and United Kingdom as a practicing architect for almost 3 years. She got her Bachelor Science of Architecture from Robert Gordon University, Scotland in 1996 and obtained her Post Graduate Diploma in Advanced Architectural Studies in 1997 from the same university. Her Masters in Advanced Architectural Studies was obtained from the same university in 2000 under the UiTM scholarship for seven months. Besides teaching Dr. Maszura has also held a number of administrative posts at faculty and university level. Dr. Maszura's contribution does not stop at faculty level, she also contributes her professional skills at university level being directly involved in community engagement and practical training visits. She actively writes articles and involves with many student activities. Her PhD research specializes on how architect, engineer and contractor (AEC) professionals would achieve high productivity efficiency by more understanding other professionals' work culture with the aid of competent technology such as Building Information Modeling (BIM) during design phase. Upon completing her doctoral studies, she was invited as the National MyAgeing Research Centre's project consultant to assist project collaboration. Her research interests also include sustainable design informatics, sustainable office architecture and environmental design. She is also an avid researcher that profound in developing new reformation for the AEC tertiary education curriculum and aligning the AEC professionals towards technological change. Currently, she is a committee member for the faculty's RIBA accreditation delegate. She published various articles in journals, proceeding and book chapters.

Cristiana Achille is an associate professor of Topography and Cartography. She graduated in Architecture at the Politecnico di Milano and obtained there a PhD in Geodetic and Topographic Sciences. She founded, with prof. Francesco Fassi, the '3D Survey Group' (Dpt. A.B.C., Politecnico di Milano) http://www.sitech-3dsurvey.polimi.it. Her research focuses on the measure and the representation of Cultural Heritage and urban context. In particular, she focuses on studying new technologies for surveying and managing data. She participates in several projects at a national and international level, where the target was a 3D accurate, complete and rapid survey of complex architectures and large monumental sites. She is Professor of Survey Techniques and a member of the National Scientific Committee CSN ICOMOS Italy – CIPA – Heritage Documentation and IIC, The International Institute for Conservation of Historic and Artistic Works - FIIC.

Kemi Adeyeye is Associate Professor in Integrated Design at the University of Bath.

Malak Al Hattab has recently earned her PhD in Civil Engineering from the American University of Beirut. Over the past five years during her PhD studies, she worked on research about Building Information Modeling (BIM), design and construction management, lean construction methods, as well as simulation. Her work has been published in international peer-reviewed journals and conferences. She plans to extend her work by integrating BIM knowledge and use to implement sustainable designs and enhance the status of the construction industry while meeting the needs of the regional trends.

Claudio Benghi studied as an architect at Politecnico di Milano in Italy because of his fascination with composition and complexity in the built environment. In time, expanding his interest in the area of information sciences, he has discovered the freedom and beauty of theoretical models, developing his career in the realm of information tools and technologies for the built environment, architecture, engineering and construction. His PhD research investigated the issues of interoperability for construction and defined a strategy to promote the uptake of networked service solutions to support it.

Patrick Erik Bradley received a Ph.D. in mathematics from Karlsruhe University, Germany, in 2002 with an emphasis in non-Archimedean geometry. Afterwards, he joined the Architectural Faculty of Karlsruhe University where he became a researcher on building stock dynamics and building information modelling. He is currently a researcher at the Institute of Photogrammetry and Remote Sensing at Karlsruhe Institute of Technology, Germany focusing on mathematical methods in data analysis and spatial modelling.

Arno Buecken studied computer science in Bonn, Germany, and Pittsburgh, PA. He received his PhD from RWTH Aachen University, where he is a group manager at the Institute for Man-Machine Interaction. His research interest focuses on the combination of remote sensing data, GIS and virtual reality.

Ricardo Codinhoto is a qualified architect with practical, teaching and research experience. He joined the University of Bath's Department of Architecture & Civil Engineering in 2015, where he holds an Associate Professor position and is the Director of Studies for the MSc Modern Building Design. He undertakes teaching and supervision at both undergraduate and postgraduate levels. Ricardo's current research interests include Building Information Modelling (BIM) applied to Design, Construction and Facilities Management, the design of healthcare facilities particularly focused on evidence-based design for wellbeing and environments for older people. He is also active in a number of research initiatives related to Design Theory, Lean Construction, Building Information Modelling Implementation and Healthcare Facilities Design, Construction and Maintenance and Project Management.

Julie Comlay is a PhD Candidate at the University of Bath. Professionally, Julie is an Architect and experienced researcher. Her research interests lie within the area of Building Information Modelling and more specifically BIM implementation for facilities management in healthcare.

P. Demian is Reader in Building Information Management at Loughborough University. He teaches design and design management and conducts research on software for design and construction, particularly BIM. His postgraduate studies were at Stanford University and undergraduate studies at Cambridge University (all in civil engineering). He is a Chartered Construction Manager, a Member of the Chartered Institute of Building and the American Society of Civil Engineers, and a Fellow of the Higher Education Academy.

Vincenzo Donato is an Architectural Engineer graduate from the University of Florence and received a double PhD degree from the University of Florence (Italy) and the Technical University of Brainwashing (Germany). Currently is a lecturer at Polytechnic of Turin and he works at the Department of Structural, Geotechnical and Building Engineering (DISEG). He carries out research and teaching activities focusing on various aspects related to "Building Information Modelling". Vincenzo has practicing experience from working as a BIM manager and as an Architect. He developed capability to work on international architectural studies, developing ideas and concepts as well standards and procedures for supporting the design of hospitals and schools using BIM. He has also been involved in the retrofitting and the renovation of existing buildings in which BIM was required. Currently his work is focused on applying BIM and diagnostic techniques for protecting historical and monumental heritage.

David Greenwood is Professor of Construction Project Management in the Faculty of Engineering and Environment, Northumbria University, UK. He is director of BIM Academy (Enterprises) Ltd., an industry-academia joint venture spin-out company, a member of the Board of Construct-IT, and former Chair of the Association of Researchers in Construction Management (ARCOM). He has published over 100 academic journal and conference papers and has authored and co-authored several textbooks. He has over twenty years of experience in consulting, training and lecturing around the world for commercial and governmental organisations and is an active promoter of better practice in the construction industry.

Martin Hoppen studied Computer Science at the University of Bonn (Germany). Since 2007, he works as a research assistant at the Institute for Man-Machine Interaction (MMI) of RWTH Aachen University (Germany). His work focuses on data modeling and data management. In 2017, he received a doctorate of RWTH for his work on data management for eRobotics applications. Currently, he continues his work at the MMI as a postdoc and group manager.

Ben Jowett is BIM Manager at Wates Construction Ltd responsible for BIM implementation and delivery for the Midlands and Northern business regions in the UK. Awarded an MSc with Distinction in Building Design Management and BIM at Northumbria University in 2017.

Mohamad Kassem is an Associate Professor in Mechanical and Construction Engineering and the Director of Enterprise and Engagement at the department of Mechanical and Construction Engineering. His qualification includes a PhD and MEng in Mechanical Engineering (Marche Polytechnic University, Italy) and an MSc in Project Management (Teesside University, UK). Mohamad's research and enterprise interest is in Digital Construction, Building Information Modelling (BIM), construction informatics, big data applications in built environment, sustainability and energy reduction, virtual reality applications in construction processes, and construction and project management. Mohamad has published more than 50 articles covering these themes in international journals and conferences. He is currently teaching aspects related to these themes to several built environment disciplines, undergraduate and post-graduate programmes within Northumbria University.

Arto Kiviniemi, Professor of Digital Architectural Design, is one of the internationally leading experts in the integrated Building Information Modelling (BIM) which he has developed both in Finland and internationally since 1996. In 1997-2002 Arto lead the Finnish national R&D programme which created the foundation for Finland's position as one of the leading countries adopting BIM. In May 2010 he moved to UK, first to the University of Salford, where he initiated the first MSc BIM programme in the UK, and in September 2013 he started in his current position at the University of Liverpool.

Nora Lombardini is Associate professor of restoration at the Politecnico di Milano. She taught architectural restoration of historic buildings at Parma, Florence and Venice universities. Her research focuses on the theory and history of restoration, analysis of the structural behaviour of historic buildings, and conservation projects aimed to the re-use of the buildings. She is involved in European projects (Tempus, Erasmus Mundus, Erasmus plus) for training at the university level. She has done workshops in Italy, but also in Ukraine, Bulgaria and China, working with local universities.

Anne Morris is an Emeritus Professor at Loughborough University. Her main interests lie in the evaluation of information services, the design of user interfaces, system usability and user information seeking behaviour. She has extensive research and consultancy experience stemming back over a period of 40 years documented in over 200 publications. A graduate of Loughborough University, Anne has degrees in Ergonomics, Information Technology, and Physics and Mathematics. Her PhD was in Information Science.

Norbert Paul graduated from the University of Karlsruhe, Germany, 1996 with a Diploma of Architecture and Urban Planning. After several years as programmer for geo-information systems and later for embedded systems he joined the Architectural Department of the Karlsruhe University in 2002 as a researcher on building information modelling (BIM). There he received the Ph.D. for his dissertation on Topological Databases. He then worked for the Geodetic Institute (GIK) at the KIT in Karlsruhe as a researcher in spatial modelling. During work he studied at the distance-learning university Fernuniversität Hagen, Germany, where he graduated 2013 as Bachelor of Computer Science. Norbert Paul than returned to the embedded systems business and now runs his own company of automated emergency power control units.

Farzad Pour Rahimian is a Senior Lecturer in Building Information Management, with expertise in the mainstream areas of BIM, including IT Integrated Design and Construction, Virtual-Reality Programming for Automation, Simulation and Optimisation of Design, Offsite-Manufactured-Construction, and Virtual Prototyping. He has extensive experience in VR/AR programming and prototyping, and was a key player on one the first and most famous game-like VR Environments for construction simulation. Farzad holds PhD degree in IT Integrated Architectural Design Studies and MSc and BSc in Architectural Engineering. He is a Fellow of Higher Education Academy (FHEA), Member of Chartered Institute of Building (MCIOB), Member of Regulatory Room of buildingSMART International, a First Class Registered Architect in Iran and member of CIB TG74, W119, and W121. He has considerable research and practice experience in BIM and IT Integrated Construction and Design (reported in more than 85 refereed articles; Cited in more than 250 publications). Farzad has been involved in supervision of several PhD studies. He is the Co-Editor of the International Journal of Architectural Research (MIT) and has been involved in various high-level committees. During his academic career, he contributed to various publicly funded research projects, on issues regarding BIM, Deign Integration and Advanced Methods of Construction.

Juergen Rossmann received his Dipl.-Ing. degree and his PhD degree from the Technical University of Dortmund. Since 2006, he is full professor and director of the Institute for Man-Machine Interaction at RWTH Aachen University, Germany. He has received several national and international scientific awards. He is member of the national academy of science and engineering in Germany.

Kirti Ruikar is an architect (CA/95/18991) with a Bachelor of Architecture degree from the University of Pune, India. She designed and managed a range of commercial and residential building projects, before enrolling onto the Doctor of Engineering (EngD) programme at Loughborough University. En-route to her doctoral research, Dr Ruikar completed an MSc in Construction Innovation and Management for which she secured a distinction. She successfully completed her doctorate studies in 2004. The focus of her research was to examine the Business Process Implications of e-Commerce in Construction Organisations. Following this she was awarded a research fellowship at the 6* Salford Centre for Research and Innovation (SCRI) in the University of Salford. She was appointed to the post of Lecturer in 2007 and is currently a Senior Lecturer in Architectural Engineering in School of Civil and Building Engineering at Loughborough University. She is a Fellow of the Higher Education Academy (FHEA). An expert in business process mapping and process management, her multidisciplinary research portfolio is reflective of her diverse interests in the fields of computer-aided design and engineering,

ICT, BIM, knowledge and information management, strategic technology management, collaborative working environments, e-readiness and electronic-business. Her research is widely published and she has over 140 publications in these fields. She has co-edited books on e-Business in Construction (Wiley-Blackwell) and Advances in Construction ICT and e-Business (Routledge) with world-leading international authors. She has co-authored the book on 'Collaborative Design Management' (Routledge) with Prof Stephen Emmitt, who is a highly regarded academic in the field. She is a committed researcher with a rich portfolio of multi-disciplinary research projects. She has supervised 13 doctoral research projects and an MPhil to successful completion. She has been a Guest Editor of the Journal of IT in Construction's special issues on, 'E-commerce in Construction' (2006), 'Technology Strategies for Collaborative Working' (2009) and 'Innovation in Construction e-Business' (2011). Since 2007 Dr Ruikar is an Associate Editor of the internationally reputed Journal of IT in Construction (ITcon); and is ITcon's (first) and only female Editor. She is a Joint Co-ordinator of CIB TG83 on e-Business in Construction (since 2011). She sits on the scientific committees of various national and international conferences.

Noha Saleeb (PhD, MSc, BArch [hons], FHEA, MCIAT) is an Associate Professor in Construction & Creative Technologies, Middlesex University UK. She is Programme Leader for the International MSc Building Information Modelling Management programme focusing on innovative digital technologies, processes for enhancement of design, information, project and construction management. She is also Programme Leader for BSc Architectural Technology programme. Her research interests lie at the intersection of architecture, IT and e-learning disciplines specialising in 3D digital architecture and Building Information Modelling. She is currently investigating implementation of Industry 4.0 technologies in design and construction, e.g. Artificial Intelligence and Big Data analytics, currently Principal Investigator for a UK International Institutional Links funded project in usage of these innovative technologies and processes for heritage assets. Initially an architect, Dr Saleeb worked for many years in industry in design, construction and onsite project coordination. She is currently providing consultancy on BIM implementation and management on live projects. Dr. Saleeb achieved many awards for her collaborative industry-academia work, knowledge dissemination, training & consultancy. Most recently: - Best European Women in Construction Award 2016 - International Business Excellence Award 2015 - Most Inspiring Tutor Award 2016 Dr. Saleeb is a board member and reviewer on several renowned journals, annual conferences and member of the Digital Working committee for several professional bodies including CIOB, BAF. Dr. Saleeb has many publications in Journals, Conferences & Books most recently: - Saleeb, N., Dafoulas, G. & Loomes, M. (2016), 'Personalisation of 3D Virtual Spaces for Enhanced Ubiquitous Learning' in Neto, F. et al. (2015)

Handbook of Research on 3-D Virtual Environments and Hypermedia for Ubiquitous Learning, Hershey, PA: IGI Global. doi:10.4018/978-1-5225-0125-1 - Saleeb, N. (2016), 'Impact of Fidelity in 3D Space Visualisation across the Construction Asset Lifecycle. Journal of 3D Information Modelling - Succar, B., Saleeb, N., Sher, W. (2016), 'Model Uses: A Requirements' Clarification Language', for Proceedings of the 40th Australasian Universities Building Education Association Conference (AUBEA 2016), 6-8 July 2016. - Dearlove, J., Saleeb, N. (2016), 'Intelligent Decision-Making System Frameworks for a Digital Plan of Work: A Theoretical Investigation For The Construction Industry', for Proceedings of the International Conference on Architectural Technology (ICAT 2016), 12-14 May 2016, Alicante, Spain - Saleeb, N. (2016), Investigation into Maturity Management for Building Information Modelling, for Proceedings of the 1st International Conference of the BIM Academic Forum, UK (BAF 2016).

Cinzia Tommasi is a PhD Student in Architecture, Built Environment and Construction Engineering. After the Master degree in Architecture at Politecnico di Milano in 2014, she was a collaborator and then a research fellow at 3D Survey Group, in A.B.C. Department, at Politecnico di Milano, until the end of 2016, when she won the Ministerial Scholarship to start the PhD program. During the year as a research fellow, the central theme of her research was the use of survey data coming from the laser scanner and photogrammetry (point cloud) in BIM processes, in particular referring to the Cultural Heritage field. She continues to develop this theme also after the beginning of the PhD program, focusing on state of the art about BIM and Cultural Heritage and making some practical tests to try to understand the issues of these methodology applied to existing buildings or sites. Simultaneously, she starts to explore the field of the valorisation, questioning the uses of the information models and the language of fruition and dissemination of Cultural Heritage.

Index

I

K

L

M

N

P

Q

R

S

T

U

V

W

Z

www.ingramcontent.com/pod-product-compliance
Lightning Source LLC
LaVergne TN
LVHW080900300926
849162LV00003B/497
* 9 7 8 1 5 2 2 5 5 6 2 5 1 *